The Human Genome

Edited by
Carina Dennis and Richard Gallagher

Foreword by
James D. Watson

nature palgrave

Acknowledgements

For invaluable input, constructive suggestions and general support during the preparation of this book, we should like to thank Barbara Cohen, Francis Collins, Elizabeth Dennis, Daniel Drell, Mark Guyer, Mark Hirst, Michael Hopkin, Ian Jones, Michael Kilborn, John MacFarlane, Joseph McInerney, Richard Nathan, Hemai Parthasarathy and Peter Wrobel. We also wish to thank Jane Ades, Barbara Izdebska, Betty Mansfield, Majo Xeridat, the National Human Genome Research Institute (NHGRI), the Department of Energy (DOE) and the Wellcome Trust for their help in illustrating the text.

Note

A version of the article which appears on pp. 4–7 was also published in *A Passion for DNA: Genes, Genomes, and Society* by James D. Watson, Cold Spring Harbor Laboratories (2001).

First published 2001 by
PALGRAVE
Houndmills, Basingstoke, Hampshire RG21 6XS and
175 Fifth Avenue, New York, N.Y. 10010
Companies and representatives throughout the world

PALGRAVE is the new global academic imprint of St. Martin's Press LLC Scholarly and Reference Division and Palgrave Publishers Ltd (formerly Macmillan Press Ltd).

ISBN 0-333-97143-4

W.H. Freeman & Company edition
0 - 7167 - 9789 - 5

This book is printed on paper suitable for recycling and made from fully managed and sustained forest sources.

A catalogue record for this book is available from the British Library.

Library of Congress Cataloging-in-Publication Data
The human genome/Carina Dennis, Richard Gallagher.
    p. cm.
    Includes bibliographical references and index.
    ISBN 0-333-97143-4
    1 Human genome. 2. Human gene mapping.
    3. Human Genome Project. I. Dennis,
    Carina II. Gallagher, Richard B.
    QH447.H835 2001
    599.93'5–dc21   2001133051

10 9 8 7 6 5 4 3
10 09 08 07 06 05 04 03 02

Produced in association with
Book Production Consultants plc, Cambridge, UK
Printed and bound in Italy

Contents

The human genome revealed

James D. Watson

Seeing the Human Genome Project's draft sequence of the human genome is highly satisfying. The way in which its 3 billion bases have been determined closely follows the course outlined more than a decade ago by the National Academy of Sciences (NAS) Committee on Mapping and Sequencing the Human Genome. Bruce Alberts, now the President of the NAS, was its chairman and I was one of its 14 other members. The predictions in our 1988 report that the human genome could be sequenced over a 15-year period for a cost of US$3 billion were more accurate than we dared guess. Two more years of work, to fill in gaps and correct mistakes, will result in an almost errorless genetic script for human existence.

That the human script would become available within our lifetimes never passed through my mind, or that of Francis Crick, when we found the double helix in 1953. Then, just learning how cells read the genetic instructions within DNA seemed a tall order. Happily, progress was faster than expected and by 1966 we knew how the genetic code utilizes groups of three DNA bases to specify the amino acid constituents of proteins – the main actors in the plays of life. Things speeded up even more after the recombinant DNA procedures of Stanley Cohen and Herb Boyer burst onto the scene in 1973. Gene cloning and manipulation metamorphosed from being dreams to becoming facts of life. Simultaneously, Fred Sanger and Walter Gilbert each developed a powerful way to determine the order of bases along DNA molecules. This meant that humans, like cells, could read the messages of genes. The way was open to ascertain the complete genetic instructions, that is to sequence the genome, of any

organism (subject to the usual constraints of money, personnel and technology).

Viral genomes were the first to be tackled, beginning with genomes of just several thousand bases. By the early 1980s, viral genomes containing more than 100 000 bases had been sequenced and bacterial genomes containing more than a million bases became realistic objectives. The completion of such genomes would tell us the number of different proteins necessary for bacterial existence. Back then I thought that the human genome, at several billion bases long, was much, much too large to take on. Soon, however, I became a strong proponent of an international Human Genome Project (HGP), believing that the large-scale mapping and sequencing resources that it would command would hasten our discovery of the genetic underpinnings of many important human diseases.

Our NAS committee wasted little time on whether or not we needed an HGP; instead, we focused on how it should be organized and financed. It seemed best to begin modestly and end with a sequencing crescendo, hopefully fuelled by much lower sequencing costs. We agreed unanimously that the first big sequencing efforts should not focus on human DNA but on DNA from the model organisms of genetics, such as baker's yeast or the fruit fly. We knew that many human genes were likely to be homologous to those of model organisms.

We proposed a 15-year-long effort, reflecting our belief that those starting the project should also be part of the finishing team. Richard Gibbs, Eric Lander, Maynard Olson, John Sulston, Bob Waterston and Jean Weissenbach all stayed the course, running increasingly larger megabase sequencing labs. Only one of our original NAS committee is no longer in science. Sadly, Dan Nathans died three years ago at the age of 70, of leukemia. During our committee deliberations, no-one proposed a shorter timeframe – technology had to be improved too much. Later, I learned that Congress likes big projects to be finished within ten years so that key initial backers are still in Washington when the achievement is celebrated. Luckily, Tom Harkin recently became that Congress rarity, a three-term Democratic Senator from Iowa, so that he and New Mexico's Republican Senator Pete Domenici will see the HGP from its commencement to its conclusion.

The improvements in technology that the HGP needed for its success materialized almost on schedule. They largely involved modifications of pre-existing methods, as opposed to great leaps forward that generate Nobel-Prize-like rewards. The current DNA sequencing machines, the workhorses of our big sequencing labs, are 1000-fold-improved descendants of the original sequencing machine put together by Mike Hunkapiller and Lloyd Smith in Lee Hood's Caltech lab. The computers and software that now compare new raw DNA sequences to pre-existing ones also do their tasks 1000 times faster than was possible when the HGP started.

A major obstacle for the correct assembly of the human genome was the vast amount of repetitive DNA (~50%), so the HGP labs decided early on to sequence DNA from known chromosomal locations. Their map-based approach, however, was challenged in May 1998 by a new company, Celera, led by Craig Venter. Celera proposed an alternative strategy whereby the genome is randomly shredded into pieces that are sequenced and then reassembled in a single process without the construction of a map, a strategy known as 'whole-genome shotgun'. Key to their approach were 300 of the new high-capacity capillary DNA sequencers that were about to be launched onto the market, as well as proprietary shotgun assembly software for use on high-powered computers. So armed, Celera promised a first draft of the human genome in only two years.

I first heard of Celera in a phone call from my former associate, Richard Roberts. Rich told me that Celera would blow the international consortium out of the water, and asked me to consider joining him on its scientific advisory board. Expecting to learn more about Celera's game plan at the soon-to-be-held Spring Genome Meeting, I reported to the National Institutes of Health (NIH) and the Wellcome Trust that Celera had marked them out for obsolescence. Later that week, Craig Venter visited NIH to tell Harold Varmus and Francis Collins that HGP's future effort might best be devoted to sequencing the mouse.

From the moment of Rich Roberts's call, I found it unthinkable that a private company should effectively control much of the human genome through key patents. This was a gene power-play that, at all costs, had to be matched. To my relief, the Wellcome Trust's immediate response was to double the budget for human genome sequencing at the Sanger Centre. Though the merits of each approach were yet to be tested, Celera's 'super shotgun' method caught the fancy of the serious press, who reported that HGP was off course. In fact, at the spring 1996 Bermuda meeting, HGP leaders had discussed Jim Weber's proposal for a low-resolution, whole-genome shotgun effort to complement the high-resolution map approach. There, Phil Green's off-the-cuff calculations, later redone and published, indicated that human DNA is too repetitive for a pure shotgun approach to assemble the genome correctly.

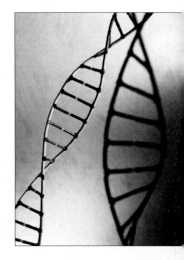

In September 1998 I returned to Washington to tell key congressional leaders that expanded federal support of the publicly funded sequencing efforts was necessary to prevent a monopoly on human genetic information. Many large pharmaceutical companies rooted for the public HGP, believing that Celera's future databases could only be validated through checking with publicly obtained sequences. To my relief, Congress increased public sequencing moneys significantly. So encouraged, HGP announced that it, like Celera, would complete a rough draft of the human genome in the spring of 2000 and, unlike Celera, it would pursue a highly accurate final product.

The February 2001 publication of drafts of the human genome by HGP and Celera represents a milestone in human history, revealing the basic features of the human genetic script. The drafts will allow us to identify most of the genes that underlie human existence. Using the genetic code to translate their message into protein products, we now have the first comprehensive overview of the molecules that make up our bodies. And it is immediately obvious that these are very similar to the molecular building blocks of other forms of life. Darwinian evolution can be increasingly described through incremental changes in underlying DNA scripts.

It is, however, unclear whether either draft is accurate enough for confident protein structure predictions. In fact proteome predictions from the two human drafts may be seriously misleading; only a virtually errorless 'gold standard' human DNA script will confidently move us into proteome waters. That so much more sequencing needs to be done, however, should in no way lessen our admiration for what both groups have accomplished.

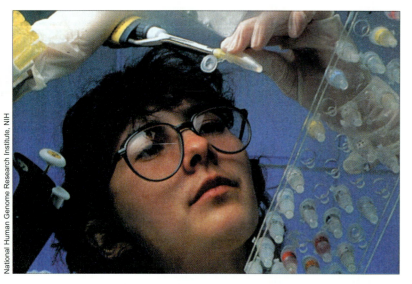

National Human Genome Research Institute, NIH

Until we saw the DNA text that underlies multicellular existence it seemed natural that increasing organismal complexity would involve corresponding increases in gene numbers. So I, and virtually all my scientific peers, were surprised last year when the number of genes of the fruit fly *Drosophila melanogaster* was found to be much lower than that of a less complex animal, the roundworm *Caenorhabditis elegans* (13 500 versus 18 500). More shocking still was the recent finding that the small mustard plant *Arabadopsis thaliana* contains many thousands more genes (~28 000) than *C. elegans*. Now we are jolted again, by the conclusion that the number of human genes is not much more than 30 000. Until a year ago, I anticipated that human existence would require 70 000 to 100 000 genes.

Why organismal complexity fails to correlate with gene numbers is not fully clear. It may be due in part to RNA splicing events which generate multiple protein products from single genes; vertebrate genes give rise to more splicing products than invertebrate genes. But the quality of respective nervous systems may be equally relevant. The roundworm, being dumber than the fruit fly, may need more specific proteins (and therefore genes) to respond to enemies or changes in its environment – the fruit fly's more advanced nervous system lets it respond to potential enemies and stresses by flying away. Plants, being totally dumb, must continually evolve new genes to respond to new enemies and climatic changes.

Many more vertebrate genomes need to be sequenced before we have a sense of how often the generation of new genes has underlain evolutionary change. We also need to know why vertebrate genomes contain so many more repetitive sequences than invertebrates. Most human repetitive sequences appear to have arisen as the result of the generation and movement of transposable genetic elements. Conceivably, many of the mutations that underlie vertebrate evolution arise from transposon movements into regulatory regions, changing gene expression patterns. The very high levels of repetitive DNA in amphibians and lungfish may reflect past needs to evolve fast for survival in their ever-changing ecological niches.

It should be possible to test the idea that changes in regulatory segments, as opposed to changes in protein-coding segments, have dominated vertebrate evolution. For example, sequence information from morphologically different breeds of dog may be informative, and hopefully funds will be made available to produce draft genomes of several breeds. How soon we shall be able to compare the chimpanzee genome with ours remains unclear. Obviously we should like to know the genetic changes that make possible the larger and more powerful human brain.

Of the many new facts emerging from the draft human genome, I am most excited by the finding that repetitive sequences are almost absent from the four clusters of homeobox genes. Unlike most functionally related human genes, the chromosomal order of homeobox genes exactly reflects their temporal expression patterns during embryonic development. In this respect, they resemble the genes of bacterial operons that are transcribed from single messenger RNA molecules; genes located at the start of bacterial operons are transcribed first by RNA polymerase molecules moving along the respective DNA. Conceivably, much of early developmental timing in humans is a reflection of the time needed for RNA polymerase molecules to transcribe the lengthy introns of homeobox genes. If so, insertions of sizeable transposable sequences into them would lethally mis-set key timing events in embryonic development.

Many, many more unanticipated observations and hypotheses will emerge as the reading of the human script extends beyond those individuals who produced it to the much larger world of interested biologists. Even the heartiest of them, however, will find themselves stretched if they take on too much. Most triumphs of the near future will probably come from focusing on the human homologues of genes of known function in one or more model organisms.

Eventually, even more important dividends will come from focusing on ourselves as human beings and making sense of the oft seemingly intractable relations between nature and nurture. There is much more to human life than interactions between its DNA script and the RNA and protein actors that carry out its instructions. The culturally derived facts and traditions that our brains pass on from one generation to the next equally affect our lives.

Our genomes, thus, can never accurately predict our futures. But we would be more than silly if we did not use their information to the fullest. The human genetic script that we are now finalizing will be regarded as the most important book ever to be read.

Remarks from the Editor of *Nature*

Philip Campbell

Occasionally, a scientific advance enraptures a particular field, sends waves of excitement through other areas of research and percolates through to the general public. The publication of the first draft and analysis of the human genome is just such an event. It provides the first meaningful look at the molecular and genetic content of what lies at the heart of every one of the trillions of cells in our bodies: the DNA that comprises our chromosomes. This book is an opportunity to experience the biological drama of the human genome, and to marvel at what it already tells us about ourselves, about our relatives in the kingdoms of life, and about those who went before us.

Often, the great breakthroughs in science appear in the pages of *Nature*. Why do researchers choose to publish in *Nature*, and what functions does it, and scientific journals in general, serve? There are several important roles. Journals offer a forum for scientists to present and debate their findings, and provide a permanent record of the progress of scientific endeavour. They offer a stamp of validation, or at least of assessment, through 'peer review', which is a thorough, formal appraisal of sub-

mitted research papers. The experts that carry out this appraisal are selected by the journal's editors, and they perform their task without recognition or financial reward. As with all research published in *Nature*, the papers on the human genome were reviewed, and greatly strengthened, by scientific peers. I salute their contribution. At *Nature*, we are also proud of the roles that our own staff play in selecting the most interesting science, coordinating its peer review, improving the clarity of its presentation and illustration, and adding complementary commentary.

The genome sequence data are made available on the Internet on a daily basis, thanks to the Human Genome Project's policy of instant accessibility. But the papers, and especially the wonderfully expressed article about the sequence that is reproduced in this book, made the meaning of all of those Cs, As, Gs and Ts clear to a huge audience. We were pleased to join in the spirit of the project and make the papers freely available to all-comers on the Internet. It was a privilege to assess the publicly funded work and to assist in making its significance as clear as possible.

Preface

For the first time in history we can read the complete set of instructions for making human beings.

Such a profound development inevitably evokes intense positive and negative reactions, sometimes both in the same individual. On the one hand, it is a pinnacle of self-knowledge, realized by a society driven by a desire to understand itself and the world around it; it is also an astonishing technical achievement. It promises practical dividends – an era of disease diagnosis, therapy and prevention that will surpass any previous development in medical science. On the other hand, the potential to 'mess with nature' through the genetic modification of plants, animals and humans induces a visceral reaction in many people. And there is a deep and widespread unease about the impact the new knowledge will have on society, indeed upon the very essence of what it means to be human.

In the words of UK Prime Minister Tony Blair, the way forward is 'to focus on the possibilities, develop them and then face up to the hard ethical and moral questions that are inevitably posed by such an extraordinary scientific discovery'. To achieve this requires a basic understanding of what a genome is and of genomics – the science of sequencing, analysing and drawing conclusions from genomes. The purpose of this book is to assist that basic understanding. It is not intended to be comprehensive. Rather, we hope it will serve as an introduction to, or a continuation of, an exploration of the human genome. We have purposefully avoided the politics and personalities involved in the project, focusing instead on the genome itself.

At the heart of the book is the scientific article describing the first assembly and analysis of our genetic code, the product of more than a decade of work by the thousands of researchers worldwide. Accompanying that article is a series of essays from leading researchers, which provides the scientific context, a sense of the excitement and a critical evaluation of the work. These essays and the scientific article are reprinted from the journal *Nature*.

To open up the Human Genome Project to a wider audience, we have written a series of introductory chapters to outline the scientific concepts and technological advances behind the project. The book begins with a definition of the genome, a description of its building blocks, the information embedded within its sequence and how each of us is a subtle genetic variation on a theme. It continues with a guide to sequencing, describing the technology that enabled scientists to sequence the genome and build the maps needed to navigate around the genome landscape. A timeline follows, tracking the project from initial stirrings to its development into a massive international mission. As a companion to the research article itself, there is an overview of the main findings that emerged from the analysis. In the final chapters, the implications of genetic information for the individual and society are considered, along with highlights from the media coverage.

This is just our first glimpse of the human genome – a so-called 'working draft'. There are still gaps to fill in and ambiguities in the sequence to resolve that will keep scientists busy until at least 2003. All the secrets of the genome are not likely to be given up for many, many years. But we have reached a landmark, the first view of the entire genome, and this is an achievement to celebrate.

Carina Dennis and Richard Gallagher
Washington and London August 2001

An owner's guide to the genome

The human genome sequence has been described as the most precious collection of information imaginable. But what exactly is the genome? What does it do? To what extent is it unique? And what do we learn from having the sequence of the genome revealed? As the owner of a unique version of the genome, you may be curious to know. In this opening chapter, we set the genome in its biological context and explain some of the concepts needed to understand the impact that the discoveries of the Human Genome Project (HGP) will have.

The big picture

Your genome is your genetic constitution. It is the information that you inherited from your parents and, in part, it directs your life. It comprises 3 billion (3 000 000 000) pieces of data in the form of deoxyribonucleic acid (DNA). The individual pieces of information are called nucleotides or bases, and these are the units of DNA. Nucleotides are linked together in tremendously long strings and their order along these strings is the DNA sequence.

What does the genome do? If you think of the body as a complex biochemical machine full of interacting chemicals, molecules and macromolecules, the genome is the blueprint for the machine. It includes the instructions for the assembly of macromolecular components; these instructions are the genes and the working components that they encode are mostly proteins. Genes influence our physical characteristics – such as eye colour, height and hair colour – as well as susceptibility to some illnesses.

However, the genome is much more than a list of component parts; it also incorporates information that controls when and where the parts should be made. Consider the miraculous early development of an embryo from a single cell to a multicellular organism – the series of transformations that brings this about requires highly coordinated expression of an enormous cast of genes. Even much more simple activities, such as fighting an infection, require elaborate patterns of gene interaction to marshal the appropriate defences. So the genome is not just a collection of genes working in isolation, but rather it encompasses the global and highly coordinated control of information to carry out a range of cellular functions. One of the great advantages of knowing the genome sequence is that we can begin to understand the regulation of cellular functions.

Your own genome is at least 99.9 per cent identical to anyone else's on the planet; compare it with that of someone closely related to you and the figure is even higher. A difference of

DNA crystals. DNA in a liquid crystalline state forms exquisite conical-shaped and fan-like structures. [Courtesy of Michael W. Davidson, Florida State University, Tallahassee]

> "I've seen a lot of exciting biology emerge over the past 40 years. But chills still ran down my spine when I first read the paper that describes the outline of our genome."
>
> *David Baltimore, California Institute of Technology*

0.1 per cent might not seem a particularly impressive amount of uniqueness, but it translates to about three million differences embedded in our genetic code, which gives plenty of scope for individuality. Some of the differences have no apparent effect, but others influence our appearance, behaviour, vulnerability to disease, and responses to medication. In essence, we are all hewn from the same genetic script, but our fine individual distinctions matter greatly.

"To see the entire sequence of a human chromosome for the first time is like seeing an ocean liner emerging out of the fog, when all you've ever seen before are rowboats."

Francis Collins, National Human Genome Research Institute

That being said, individuals cannot be reduced to their genetic characteristics; we are much more than simply the products of our genomes. Issues of what we share and how we differ, and of what our genes do and do not define, are complex. The Universal Declaration on the Human Genome and Human Rights, by the United Nations Educational, Scientific and Cultural Organization, provides a wonderful expression of this concept. In calling for respect for human uniqueness and diversity, the declaration states that: 'The human genome underlies the fundamental unity of all members of the human family, as well as the recognition of their inherent dignity and diversity. In a symbolic sense, it is the heritage of humanity.'

What are the practical benefits of determining the sequence of the human genome? For scientists, they are enormous and immediate. Virtually every field of biology, from biochemistry to behavioural psychology, palaeontology to parasitology, conservation to cancer research, will gain new insight. One application of the sequence will be to open a window on our history, providing a biological scroll of how our ancestors dispersed and settled around the world.

Humans: bipedal primate mammals distinguished by a highly developed brain and by an erect body carriage that frees the hands for manipulation. Vital statistics include an average adult weight range of 36–95 kg; average height of 1.7 m; and an average lifespan of 75 years.
[The cover image, by Eric Lander, was created by Runaway Technology, Inc. (www.photomosaic.com) using PhotoMosaic by Robert Silvers from original artwork by Darryl Leja. It is used courtesy of the Whitehead Institute for Biomedical Research.]

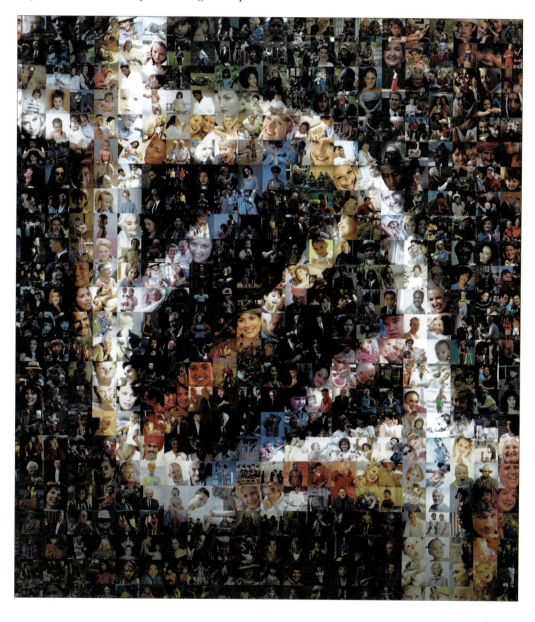

The benefits from the sequence will extend far beyond the research community, to revolutionize medicine. Our genetic make-up influences our susceptibility to disease. Most common human diseases, such as cardiovascular disease, rheumatoid arthritis and diabetes, involve several genes and are also influenced by environmental factors. The genetic basis of these complex diseases, which are known as 'multifactorial' or 'polygenic', is difficult to decipher. How will genomics help? In addition to identifying all the genes, researchers have been cataloguing variation within the genome, pinpointing the locations and types of genetic differences between individuals. This will help to identify genetic profiles that are prevalent in complex diseases. Not only will this lead to early detection and treatment of disease, but it will also propel us towards a new era of prevention, which, as the saying goes, is better than cure. Knowing the sequence of the genome and how it varies between individuals will also shed light on why some people with a particular disease respond better than others to drug treatment. And it will reveal new targets for the development of medicines.

The sequencing of the genome is only the first step on a long journey to full understanding of the biology of humans. However, although there is still much to learn, we are already gaining great insight into the nature of the genome and what makes us different from other organisms.

Zooming in on the genome

Before plunging into the details of our genome sequence, it helps to have a sense of biological scale. Here we start with the human body and break it down into progressively smaller components.

The first level of division is the ten interdependent systems that make up the body. These are the skeletal, muscular, circulatory, nervous, respiratory, digestive, excretory, endocrine, reproductive and immune systems. Each of these is composed of a network of interacting tissues and organs which, in turn, are made up of a unique composition of cell types.

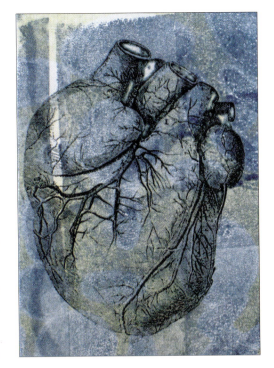

The heart is the pump of the circulatory system. Like all organs, it has a unique composition and function. It contains a wide range of specialized cell types that interact to provide structure and function.

"Let us be in no doubt about what we are witnessing today: a revolution in medical science whose implications far surpass even the discovery of antibiotics, the first great technological triumph of the 21st century."

UK Prime Minister Tony Blair

Cells are the basic units of all forms of life. The number of cells in an organism can range from one to trillions. The simplest forms of life are free-living, single-celled organisms such as bacteria, whereas humans are estimated to be made up of some 75 trillion (75 000 000 000 000) cells. Cells can vary greatly in size: the typical human cell is around 20 micrometres (0.00002 m) in diameter, whereas some nerve cells are over a metre in length. Although all cells of the human body have a similar basic structure, they display tremendous diversity. There are more than 200 different types of human cells, all varying in appearance, lifespan and function. And yet each cell of an individual contains the same genetic information. It is the selective expression of that information, through the switching on and off of genes, that

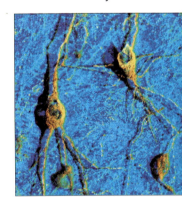

All living things are made of cells. Shown here is a confocal microscopy image of the brain showing the nerve cells in yellow. [Medical Microscopy Science / Wellcome Photo Library]

"The availability of genome sequence is just the beginning. Scientists now want to understand the genes and the role they play in the prevention, diagnosis and treatment of disease."

Randy Scott, President of Incyte Genomics

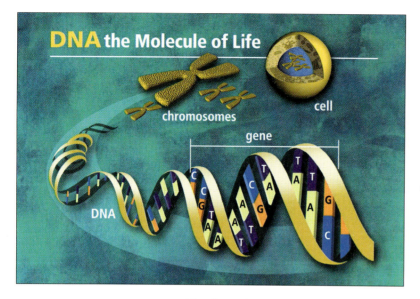

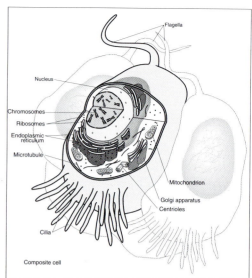

Composite cell

Inside the cell. Above right: some of the many organelles of a cell, each of which has its own distinct functions (NHGRI). DNA is found within the nucleus, tightly packaged into chromosomes, as illustrated above. [NHGRI, NIH]

provides the variety. An analogy to the genome would be a well-stocked kitchen; just as different combinations of ingredients can be used to concoct a host of different dishes, so various patterns of gene expression can produce a variety of cell types.

Each cell is further subdivided into structures called organelles (meaning 'small organs'). They are dedicated to specific tasks

inside the cell; for example, mitochondria are the power generators, ribosomes are protein factories, lysosomes act as waste-disposal units, and the endoplasmic reticulum labels, sorts and transports molecules. The organelle that concerns us here is the nucleus, which could be considered the headquarters of the cell – the nucleus houses the DNA.

Inside the nucleus the long, slender threads

A spectral karyotype of human chromosomes

Chromosomes (meaning 'coloured bodies') can be readily stained with certain dyes. The differences in size and banding pattern allow the chromosomes to be distinguished from each other, an analysis called karyotyping. Here, each chromosome is labelled with a different colour, a technique that is useful for identifying chromosome abnormalities. Some human diseases are caused by major chromosomal abnormalities, including missing or extra copies of chromosomes or breaking and rejoining of parts of chromosomes (called translocations). For example, Down syndrome results from a third copy of chromosome 21. [NHGRI, NIH]

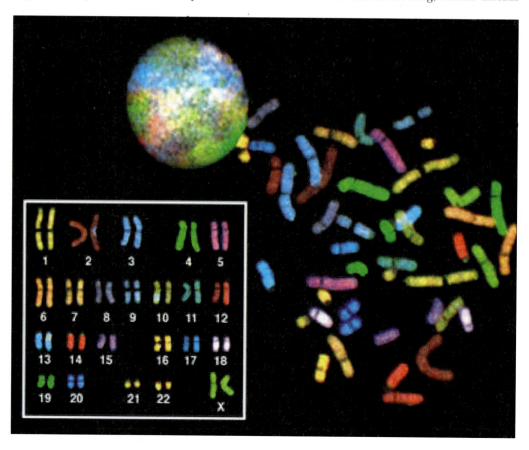

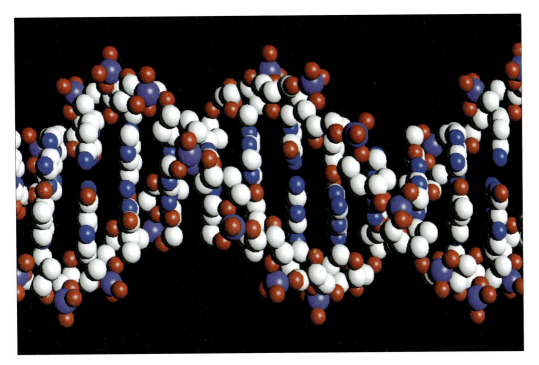

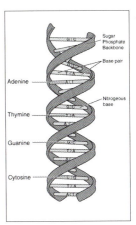

of DNA are tightly coiled into distinct, rod-like structures called chromosomes. Each species packages its DNA into a particular number of chromosomes, of characteristic size and shape. A mosquito has just 6, a pea plant makes do with 14, whereas some species of fern boast over 1000. Cats and dogs have 38 and 78 respectively, and chimpanzees have 48. Humans have 46 chromosomes arranged in 23 pairs, 22 of which are called autosomes. They are numbered 1–22, in decreasing order of size (except for chromosome 21, which is actually smaller than chromosome 22). The remaining pair is the sex chromosomes. Normal males have an X and a Y chromosome, whereas normal females have two X chromosomes. Y is the smallest chromosome of the human genome, being little more than a stump.

Defining DNA

The basic unit of DNA – the nucleotide – is composed of a sugar molecule, a phosphate molecule and a nitrogenous base. The base is the information-carrying element and exists in four different versions: adenine, cytosine, guanine and thymine. These are abbreviated to A, C, G and T, the four letters of the genetic alphabet.

Strings of nucleotides are joined together to form an incredibly long strand of DNA; unwound and joined together, the DNA from a single cell would be almost 2 metres long but only 50 trillionths of a centimetre wide. In cells, DNA molecules consist of two strands that wrap around each other to resemble a twisted ladder, the famous double helix. The sides of the ladder are made of sugar and phosphate molecules, and the 'rungs' are formed by pairs of bases, one on each strand. A remarkable feature of DNA is the way in which the pairs of bases are held together: A only pairs with T, and G with C. These pairings are termed 'base pairs'.

The linear order of the bases along a DNA strand represents its sequence. The genome size of an organism is usually stated as the total number of base pairs; the human genome contains roughly 3 billion. There is little correlation between the complexity of an organism and the size of its genome. Although the human genome contains at least 200 times more DNA than that of yeast, and is comparable to that of frogs and sharks, it is dwarfed by the genome of the newt, which has 15 billion base pairs and, extraordinarily, by that of a single-celled micro-organism, *Amoeba dubia*, which is 200 times bigger than the human genome.

The slice of life

Throughout life, cells are constantly being replaced and replenished. This is possible because most cells are able to replicate. Before a cell divides, it first copies its genome so that identical genetic material can be distributed to each of its daughter cells.

Structure of DNA. The double-helical structure of DNA is shown above left [c. Pete Artymiuk/Wellcome Picture Library]. Above, base pairing is illustrated. The binding of two nucleotides forms a base pair. Cytosine and guanine are bound together by three hydrogen bonds, whereas adenine and thymine are bound by two hydrogen bonds. Therefore the link between cytosine and guanine is much stronger than the link between adenine and thymine. [NHGRI, NIH]

►18

DNA: an icon

Photomicrograph of high-density liquid crystalline DNA. [Courtesy of Michael W. Davidson, Florida State University, Tallahassee]

The history of how DNA was discovered, its importance recognized, and its structure revealed is extraordinary. It has filled the pages of entire books, recounting the fascinating discoveries and equally fascinating people behind the science. To appreciate how DNA went from being considered a 'boring molecule' to being hailed as a modern-day icon, it is worth recapping a few of the highlights along the way.

In a less-than-auspicious beginning, DNA was first identified in cells of pus from discarded surgical bandages. This was in 1869, when the Swiss biochemist Friedrich Miescher described a substance that could not be broken down by protein-digesting enzymes and therefore 'cannot belong among any of the protein substances known hitherto'. Because it came from the nucleus, Miescher named it nuclein, but it was later found to have acidic properties and was renamed nucleic acid. There are two varieties of nucleic acid which, depending on the presence of one of two types of sugar – ribose and deoxyribose – are called ribonucleic acid (RNA) and deoxyribonucleic acid (DNA).

Clues that DNA can transmit information came from studies of the bacterium pneumococcus, the agent of pneumonia, which was a widespread killer at the time. In the 1920s, Frederick Griffiths described two strains, a virulent one that was lethal when injected into mice, and a non-virulent one. Neither the non-virulent strain nor a preparation of the virulent strain that had been killed by heating was lethal. But when the heat-killed virulent bacteria were injected together with the untreated non-virulent strain, the mice died. Somehow, the dead virulent strain had conferred lethality on the non-virulent strain. In 1943, Oswald Avery and colleagues revealed DNA to be the transforming factor that was at work. Using about 20 gallons (76 litres) of bacteria, the team purified DNA that was free from all other substances, and showed that it alone was responsible for the transformation.

At the time, the work received little attention. DNA was thought to be too simple a molecule, especially in comparison to proteins, to contain all of the genetic information for an organism. Definitive evidence that DNA is the hereditary material was finally provided by Alfred Hershey and Martha Chase in 1952. They carried out experiments using bacteriophages (literally meaning 'eaters of bacteria'), which are viruses that infect and seize control of bacteria, hijacking the cell's machinery to produce more virus particles. Bacteriophages contain both protein and DNA, but which of these mediates the takeover? Hershey and Chase labelled two cultures of bacteriophages: one with radioactive phosphorus, which tags DNA; and one with radioactive sulphur, which tags proteins. They then infected different colonies of bacteria with the different bacteriophage populations, and tracked the location of the radioactive material. Radioactive sulphur (and therefore the protein) remained outside the bacteria, whereas radioactive phosphorus (and therefore the DNA) was on the inside. Furthermore, newly produced bacteriophages contained radioactive phosphorus. These results firmly established DNA as the material of heredity.

The acceptance of DNA as the chemical basis of heredity inspired many scientists to attempt to discover its structure. One of these was Rosalind Franklin, a chemist at University College London. She used X-ray crystallography, a technique in which an X-ray beam is fired at a crystal of the substance of interest and the resulting scattering of the X-rays is used to build up a picture of the structure. Franklin and her colleague Maurice Wilkins obtained great insight into the structure of DNA, notably its helical nature. But it was James Watson and Francis Crick, two maverick intellectuals working in Cambridge, UK, who finally uncovered the elusive structure of DNA.

It is one of the most frequently told stories of science, not least because, as later recounted by Watson, it is a tale of boundless ambition, impatience with authority and disdain for conventional wisdom. The pair's first attempt at solving the structure (at the end of 1951) went badly wrong; because of a muddled

No. 4356 April 25, 1953   N A T U R E   737

equipment, and to Dr. G. E. R. Deacon and the captain and officers of R.R.S. *Discovery II* for their part in making the observations.

[1] Young, F. B., Gerrard, H., and Jevons, W., *Phil. Mag.*, **40**, 149 (1920).
[2] Longuet-Higgins, M. S., *Mon. Not. Roy. Astro. Soc., Geophys. Supp.*, **5**, 285 (1949).
[3] Von Arx, W. S., *Woods Hole Papers in Phys. Oceanog. Meteor.*, **11** (3) (1950).
[4] Ekman, V. W., *Arkiv. Mat. Astron. Fysik.* (Stockholm), **2** (11) (1905).

MOLECULAR STRUCTURE OF NUCLEIC ACIDS

A Structure for Deoxyribose Nucleic Acid

WE wish to suggest a structure for the salt of deoxyribose nucleic acid (D.N.A.). This structure has novel features which are of considerable biological interest.

A structure for nucleic acid has already been proposed by Pauling and Corey[1]. They kindly made their manuscript available to us in advance of publication. Their model consists of three intertwined chains, with the phosphates near the fibre axis, and the bases on the outside. In our opinion, this structure is unsatisfactory for two reasons : (1) We believe that the material which gives the X-ray diagrams is the salt, not the free acid. Without the acidic hydrogen atoms it is not clear what forces would hold the structure together, especially as the negatively charged phosphates near the axis will repel each other. (2) Some of the van der Waals distances appear to be too small.

Another three-chain structure has also been suggested by Fraser (in the press). In his model the phosphates are on the outside and the bases on the inside, linked together by hydrogen bonds. This structure as described is rather ill-defined, and for this reason we shall not comment on it.

We wish to put forward a radically different structure for the salt of deoxyribose nucleic acid. This structure has two helical chains each coiled round the same axis (see diagram). We have made the usual chemical assumptions, namely, that each chain consists of phosphate diester groups joining β-D-deoxyribofuranose residues with 3′,5′ linkages. The two chains (but not their bases) are related by a dyad perpendicular to the fibre axis. Both chains follow right-handed helices, but owing to the dyad the sequences of the atoms in the two chains run in opposite directions. Each chain loosely resembles Furberg's[2] model No. 1 ; that is, the bases are on the inside of the helix and the phosphates on the outside. The configuration of the sugar and the atoms near it is close to Furberg's 'standard configuration', the sugar being roughly perpendicular to the attached base. There

This figure is purely diagrammatic. The two ribbons symbolize the two phosphate—sugar chains, and the horizontal rods the pairs of bases holding the chains together. The vertical line marks the fibre axis

is a residue on each chain every 3·4 A. in the z-direction. We have assumed an angle of 36° between adjacent residues in the same chain, so that the structure repeats after 10 residues on each chain, that is, after 34 A. The distance of a phosphorus atom from the fibre axis is 10 A. As the phosphates are on the outside, cations have easy access to them.

The structure is an open one, and its water content is rather high. At lower water contents we would expect the bases to tilt so that the structure could become more compact.

The novel feature of the structure is the manner in which the two chains are held together by the purine and pyrimidine bases. The planes of the bases are perpendicular to the fibre axis. They are joined together in pairs, a single base from one chain being hydrogen-bonded to a single base from the other chain, so that the two lie side by side with identical z-co-ordinates. One of the pair must be a purine and the other a pyrimidine for bonding to occur. The hydrogen bonds are made as follows : purine position 1 to pyrimidine position 1 ; purine position 6 to pyrimidine position 6.

If it is assumed that the bases only occur in the structure in the most plausible tautomeric forms (that is, with the keto rather than the enol configurations) it is found that only specific pairs of bases can bond together. These pairs are : adenine (purine) with thymine (pyrimidine), and guanine (purine) with cytosine (pyrimidine).

In other words, if an adenine forms one member of a pair, on either chain, then on these assumptions the other member must be thymine ; similarly for guanine and cytosine. The sequence of bases on a single chain does not appear to be restricted in any way. However, if only specific pairs of bases can be formed, it follows that if the sequence of bases on one chain is given, then the sequence on the other chain is automatically determined.

It has been found experimentally[3,4] that the ratio of the amounts of adenine to thymine, and the ratio of guanine to cytosine, are always very close to unity for deoxyribose nucleic acid.

It is probably impossible to build this structure with a ribose sugar in place of the deoxyribose, as the extra oxygen atom would make too close a van der Waals contact.

The previously published X-ray data[5,6] on deoxyribose nucleic acid are insufficient for a rigorous test of our structure. So far as we can tell, it is roughly compatible with the experimental data, but it must be regarded as unproved until it has been checked against more exact results. Some of these are given in the following communications. We were not aware of the details of the results presented there when we devised our structure, which rests mainly though not entirely on published experimental data and stereochemical arguments.

It has not escaped our notice that the specific pairing we have postulated immediately suggests a possible copying mechanism for the genetic material.

Full details of the structure, including the conditions assumed in building it, together with a set of co-ordinates for the atoms, will be published elsewhere.

We are much indebted to Dr. Jerry Donohue for constant advice and criticism, especially on interatomic distances. We have also been stimulated by a knowledge of the general nature of the unpublished experimental results and ideas of Dr. M. H. F. Wilkins, Dr. R. E. Franklin and their co-workers at

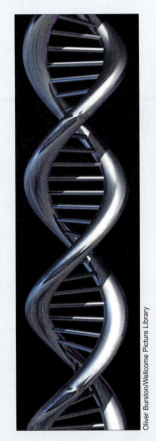

Oliver Burston/Wellcome Picture Library

Watson and Crick's paper published in *Nature* in April 1953.

King's College, London. One of us (J. D. W.) has been aided by a fellowship from the National Foundation for Infantile Paralysis.

J. D. WATSON
F. H. C. CRICK
Medical Research Council Unit for the
Study of the Molecular Structure of
Biological Systems,
Cavendish Laboratory, Cambridge.
April 2.

[1] Pauling, L., and Corey, R. B., *Nature*, **171**, 346 (1953) ; *Proc. U.S. Nat. Acad. Sci.*, **39**, 84 (1953).
[2] Furberg, S., *Acta Chem. Scand.*, **6**, 634 (1952).
[3] Chargaff, E., for references see Zamenhof, S., Brawerman, G., and Chargaff, E., *Biochim. et Biophys. Acta*, **9**, 402 (1952).
[4] Wyatt, G. R., *J. Gen. Physiol.*, **36**, 201 (1952).
[5] Astbury, W. T., Symp. Soc. Exp. Biol. 1, Nucleic Acid, 66 (Camb. Univ. Press, 1947).
[6] Wilkins, M. H. F., and Randall, J. T., *Biochim. et Biophys. Acta*, **10**, 192 (1953).

N A T U R E   April 25, 1

Molecular Structure of Deoxypentose Nucleic Acids

WHILE the biological properties of deoxypentose

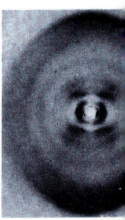

James Watson (left) and Francis Crick (right) with their model of the DNA double helix. [A. Barrington Brown/Science Photo Library]

to others, Watson and Crick made another attempt. Watson was able to examine the X-ray evidence from Franklin and Wilkins in January 1953, at which point he and Crick went into a frenzy of model building, using large, three-dimensional models that were one of the keys to their success. On 7 March 1953 they were ready to announce to fellow patrons of the Eagle pub in Cambridge that they had found 'the secret of life'.

'We wish to suggest a structure for the salt of deoxyribose nucleic acid (D.N.A.),' they announced in an article in *Nature* on 25 April 1953, continuing: 'This structure has novel features which are of considerable biological interest.' The paper, a model of clarity, precision and understatement, is reproduced on the previous page. What Watson and Crick suggested was a winding double helix in which the nucleotide bases on each strand are interlocked (A with T, and C with G), holding the two strands together. Each DNA molecule in the double helix thus forms a template for the other, immediately suggesting how DNA would replicate itself.

The model had great beauty and simplicity, and fitted the experimental data available, so it was soon accepted as correct. Max Delbruck, a theoretical physicist who became a principal figure in genetics, wrote: 'It might be said that Watson and Crick's discovery of the DNA double helix in 1953 did for biology what many physicists hoped in vain could be done for atomic physics: it solved all the mysteries in terms of classical models and theories, without forcing us to abandon our intuitive notions about truth and reality.'

For this achievement, Watson, Crick and Wilkins won the 1962 Nobel Prize for physiology/medicine. Franklin, whose work greatly contributed to the discovery, had died of cancer several years earlier at the age of 37.

recollection of a lecture that Franklin had given, they came up with a three-chain model with the backbone on the inside. Undaunted, and despite being instructed by the head of their laboratory to leave DNA structure

---

What do an interior design company, a group of musicians, an electronic magazine and a nightclub in San Francisco have in common? DNA! The double helix has taken its place as a modern icon. Some of its many uses in science, art, leisure and advertising are illustrated on the facing page. Here, in their own words, is why this disparate collection of entities chose DNA as a symbol.

"We used the name and DNA image as being the ultimate building blocks and blueprint in design that, hopefully, reflects the way we work. The double helix has a strong simple visual impact, but is complex and detailed in the way that it is generated, a perfect representation of our company. As an image it is open to lots of opportunities for us to express our creative abilities."

Chris Page, DNA Design
(http://subscriber.scoot.co.uk/dna_design/)
A pair of Christmas cards from DNA Design is shown.

"The name of our electronic band 'Freaky DNA' references DNA as a way to symbolize the dichotomy which is often assumed between the

ideals of the scientific and the artistic. Anything which has its origins in freaky DNA is likely to evolve into something new and unexpected, sometimes mutated and sometimes beautiful but always with a life of its own, much like our goals when we create music."

Leonard Paul (http://www.sfu.ca/~leonardp/)

"What I had in mind was a magazine which would explore the unique and/or the unknown, and of course with the intention of an alternative perception of things around us. And that was exactly DNA: something unique, a code man has not yet broken, a new perspective leading us to another dimension!

DNA magazine tries to include everything an alien from another world would like to know about this one. In other words, our essence."

George Drakakis
(http://www.dnamag.gr/index.htm)

"DNA is the fundamental building block of all life on earth, and I wanted the club to be the fundamental building block of club life in San Francisco. Even in 1985 the term DNA and the double helix symbol were widely recognized, and 'not too scary' – for the general public. Finally I wanted a name that was inviting – 'a place where your DNA can lounge'.

Barry Synoground, DNA Lounge, San Francisco (http://www.dnalounge.com)

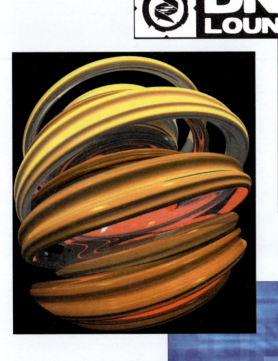

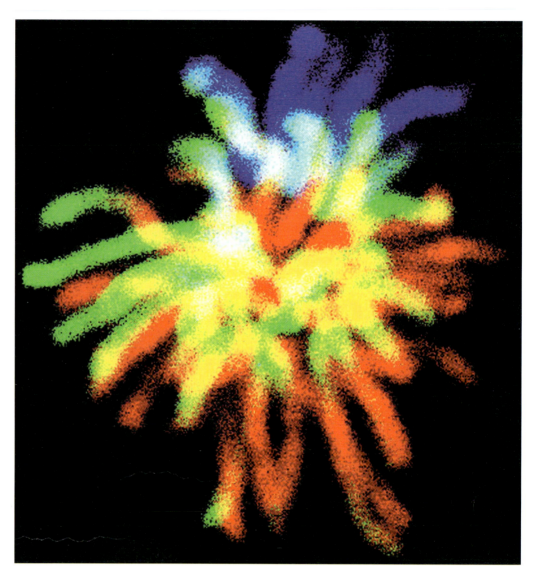

Cell division. Before cells divide, they copy their chromosomes. A copy of each chromosome is then pulled to either end of the cell so that, upon division, each progeny cell receives the full complement of chromosomes. This picture shows a cell about to divide, with its chromosomes stained with a dye that binds specifically to DNA. [Rachel Errington, Sally Davies and Paul J. Smith/ Wellcome Trust Medical Photographic Library]

The process of duplicating and dividing the genetic material of a cell is called mitosis. It happens as follows: the parent cell makes copies of all its chromosomes, which line up along the midline of the cell and are then pulled apart, so that each end of the cell contains a copy of every chromosome. A membrane then forms through the middle of the cell, dividing it in two, with each daughter cell containing an identical and complete set of chromosomes.

How are the chromosomes copied? Existing DNA molecules are used as templates to copy new ones, a process known as DNA replication. The double helix along the stretch of DNA to be copied is unwound and the weak bonds between the base pairs break, allowing the strands to separate. A DNA-copying molecule, called DNA polymerase, then binds to each strand and moves along it. DNA polymerase makes an exact replica of the old strand that used to be paired with the template strand. It takes new bases, which float freely inside the nucleus, and inserts them into the growing strand, strictly obeying the base-pairing rules. A is only inserted opposite T and C opposite G, ensuring that the new strand is an exact copy of the old one. Once the copying is complete, the strands rewind and the helices reform, resulting in two double-stranded molecules of DNA (each containing one original strand and one new strand) which are divided between the two daughter cells.

From genes to proteins

The gene is the fundamental unit of inheritance. It is a stretch of sequence in a specific position on a DNA strand that carries the instructions for making a particular protein. Proteins are the

building blocks of life; they provide the structural architecture of cells and tissues, enzymes for essential biochemical reactions, and signalling molecules to coordinate cellular activities. When a cell is making a protein, the gene is said to be 'expressed', 'active' or 'switched on'. As discussed earlier, all cells in an organism contain the same genes but the subset of genes that are expressed defines the physical nature of a given cell as well as its function in the body.

In more complicated organisms than bacteria, genes are divided into sections that code for proteins, called exons (meaning 'expressed sequences'), interrupted by non-coding spacers

> ## "The gene is by far the most sophisticated program around."
>
> *Bill Gates, CEO Microsoft*

called introns (meaning 'intervening sequences'). Human genes vary greatly in length; whereas the average protein-coding sequence of a gene is about 1000 to 2000 base pairs, long stretches of non-coding sequence interspersed between exons can extend the boundaries by 20 000–100 000 base pairs. The largest known human gene, which encodes dystrophin (an important protein in the scaffolding of muscle cells), is 2.4 million base pairs long, of which only 14 000 actually code for the protein.

A single gene can be used by a cell to make more than one protein. This can be done by splicing together different combinations of exons within a gene. Alternatively, the protein made by a gene can be modified after it is produced, by adding different chemical groups that change its properties. So there are many more proteins than genes, but no-one knows exactly how many yet. There are now thought to be 30 000–40 000 human genes, and some scientists estimate that there could be more than 120 000 different proteins in the human body.

Proteins are made up of long chains of subunits called amino acids, of which there are 20 different kinds. How do the 4 letters of the DNA alphabet translate to the 20 different building blocks of proteins? Like many good things, the genetic code happens in threes. Each specific sequence of three DNA bases (called a 'codon') codes for one amino acid. For example, the base sequence GCA codes for the amino acid alanine, and AGA codes for arginine. The genetic code is thus a series of 3-base codons

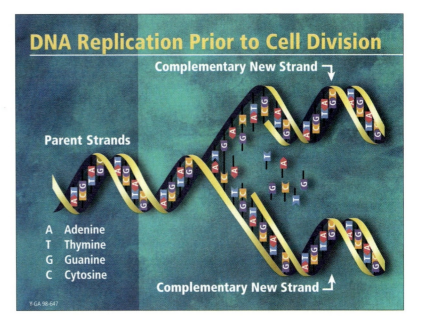

DNA Replication Prior to Cell Division

Complementary New Strand

Parent Strands

A Adenine
T Thymine
G Guanine
C Cytosine

Complementary New Strand

Y-GA 98-647

DNA replication. To replicate before cell division, the DNA double helix unwinds and each strand acts as a template for making a new complementary strand. A new strand is the mirror image of the template strand according to the rules of base pairing: A with T, and G with C. This results in two DNA molecules (each consisting of one old and one new DNA strand), whose sequences are identical to those of the original DNA. [DOE Human Genome Program]

that tell the cell's protein-synthesizing machinery the order in which to construct a string of amino acids to make a particular protein.

The genetic code, like the English language, has a bit of flexibility in its usage – there are several ways to say the same thing. There are 64 possible codons ($4 \times 4 \times 4$) but only 20 amino acids, so more than 1 codon may code for a single amino acid. For example, CGC, CGA and CGG all code for arginine.

The information for making proteins is not handed directly from the DNA to the cell's protein manufacturers. Rather, there is an intermediary called messenger ribonucleic acid (mRNA). RNA is very similar to a single strand of DNA, except that it has a different base in place of thymine, called uracil (abbreviated to U). The RNA molecule is a copy of the protein-coding information of a gene, and is copied

> ## "Most of biology happens at the protein level, not the DNA level."
>
> *Craig Venter, President of Celera Genomics*

from the DNA template by a process known as transcription. Think of it like a photocopier – keeping the master copy safe but reeling out one, ten or a thousand copies of different genes in different cells. The mRNA is then shuttled out of the nucleus to special organelles (called the rough endoplasmic reticulum) where the protein-making machines, which are termed ribosomes, are located. The protein-synthesizing

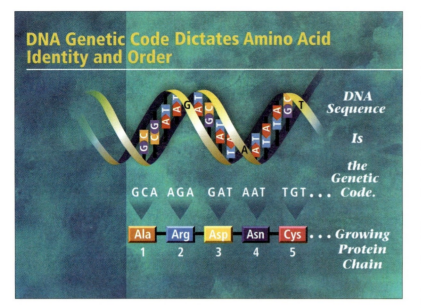

DNA Genetic Code Dictates Amino Acid Identity and Order

DNA Sequence Is the Genetic Code.

GCA AGA GAT AAT TGT...

Ala Arg Asp Asn Cys ... Growing Protein Chain
1   2   3   4   5

The genetic code. The language of DNA and its intermediary message, mRNA, consists of just 4 characters, which are translated to the 20-character language of proteins. The genetic code consists of a linear series of nucleotides read three at a time – these triplets are called 'codons'. Each codon specifies a particular amino acid to be added to the end of the growing protein chain. This process of converting the genetic code into protein is called 'translation'. [DOE Human Genome Program]

machinery translates the codons of the RNA strand into a string of amino acids. This progression of DNA to RNA to protein is known as the central dogma of genetics.

The genetic language is universal; our DNA speaks the same language as the DNA in a plant or a fly. There are slight differences in dialect, but our cells would have no problem in making a protein encoded by a gene from just about any organism on Earth! This is the reason why genetic modification works so well and why the ability to transfer genes between species is now routine in research. A further practical benefit is that if we know the genes of one organism, we can use this information to search for genes in another.

Less than 2 per cent of the human genome is made up of protein-coding sequences. To think of it another way, this means that genetic instructions take up just a few centimetres of the 2-metre strand of DNA that is packed inside every cell. The remainder has been labelled as 'junk' DNA. The function of these vast tracts of non-coding sequence is still largely obscure, but they do seem to contain instructions that help to control what proteins are expressed where, and when.

Personal genomes

Clearly, we are all unique at many levels, including the level of the genome. To understand how we differ from each other, we need to understand how a parent's genome is transmitted to his or her children. Genetic material is passed to the next generation through the sex cells,

sperm and egg. Whereas most cells of the human body possess pairs of each of the 23 chromosomes, the sex cells contain only a single copy. When an egg and sperm unite during fertilization, the resulting embryo inherits half a copy of each parent's genome to restore the normal number of chromosomes.

Sex cells (or gametes) are created by meiosis, a special kind of cell division that only happens in the egg-producing cells of the ovaries and the sperm-producing cells of the testes. The chromosomes are copied as in normal cell division (mitosis), but the cells divide twice rather than just once. This produces four sex cells, each of which contains only one set of the 23 chromosomes.

The chromosomes inherited are not simply standard copies from our parents. Instead, during the first round of division, the pairs of homologous chromosomes swap pieces of their genetic material. Pairs of chromosomes come together, break at identical points along their length, exchange equivalent pieces of DNA,

> "Man with all his noble qualities ... still bears in his bodily frame the indelible stamp of his lowly origin."
>
> *Charles Darwin 1809–82, natural historian;* The Descent of Man *(1871), closing words*

and then rejoin. This shuffling is known as 'recombination', and it produces chromosomes with new combinations of genes – and therefore traits – that may not have been present in either parent. Children thus inherit 'patchwork' chromosomes, consisting of alternate portions of the chromosomes that originated in their grandparents.

Most differences between individual human genomes are very small – a few bases missing here, a few extra inserted there. The overwhelming majority of variation has no effect, as it occurs in parts of the genome that do not code for genes, but occasionally these small changes give rise to altered proteins. The most common variations are single-base substitutions, known as 'single nucleotide polymorphisms' or SNPs (pronounced 'snips'). An example of a SNP that could affect the structure of a protein is a change from A to C in the sequence CAT. CAT encodes the amino acid histidine, whereas CCT gives proline.

How do such changes in the human genome sequence come about? Well, nobody – and nothing in biology – is perfect. Although the cell generally copies its DNA accurately, occasionally a wrong base is inserted into the new DNA sequence, or some bases are inadvertently skipped or added. This slip-up in DNA copying can occur in the sex-producing cells, and hence be inherited by a child and passed down through subsequent generations. Alternatively, genetic changes can occur during division of normal cells of the body and are not passed on to the next generation – these are known as 'somatic' changes.

Our genomes are therefore a combination of old and new changes. Cumulative alterations can be passed down through the generations, which is why we are genetically more similar to our relatives. Changes that occurred in distant human ancestors are likely to be present in many groups of people around the world, whereas recent changes will be found in more localized populations.

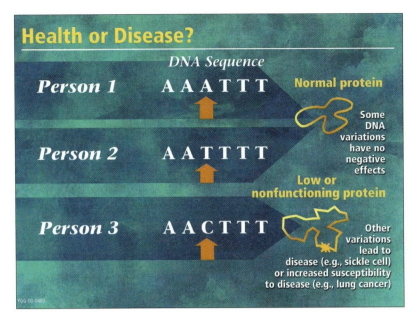

Health or Disease?

DNA Sequence

Person 1 A A A T T — **Normal protein**

Person 2 A A T T T

Person 3 A A C T T

Some DNA variations have no negative effects

Low or nonfunctioning protein

Other variations lead to disease (e.g., sickle cell) or increased susceptibility to disease (e.g., lung cancer)

YGG-00-0480

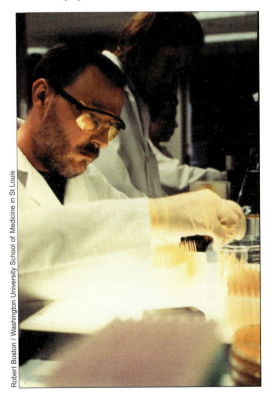

Robert Boston / Washington University School of Medicine in St Louis

Genetic variation and disease

Genetic variation is a double-edged sword. Changes in the genome sequence create diversity and allow any population, including humans, to have a broader range of attributes that help it adapt to changing environments. Some genetic changes, or 'mutations', however, can be disadvantageous in that they cause disease.

Errors in single genes are responsible for more than 4000 hereditary diseases, including cystic fibrosis, Tay-Sachs disease, sickle-cell anaemia and Duchenne muscular dystrophy. Most hereditary diseases are rare, but collectively they affect a substantial proportion of the population. You probably know or have heard about someone who is affected by an inherited genetic disease. Stephen Hawking, one of the world's great cosmologists, suffers from amyotrophic lateral sclerosis (ALS), the hereditary disease that took the lives of baseball player Lou Gehrig and actor David Niven. Huntington disease claimed the life of songwriter Woody Guthrie. But genetic disease takes its heaviest toll among the young, causing a substantial percentage of all infant mortalities.

Our chromosomes exist in pairs and we have two copies of most genes (the exception being some genes on the sex chromosomes; males

Genetic variation. Most variation in the human genome arises from substitutions of individual nucleotides, called single-nucleotide polymorphisms (SNPs), which are like typographical errors. In the example illustrated here, A, C or T can occupy the same position in the genome sequence. Fortunately, most of the genetic changes (or 'mutations') are harmless because they do not occur in parts of the DNA that contain instructions for making a protein. Sometimes, however, a mutation in a gene can affect the structure or function of the protein encoded by the gene, and this can alter how cells behave in the body and lead to disease. [DOE Human Genome Program]

"Now, for the first time, we have an historical anthology of ourselves, some of it passed down for a billion years. We're just learning how to read the story, and it's sure to enthral us for decades to come."

Eric Lander, Whitehead Institute

have a single X chromosome so they will have only one copy of some X-linked genes where there is no equivalent counterpart on the Y chromosome). Alternative forms of our genes are called 'alleles', and a single allele of each gene is inherited separately from each parent. Different alleles produce variation in inherited characteristics such as hair colour and blood type.

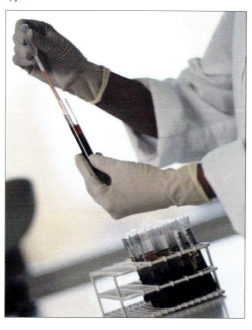

"It would surprise me enormously if in 20 years the treatment of cancer had not been transformed."

Mike Stratton, Cancer Genome Project

When a mutation inactivates one copy of a gene, the second copy can sometimes act as a back-up and compensate for the deficiency. So some genetic diseases do not develop unless a person has two defective copies of the gene – such diseases are termed 'recessive'. Cystic fibrosis is an example of a recessive trait; it arises from a mutation in the gene encoding a protein that serves as a channel through which chloride ions enter and leave cells. About 1 in 20 people carries the cystic-fibrosis mutation in one copy of the gene. It is only when two defective copies

are inherited that the disease develops. For other conditions, however, a change in just one copy of a gene is enough to cause disease. This form of inheritance is called 'dominant' because the flawed copy dominates the normal copy of the gene. An example is Huntington disease, which is caused by an unusual repetition of a triplet of bases, CAG, in the gene that encodes the protein huntingtin (the normal function of which is still unclear). In unaffected people CAG may be repeated 11–34 times, but in an affected patient the triplet is repeated many more times, sometimes more than 80. This results in the production of an abnormal huntingtin protein, which disrupts the function of nerve cells.

The examples given above are inherited genetic diseases, in which a genetic mutation is passed from parents to offspring through the parents' sex cells. Genetic changes in somatic cells – that is, normal cells of the body that are not passed on to the next generation – can also cause disease. Cancer, for the most part, is caused by somatic mutations that result in uncontrolled cell growth and cell division. Cancer typically arises from the accumulation of mutations in not one but several genes. It starts with a single mutation in a cell that is transmitted to all its daughter cells through mitosis. Some divisions later, a second mutation occurs which is also transmitted to daughter cells, and so on. Cancer-causing mutations typically give a cell a 'growth advantage' over normal cells, enabling it to divide more quickly. Thus, the disease often arises from mutations in three types of gene: 'oncogenes', which promote cell growth and division; 'tumour-suppressor genes', which keep cell division in check; and DNA-repair genes, which fix damaged genes.

Knowing the complete sequence of human DNA has transformed the process of identifying genes that are responsible for hereditary diseases. And now that the catalogue of human variation is rapidly being compiled, important studies are being undertaken to link the common complex diseases with underlying genome variation. The revolution is under way.

"This is the outstanding achievement, not only of our lifetime, but in terms of human history. I say this because the Human Genome Project does have the potential to impact on the life of every person on this planet."

Michael Dexter, Wellcome Trust

The art of DNA sequencing

The Human Genome Project (HGP) has often been compared to the space programme. Both have attracted much public and media attention; and both have been characterized by curiosity, a spirit of adventure and a high degree of interdisciplinary collaboration, as well as by competition and controversy. Their respective end products – the sequencing of the human genome and the landing of people on the moon – are landmark achievements in science and technology. Just as our admiration for space exploration is deepened by an awareness of the hurdles that were overcome, so an understanding of the technical challenges of genome sequencing gives a better appreciation of the venture into our molecular selves. This chapter explains how the human genome was sequenced, recounts the historic breakthroughs that made it possible and describes the technical developments that accelerated the project towards its goal.

The recipe for genome sequencing

DNA sequencing is the term used to describe the laboratory process of reading the order of the four letters of the genetic alphabet (A, C, G and T) along a strand of DNA. Sequencing a genome therefore means reading every single letter of an organism's DNA. Let's start with an overview of how a genome is sequenced:

Step 1 – select your organism.

Step 2 – isolate the DNA from cells and prepare large samples of high quality.

Step 3 – cut the purified DNA at random into manageably sized, overlapping pieces.

Step 4 – insert the DNA pieces into packages for the production of limitless copies (or 'clones').

Step 5 – read the order of bases for each DNA piece.

Step 6 – determine the overlap of each piece and assemble the sequences to give the final genome sequence.

Reading the genetic script

There are a couple of different approaches to DNA sequencing, the most popular of which is the 'chain-termination method'. This takes advantage of a naturally occurring DNA polymerase, a protein that plays a central role in replicating the genetic information of a cell. This process is essential as it ensures that every time a cell divides, both new cells contain their own set of DNA. DNA polymerase molecules attach to both strands and use them as templates to string together new, complementary strands from single nucleotides. The result is two sets of double-stranded DNA – one set for each new cell. This DNA-copying process is astonishingly accurate and the strict rules of base pairing – that A pairs only with T, and G with C – ensure that two exact copies are replicated from the parent cell.

To sequence DNA, this copying process is re-created in the laboratory, with a twist. Along with all of the ingredients needed to make new DNA – which include a DNA template to be copied, DNA polymerase and single A, C, G and T nucleotides – modified versions of the nucleotides known as 'terminator nucleotides' are also included. DNA polymerase can add terminator nucleotides to the new growing strand of DNA, after which no further nucleotides can be added, so the polymerase stops in its tracks. As the addition of a terminator nucleotide (instead of a normal nucleotide) is a random process, the result is a collection of incompletely copied fragments that all start at the same point but differ in length by a single base, depending on how far the polymerase got

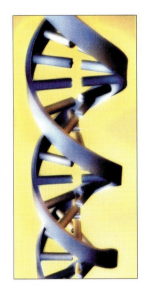

How to sequence DNA

a) DNA polymerase copies a strand of DNA.

b) The insertion of a terminator base into the growing strand halts the copying process. This is a random event that results in a series of fragments of different lengths, depending on the base at which the copying stopped. The fragments are separated by size by running them through a gel matrix, with the shortest fragments at the bottom, largest at the top.

c) The terminators are labelled with different fluorescent dyes, so each fragment will fluoresce a particular colour depending on whether it ends with a A, C, G or T.

d) The sequence is 'read' by a computer. It generates a 'sequence trace', as shown here, with the coloured hillocks corresponding to fluorescent bands read from bottom to top of one lane of the gel. The computer translates these fluorescent signals to DNA sequence as illustrated across the top of the plot.

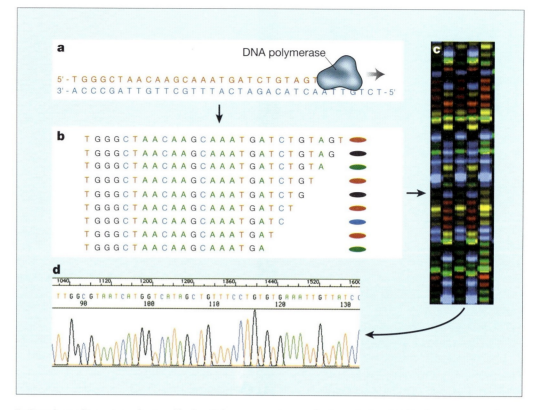

before inserting a terminator. Each of the terminator bases can be labelled with differently coloured dyes (for example, green, blue, yellow or red), which is important for distinguishing the fragments later.

Fragments of DNA can be separated according to their size. This is achieved by propelling the DNA through either a very thin sheet or a tiny capillary of acrylamide gel, a substance that is like firm jelly (or Jell-O). After the DNA samples are put onto the gel, an electric current is applied and the DNA fragments migrate towards the positive pole because the DNA itself has a negative charge – this process is known as gel electrophoresis. The gel is a densely packed meshwork through which the shorter fragments can move faster than the longer ones. After a few hours, this creates a ladder of DNA fragments spaced out along the gel, with the shortest at the bottom. Moving up the gel, each fragment is bigger than the previous one by just a single base.

How does this ladder of fragments translate into the sequence? Each fragment is tagged at its end with one of the four terminators labelled with different dyes. Gels used for sequencing are mounted onto specialized equipment and, as the fragments move through the gel, they pass a laser beam at a set point, which excites the dye, causing it to fluoresce in its particular colour. This fluorescent signal is captured by detectors and displayed as a coloured band on a computer monitor – the pattern thus looks like a 'ladder' of red, green, yellow and blue bands. The computer interprets the two pieces of information that it has for each band: the position (which indicates its length) and the colour (which identifies the base at the end of that fragment) and converts this information into the order of bases of the sequence.

Gel electrophoresis is limited by the number of bands that it can clearly resolve. For any sequencing reaction only about 500–800 bases can be read, so the DNA must first be cut into tiny pieces to be sequenced.

Getting automated – fast, faster, fastest

Technical wizardry, combined with intensive automation, has transformed the pace of DNA sequencing. Robots and computers are now used in every aspect of the process: performing the sequencing reactions, transferring the reactions onto gels, separating the fragments by electrophoresis, and reading the order of bases. Work that once took many days and many people can now be done in minutes, and a few staff members can keep the whole operation ticking 24 hours a day.

This is a far cry from the mid-1980s, when state-of-the-art laboratories could sequence only about 500 bases a day. Then Leroy Hood and his colleagues began to automate the process – their first sequencing machine could read up to 15 000 bases in a single day. The DNA-sequencing machines of today are even more powerful, and some can accurately sequence up to 400 000 bases per day. This has led to spectacu-

Wellcome Photo Library

The origins of sequencing

Fred Sanger (left) and Walter Gilbert (below) shared the 1980 Nobel Prize for chemistry for their independently developed methods of sequencing DNA. For Sanger this was a second Nobel Prize, a feat achieved by only four people; his first was awarded in 1958 for work on the structure of proteins, in particular insulin. Sanger's chain-termination method (also known as the 'dideoxy' method after the chemistry of the terminator bases used in the reaction) is the most widely used today. Gilbert, together with Allan Maxam, developed an alternative strategy that uses chemical agents to cleave DNA specifically at A, C, G or T bases. The fragments, which are radioactively labelled, are separated by size on gels and exposed to film, and the resulting ladder is used to determine the sequence.

The Sanger Centre, near Cambridge, UK, is named after Fred Sanger. It is one of the largest sequencing centres in the world and deciphered roughly one-third of the human genome.

Walter Gilbert is notable for his entrepreneurial spirit, having launched several biotechnology companies. He also proposed the sequencing of the human genome as a commercial venture – more than ten years before the company Celera Genomics pursued this very goal.

lar progress: researchers of the HGP sequenced 90 per cent of the human genome in just 15 months.

The technology revolution triggered dramatic growth. Over the past decade the amount of

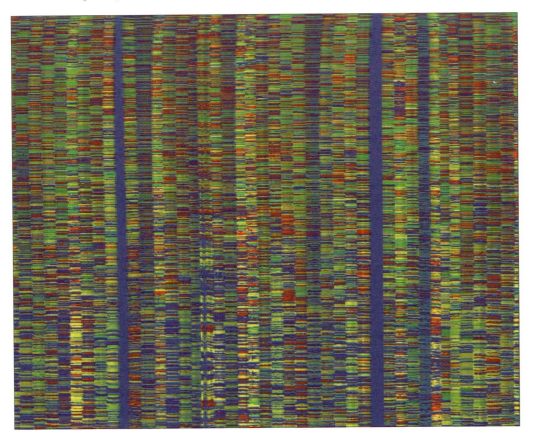

The sequence rainbow. The output from an automated DNA sequencing machine used to determine the complete human genome sequence. Each vertical lane shows the sequence of bases in a given stretch of DNA. Each of the four different bases is labelled with one of the four coloured dyes. The order of the bases is analysed by a computer and assembled to give the continuous base sequence for a segment of the genome. [Sanger Centre/Wellcome Photo Library]

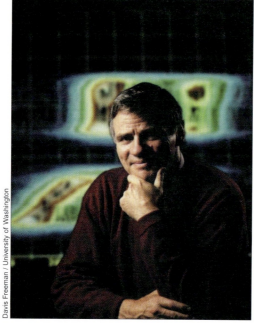
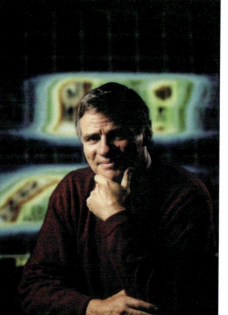

Davis Freeman / University of Washington

Those magnificent men and their sequencing machines

The first automated DNA-sequencing machine was invented by Leroy Hood (above), Lloyd Smith and Mike Hunkapiller (below) in 1985. They also improved Sanger's sequencing method by replacing unstable and hazardous radioisotope tags with fluorescent dyes of different colours to tag each of the four DNA terminator bases. Hood and Hunkapiller (and several others) set up Applied Biosystems Incorporated to market the gene-sequencing machine the following year.

In 1998, Hunkapiller's group at PE Biosystems developed a new sequencer, the ABI Prism 3700 DNA Analyzer, which was faster and even more automated than previous sequencers.

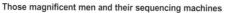
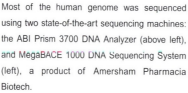
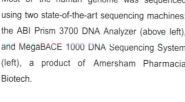

[Both images: Dr Eric Green National Human Genome Research Institute, NIH]

Most of the human genome was sequenced using two state-of-the-art sequencing machines: the ABI Prism 3700 DNA Analyzer (above left), and MegaBACE 1000 DNA Sequencing System (left), a product of Amersham Pharmacia Biotech.

sequence data collected worldwide has doubled every 18 months, while the cost of sequencing has halved. Thus sequencing volume is increasing, and the cost of sequencing is coming down by a factor of 10 every five years. This is reminiscent of the doubling of microprocessor speed every 18 months, which in the computing industry is known as Moore's law (named after Gordon Moore, co-founder of Intel), despite the fact that it's an observation rather than an actual law. The cost of sequencing has come down, from more than US\$10 per base in the late 1980s to about 10 cents per base in 2001. These trends have not just accelerated the sequencing of the human genome; they have also fuelled genome projects for a wide range of animal, plant and microbial species.

Inside the molecular biologist's toolbox

Sequencing requires short stretches of DNA, so genomes must be cut into very small pieces, which are then copied in sufficient amounts for sequencing. Here we look at the tools required to chop up the genome and copy the pieces.

As scissors to clip DNA molecules at specific sites, scientists use proteins called 'restriction enzymes'. These proteins are produced naturally by bacteria as a defence against viral infection. Bacterial DNA is protected by a process known as methylation so that only foreign (viral) DNA is affected – chopping up the viral genome halts the infection. More than 3000 such enzymes have been isolated from different bacteria, each of which recognizes short stretches of DNA, typically 4–8 bases long. By using combinations of restriction enzymes, scientists can tailor the size of the DNA fragments that they desire.

The next step is to produce the large quantities of genomic fragments required for sequencing. This involves making copies (or 'clones') of the fragments. Most restriction enzymes don't leave nice, neat edges as they slice through double-stranded DNA. Instead, a few unpaired bases trail from one of the strands; the sequence of this overhang is specific to the particular restriction enzyme used. These bases can attach to complementary overhangs on other DNA molecules cut by the same enzyme, like flaps of Velcro sticking together. This allows the creation of new combinations, known as 'recombinant' molecules. An enzyme that welds the ends together – a so-called 'DNA ligase' – is used to prevent the recombinant molecules from becoming unfastened.

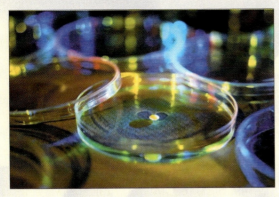

Tools of the trade. In 1970, Hamilton Smith purified the first restriction enzyme, a molecule that cuts DNA at specific sites. Smith shared a Nobel Prize in 1978 with Werner Arber and Daniel Nathans for the discovery of restriction enzymes and their application to molecular biology. Their work added 'molecular scissors' to the toolkit for manipulating DNA and paved the way for the development of genetic engineering.

In 1972, Paul Berg and Herbert Boyer cut and pasted together two DNA strands to create the first recombinant DNA molecule. The following year, Boyer, together with Stanley Cohen, created the first recombinant organism by splicing together sections of viral and bacterial DNA and transferring the resulting recombinant DNA molecule into a bacterial host. On the heels of this discovery, they cloned the first animal gene. They fused a frog gene that encodes ribosomal RNA (part of the cell's protein-synthesizing machinery) to DNA from a bacterium and put the recombinant DNA molecule back into a bacterial cell, where it produced the frog ribosomal RNA.

In 1977, a human gene was cloned by the first genetic engineering company, Genentech, which was co-founded by Herbert Boyer. The gene encoding somatostatin (an inhibitor of human growth hormone) was inserted into bacteria, which began making somatostatin. This was the first time a human protein had ever been produced by a living creature outside the human body. The following year, Genentech cloned human insulin which, when marketed in 1992, became the first recombinant-DNA drug.

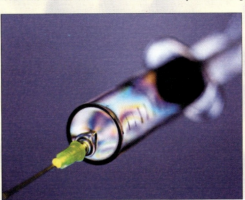

This approach is used to insert (or 'splice') the genomic DNA of interest into bacterial plasmids. Plasmids are tiny rings of DNA found inside bacteria that are copied independently of the bacterium's main genome. They can be manipulated in the laboratory and are easily isolated. The recombinant plasmid, bearing the inserted genomic DNA, is returned to its bacterial host, where it is copied as the bacterium divides. Thus, the bacteria can be thought of as tiny biological 'factories', making endless clones of the genomic DNA fragment.

The creation of recombinant molecules and their cloning into micro-organisms are much used techniques in molecular biology. The cloning of a DNA fragment should not be confused with the cloning of animals, such as Dolly, the famous cloned sheep. Cloning in molecular biology is simply the use of micro-organisms to reproduce limitless quantities of a DNA fragment – a sort of biological photocopying. The organisms used are usually bacteria, particularly the species *Escherichia coli*, which is a favourite workhorse of biologists.

Plasmids can comfortably carry small inserts of DNA but, as we shall see, bigger fragments are required for some purposes. To hold these, another type of package (or vector) called an artificial chromosome can be used. These can be inserted into bacteria or yeast, where they are replicated in the same way as normal chromosomes and are therefore termed bacterial or yeast artificial chromosomes (BACs or YACs) respectively. Inserts of more than 1 000 000 base pairs of genomic DNA can be carried by YACs, but these can be unstable, so BACs that hold up to 300 000 base pairs are more commonly used.

To ensure that the whole genome of the organism is sequenced, researchers produce DNA libraries. A DNA library is a collection of individually cloned fragments which together constitute the entire genome of the organism. These have a role analogous to that of conventional book libraries – storing an entire collection of information in a convenient, well-catalogued set of packages as a resource that can be used by anyone in the community.

Constructing an overlapping clone library. A collection of clones of chromosomal DNA, called a library, has no obvious order indicating the original positions of the cloned pieces on the uncut chromosome. To establish the order of cloned fragments, libraries of clones containing partly overlapping regions must be constructed. These clone libraries are ordered by dividing the inserts into smaller fragments and determining which clones share common DNA sequences.

Importantly, because of the way in which the cloned fragments are originally generated, many of the clones are partially overlapping so that the information is found in many entries.

Sequencing many overlapping entries might sound like a waste of time, but it actually serves two purposes: it ensures high accuracy in the final sequence, and the overlaps allow the fragment information to be reassembled into a single, complete set of information that represents the entire genome.

Piecing together the genome puzzle

Sequencing a genome is like doing an enormous jigsaw puzzle. First, the genome is cut into small pieces; next, the individual pieces are sequenced; and, finally, the pieces are put back together in the correct order. As noted above, the reason that the pieces can be put back together is that they are partially overlapping, allowing one fragment to be exactly fitted to the correct neighbour, just as the unique curves of jigsaw-puzzle pieces allow only one possible assembly of the puzzle.

The human genome poses a particularly daunting brainteaser. At roughly 3.2 gigabases (3 200 000 000 bases) it is 25 times larger than any previously sequenced genome; indeed, it is bigger than the combined number of bases sequenced in all previous genome projects by a factor of 8. Moreover, the abundance of DNA sequences that are repeated many times over – comprising over half of the genome – also poses a significant technical challenge, as one region could easily be mistaken for another when the sequence is assembled.

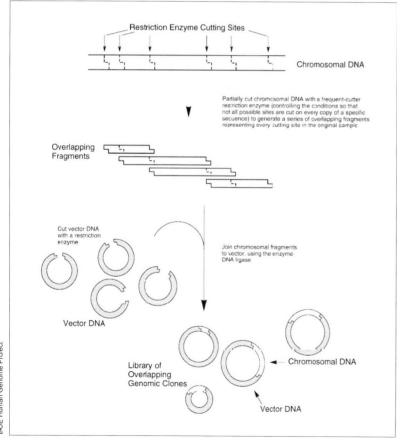

➤32

Making maps

Archival photograph by Mr Steve Nicklas, NOS, NGS / Treasures of the NOAA Library Collection

Genome maps are constructed by determining the order of thousands of landmarks scattered across the DNA, and are usually needed before embarking on the sequencing of the genome. To explain mapping, let's try an analogy. Suppose that sequencing of the human genome is like flying across a continent, taking photographs that each cover 10 square miles. Assembly is equivalent to piecing these photographs together to get an accurate picture of the continent as a whole. Some landscapes and cityscapes would be easy to piece together because they contain recognizable landmarks – think of the Eiffel Tower, the Grand Canyon or the Great Pyramids – but others would be extremely difficult, such as deserts, featureless farmland or miles of suburbia. But imagine that before taking the photographs, every sector was labelled with a distinctive landmark. Now entire continents could be pieced together relatively easily. Different types of landmarks (or markers) can be used in genome mapping, but they must be unique sites that identify specific coordinates on a chromosome.

Aiding the map-making process, the human genome already naturally exists in smaller segments, namely the chromosomes. We have 22 pairs plus either two X chromosomes (in the case of females) or an X and a Y (for males). Each chromosome can be broken down into smaller maps that vary in the level of resolution, the type of marker used or the way in which the distances between markers are measured. Genome maps can be broadly classified as either genetic or physical.

During the formation of the sex cells, pairs of chromosomes break at specific sites and exchange equivalent pieces of DNA. If two sites on the DNA are close together, then the chance of their being separated and reshuffled during the formation of the sex cells is less than if they were far apart. This provides the conceptual basis for genetic (or linkage) maps. Such maps determine the order of specific DNA markers along a chromosome by examining how they are inherited through generations. Genetic maps only give the relative positions of markers, not their physical locations in the genome or actual distances, but their great power is in identifying genes that underlie physiological or morphological traits. They have proved extremely useful in the identification of important disease genes, including those involved in cystic fibrosis and sickle-cell anaemia.

Physical maps, on the other hand, determine the precise physical locations of landmarks in the genome; the distance from one landmark to the next is measured in base pairs. Some landmarks are variations in the DNA sequence that can be distinguished by whether they are cleaved by specific restriction enzymes – such landmarks are called 'restriction-fragment-length polymorphisms' (RFLPs). Other landmarks are short, repeated sequences that vary in the number of repeated units, and are known as 'variable number of tandem repeats' (VNTRs). Modern maps typically use short sequences of about 200–500 base pairs that occur only once in the genome; these are called 'sequence-tagged sites' (STSs).

The Human Genome Project

What is the Human Genome Project?

The Human Genome Project (HGP) is an international research programme that was set up to characterize the genomes of humans and other organisms; to develop the new technology needed to do so; and to address the ethical, legal and social implications of this new information. All the data produced are widely and freely available.

When did it start? The possibility of sequencing the human genome was first discussed in the mid-1980s. The HGP officially started in 1990 with a 15-year plan to map and sequence the human genome.

What was the plan? The HGP was launched with the aim of developing genomic tools and resources, and of mapping and sequencing the human and other genomes. Throughout the project, maps of human chromosomes have been made; sequencing technology has been improved; computational tools have been designed; and strategies for collecting, analysing and storing data have been developed. The study of the effect on society of large amounts of new genetic knowledge and technology, and the recommendation of policy to maximize the benefits and minimize the risks, have been an integral part of the HGP since its inception.

Initial efforts were focused on the genomes of laboratory organisms such as yeast (*Saccharomyces cerevisiae*) and the

Who is sequencing the human genome?

The group of scientists within the HGP that is sequencing the human genome is known as the International Human Genome Sequencing Consortium (IHGSC). It is composed of more than 2000 scientists at 20 institutions in 6 countries. The institutes are listed below in order of total genomic sequence contributed, with the first five centres producing the bulk (about 85%) of the sequence.

1. Whitehead Institute for Biomedical Research, Center for Genome Research, Cambridge, Massachusetts, USA
2. The Sanger Centre, Cambridge, UK
3. Washington University Genome Sequencing Center, St Louis, Missouri, USA
4. US Department of Energy Joint Genome Institute, Walnut Creek, California, USA
5. Baylor College of Medicine Human Genome Sequencing Center, Houston, Texas, USA
6. RIKEN Genomic Sciences Center, Yokohama, Japan
7. Genoscope and CNRS UMR-8030, Evry, France
8. GTC Sequencing Center, Waltham, Massachusetts, USA
9. Department of Genome Analysis, Institute of Molecular Biotechnology, Jena, Germany
10. Beijing Genomics Institute/Human Genome Center, Beijing, China
11. Multimegabase Sequencing Center, Institute for Systems Biology, Seattle, Washington, USA
12. Stanford Genome Technology Center, Stanford, California, USA
13. Stanford Human Genome Center, Stanford, California, USA
14. University of Washington Genome Center, Seattle, Washington, USA
15. Department of Molecular Biology, Keio University School of Medicine, Tokyo, Japan
16. University of Texas Southwestern Medical Center, Dallas, Texas, USA
17. University of Oklahoma Advanced Center for Genome Technology, Norman, Oklahoma, USA
18. Max Planck Institute for Molecular Genetics, Berlin, Germany
19. Cold Spring Harbor Laboratory, Lita

What is the 'working draft'?

The working draft is an intermediate stage in the generation of the true goal of the HGP, a high-quality, 'finished' sequence of the human genome. In the working draft, each letter of the genetic code has been read at least four to five times (4–5× coverage). Although the draft sequence is very useful, it still has some gaps and ambiguities. The ultimate goal of the HGP is to produce a completely finished sequence, with no gaps and at least 99.99 per cent accuracy (which will require about 9× coverage). A finished, high-quality sequence is expected to be completed by the end of year 2003, two years earlier than originally planned.

Why 'finish' the sequence? The working draft has allowed a first view of the whole human genome. But a completely accurate, finished sequence is needed to compile a comprehensive inventory of genes and to identify the sequences involved in control mechanisms, which are embedded in the vast stretches of non-coding regions between genes.

An accurate reference sequence will also help to determine the variation that exists between the DNA of different individuals, how the sequence varies in distinct populations around the world, and how our DNA is different from that of our ancestors and closest relatives in the animal kingdom.

How fast do HGP scientists sequence? Twenty years ago, deciphering 12 000 bases would have taken a year or more. Three years ago, when pilot sequencing projects to evaluate the feasibility of human DNA sequencing were initiated, sequencing 12 000 bases took 20 minutes. Today, the consortium churns out 1000 bases of raw sequence per second, or 12 000 bases of 'working draft' per minute.

In 1990, when the HGP began, DNA sequencing probably cost at least US$10 per base. The process has been streamlined and automated to read sequences faster and more cheaply, reducing the cost to roughly 10 cents per base.

Whose DNA is being sequenced? The human reference sequence does not correspond to any one person's genome. The sequence is derived from the DNA of a large number of volunteers, who responded to local public advertisements near the laboratories where the DNA 'libraries' were being prepared. Researchers collected blood (male and female) or sperm samples (male only, obviously) from a large number of donors from diverse populations. About five to ten times as many volunteers donated samples as were eventually used and all identification was removed before samples were selected, so neither donors nor scientists know whose DNA has been sequenced.

Who has access to the human DNA sequence? Anyone who can log on to the Internet has full access to the sequence data held in public databases. Sequence assemblies of 1000–2000 bases are deposited into the databases within 24 hours of completion, and the data can be accessed without any restriction. The code of practice for rapid data release was agreed upon at a meeting of the HGP held in Bermuda in February 1996 and is known as the 'Bermuda principles'.

roundworm (*Caenorhabditis elegans*). These projects helped to refine DNA-sequencing technology and the conceptual strategies needed to tackle more complex genomes, and at the same time provided crucial information for interpreting the human sequence. Only a very small amount – as little as 1.5 per cent – of the human genome encodes proteins. Comparing human DNA with that of other organisms helps to identify human genes amongst the sea of sequences and also provides clues as to their function.

In 1996 a series of pilot sequencing projects was initiated to test the feasibility of deciphering the human genome. Within a couple of years, these efforts were deemed successful, and the large-scale (or 'high-throughput') effort to sequence the human genome was launched in March 1999. Around this time, the HGP adopted a strategy that would first generate a 'working draft', so that scientists would have useful data covering the entire genome as soon as possible, rather than having to wait for the sequence to be 'finished'.

With the infrastructure and resources in place, productivity skyrocketed, and more than 90 per cent of the sequence was produced in just 15 months. In February 2001, HGP scientists published the sequence, assembly and analysis of the working draft of the human genome.

Annenberg Hazen Genome Center, New York, USA

20. GBF German Research Centre for Biotechnology, Braunschweig, Germany

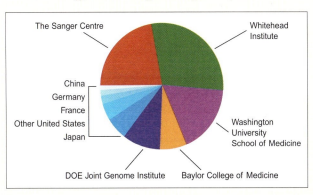

The Sanger Centre — Whitehead Institute — China — Germany — France — Other United States — Japan — DOE Joint Genome Institute — Baylor College of Medicine — Washington University School of Medicine

How much does it cost, and who pays?
About US$300 million has been spent world-wide to generate the draft sequence. The 15-year HGP effort has a total projected funding of US$3 billion (1990–2005), which encompasses a wide range of scientific activities related to genomics – the sequencing of the human genome represents only a fraction of the overall effort. Investment has been made in developing the technology and creating the resources to make large-scale sequencing feasible, allowing the sequencing of genomes of other organisms, the training of scientists, and the examination of the societal implications of genome research.

The project is funded by grants from government agencies and public charities in the various countries. These include the US National Institutes of Health (NIH), the US Department of Energy (DOE), the Wellcome Trust in the UK and the UK Medical Research Council, as well as agencies in Japan, France, Germany and China.

Where is it stored? There are international public nucleotide-sequence databases in the United States (GenBank), Europe (the European Bioinformatics Institute) and Japan (the DNA Database of Japan). These databases exchange information daily, so that all sequence information is available from any of the three.

As well as sequence data for the human genome, these databases contain the complete genome sequences of 'model' organisms (those most frequently used in the laboratory) such as yeast, worm, fruit fly and many microbes, as well as DNA sequences from other genomes that are currently being sequenced. Scientists in their thousands access the databases every day.

Links to the databases as well as other human-genome resource sites can be found at:

http://www.ensembl.org/genome/central
and
http://www.ncbi.nlm.nih.gov/genome/central

Also see Box 2 on page 88

There are two basic approaches to solving the puzzle of genome sequencing. One is the 'map-based' or 'clone-by-clone' method. This is a hierarchical process in which the genome is broken up into progressively smaller segments. The sequence of the smallest pieces is determined and the pieces are then reassembled into progressively larger segments until the entire genome has been pieced back together. The first set of large segments is used to build a physical map, which determines the origin of each segment within the genome and the position of each of the segments relative to all the others. Then each of these segments is diced into small,

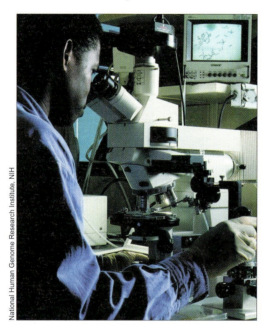

National Human Genome Research Institute, NIH

overlapping pieces and sequenced, a process that has been termed 'shotgun sequencing'. Once all the sequences have been determined, the overlaps are used to reconstruct the continuous sequence of the large segment and these are then assembled to give the whole genome. The overall approach is also known as the hierarchical shotgun method.

An alternative strategy is to shatter the entire genome and sequence all of the tiny pieces directly, completely bypassing the construction of a physical map. Computer programs assemble the millions of sequenced fragments into continuous stretches, ultimately reassembling the full genome. These programs consist of a staggered series of mathematical steps that sort, edit and assemble fragments, and are typically designed so that the easier steps are done first, followed by the harder moves. This

strategy is called the whole-genome shotgun method.

How do the two methods compare, using the jigsaw-puzzle analogy? Blank side up, all the pieces look roughly the same. Turning them over (sequencing) shows you the picture on each piece. After this, they can be assembled in two different ways. One is by comparing the pieces with the picture on the box; for example by gathering and piecing together all the blue and white bits that make up the sky. This is equivalent to a map-based method, with the picture serving the function of the map. The alternative is to take each piece individually and figure out which other pieces or clusters it fits with. This is analogous to the whole-genome shotgun method.

Both approaches have been successfully. A map-based method was used to sequence the first eukaryotic genome, the baker's yeast *Saccharomyces cerevisiae*, and for the first animal genome, the roundworm *Caenorhabditis elegans*. The first free-living microbe to be sequenced, the bacterium *Haemophilus influenzae*, was sequenced using a whole-genome shotgun method, as was the genome of the fruit fly *Drosophila melanogaster*.

The two methods each have their advantages and disadvantages. The map-based approach is reliable, and the maps have other uses besides sequence assembly. Much of the sequencing technology in use today was developed in laboratories that used this method in the late 1980s and 1990s, providing a valuable foundation for much of the work on the human genome. But the construction of maps is time consuming, particularly for a genome the size of a human's. On the other hand, whole-genome shotgunning is (potentially) very fast, but the reconstruction of a genome from so many tiny pieces poses obvious challenges.

Which method was used on the human genome? The answer is both. Researchers of the HGP, on which we concentrate here, primarily used the map-based approach, constructing a physical map of the whole genome that was used to assemble the sequence. However, whole-genome shotgun data were an important aspect of their strategy. And whole-genome shotgunning was the primary technique used by the company Celera Genomics, although its assembly also incorporated map-based data. In other words, as with solving jigsaw puzzles, a mixed strategy can often be best.

Celera Genomics

WHAT IS CELERA?

As indicated by its name (meaning 'swift' in Latin) and its slogan 'Discovery can't wait', Celera is very much about speed. Launched in 1998 in Rockville, Maryland, with J. Craig Venter at the helm, the company set out with an ambitious plan to sequence the human genome in just three years.

Celera's first major project was the sequencing of the fruit fly *Drosophila melanogaster* genome. Upon successful completion, sequencing of the human genome began in September 1999. The sequence was generated over nine months and, together with the sequence data made freely available by the Human Genome Project (HGP), Celera assembled the genome. In a joint announcement on 26 June 2000, the HGP and Celera celebrated their respective first assemblies of the human genome. Celera went on to publish its initial analysis in the 16 February 2001 issue of *Science*, the same week that the public effort reported their findings in *Nature*.

WHAT WAS CELERA'S APPROACH?

Celera used whole-genome shotgunning (WGS) to sequence the human genome. Chromosomes were randomly sheared into millions of pieces of 2 000 and 10 000 base pairs in length. The pieces were inserted into plasmid vectors (to create DNA 'libraries') and propagated in *E. coli* to produce millions of copies of each fragment. A key feature of Celera's approach is that both ends of each fragment of DNA are sequenced (called 'paired-end sequencing') to help determine the order and orientation of the fragments during assembly.

Combining Celera's sequence and the HGP's data, computer algorithms assembled the millions of sequenced fragments into a continuous stretch of the human genome. Two methods of

assembly were used: a whole-genome assembly algorithm and compartmentalized shotgun assembly (CSA), whereby the genome data are divided into segments or compartments, which are then assembled. The CSA assembly was used for the annotation reported in *Science*. Comparing the two assemblies served as an accuracy check of the assembly process.

WHOM DID CELERA SEQUENCE?

To sequence the human genome, Celera used DNA samples collected from five donors who identified themselves as Hispanic, Asian, African-American or Caucasian. The five people (two men and three women) whose genomes were used were selected from a pool of twenty-one individuals, who volunteered in response to newspaper advertisements and outreach efforts.

SEQUENCING POWER, COMPUTING POWER

The speed of Celera's genomic sequencing is dependent upon a powerful, highly automated sequencing machine, called ABI Prism 3700 DNA Analyzer. With 300 of these sequencers running day and night, it isn't surprising that Celera clocked up an electricity bill of around US$1 million a year!

Computers played a huge role in the genome assembly. Celera relied on high-performance computer technology to manage the more than 80 terabytes of data and to perform what are believed to be some of the most complex computations in the history of supercomputing. The calculation to perform the initial assembly involved 500 million trillion sequence comparisons requiring over 20 000 CPU (central processor unit) hours on Celera's supercomputer. Celera's final assembly computations required 64 gigabytes of memory.

Assembling the human genome

The members of the HGP created a high-resolution map to help to assemble the genome sequence. They derived the physical map using the following strategy. Human DNA was cut into fragments of 100 000–200 000 base pairs using restriction enzymes, after which the fragments were cloned into bacterial artificial chromosomes (BACs) and inserted into bacteria, which stored and replicated the human DNA so that it could be prepared in sufficient quantities for sequencing. A collection of clones containing an entire genome is called a BAC library – it

took about 20 000 BAC clones to store the human genome.

The BAC clones were then 'fingerprinted'; that is, they were cut with a restriction enzyme to generate a unique pattern of fragments. By comparing the fingerprints of different clones, it was possible to identify ones that overlap with each other (because they contain a similar subset of fragments) and thus to order the BAC clones. Each BAC clone was also 'mapped' to determine where its human DNA came from within the set of human chromosomes. This was done by looking for recognizable

Genome Gallery

A selection of notable genomes that have been sequenced.

Φ X 174
(1977) 5386 bp
First genome sequenced, a bacteriophage

Haemophilus influenzae
(1995) 1 830 000 bp
First genome of a free-living organism

Mycoplasma genitalium
(1995) 580 000 bp
Smallest genome of any free-living organism

David Scharf / Science Photo Library

Saccharomyces cerevisiae
(1996) 12 100 000 bp
First genome of a 'eukaryotic' (nucleus-containing) organism, the yeast used by brewers and bakers

Methanococcus jannaschii
(1996) 1 660 000 bp
The first genome from the third kingdom, *Archae*, which comprises microbes that live in harsh environments, for example thermal springs

Escherichia coli
(1997) 4 670 000 bp
Workhorse bacterium for biologists

Helicobacter pylori
(1997) 1 660 000 bp
Bacterium associated with gastric disease

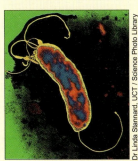

Dr Linda Stannard, UCT / Science Photo Library

Genome Gallery

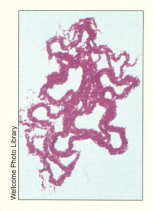

Mycobacterium tuberculosis
(1998) 4 400 000 bp
Cause of the disease
tuberculosis

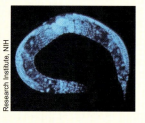

Caenorhabditis elegans
(1998) 97 000 000 bp
The first genome sequence
of an animal, the roundworm

Deinococcus radiodurans
(1999) 2 600 000 bp Highly
radiation-resistant bacterium

First human chromosomes
(1999 and **2000)**
Chromosomes 22,
48 000 000 bp and 21,
45 000 000 bp respectively

Drosophila melanogaster
(2000) 180 000 000 bp
Fruit fly, an important laboratory organism in genetics

genomic landmarks in each BAC clone. Integration of all this information generated a physical map of the genome, from which the precise genomic location of the sequence from each clone, and its spatial relation to the human DNA in other BAC clones, could be determined.

For sequencing, each BAC clone was cut into smaller, overlapping fragments of about 2000 base pairs in length. These were inserted into plasmids and a sequencing reaction was performed at each end of the plasmid clones. The full sequence of the BAC clone was derived by piecing together the sequences from all these tiny fragments, first to form longer-sequence 'contigs' (short for 'contiguous sequences'), and eventually to obtain a single-sequence contig representing the entire human fragment carried by the BAC in question. As the order and overlap of BAC clones were known from the physical map, even longer stretches of sequence could then be assembled. Ultimately, the contigs were fitted together to assemble whole chromosomes and eventually the entire sequence of the human genome.

Going from a sequence of 500 base pairs produced by a sequencing machine to a genome of 3 200 000 000 base pairs is a mind-boggling leap that requires huge computing power. Literally millions of sequence reads are generated, and the data are then fed into 'assembler' programs, which identify and merge overlaps in the sequence.

Quality control is obviously very important. Software has been developed to monitor the quality of the raw sequence data and to assess the likelihood that the base identified by the 'base-calling' program is the correct one. Sequence reads that contain a lot of errors or ambiguities are weeded out before they are fed into the assembly programs. PHRED (pronounced 'Fred') is the standard software program used to assign quality values to each base read by the sequencing machine. The standard computer program that then assembles overlapping sequences is called PHRAP (pronounced 'Frap'). Both of these programs were developed by Phil Green and colleagues at the University of Washington.

Although assembly computer programs can sift through conflicting information and make predictions about which sequence is likely to be correct, the assembly and final polishing are done by humans. Experts, known as finishers, painstakingly identify every gap and ambiguity,

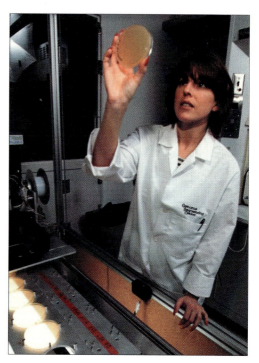

A scientist examines the growth of bacterial colonies that can be used as 'factories' for making many copies of human genomic fragments. [Robert Boston / Washington University School of Medicine in St Louis]

and work out what experiments need to be done to fill in the missing pieces and resolve the discrepancies. The genome sequence must be read many times over to be sure that no base has been missed or misread by the sequencing machines. Some stretches are easy to read and need to be sequenced only a few times to determine the order of the bases accurately. For other regions, the sequence must be read many times to generate a high-quality sequence. The number of times that a genome sequence is re-read is called the 'depth of coverage'.

The goal of the HGP is to generate a high-quality 'finished' sequence for the human genome. The agreed standard for the finished sequence is that each letter should be designated as 99.99 per cent accurate, which means there will be no more than 1 error per 10 000 bases. To achieve this, each base must be read an average of nine times (that is, it will have $9\times$ depth of coverage). This level of accuracy is needed to recognize all genes and their regulatory components, and to detect variation between different human genomes.

The human genome sequence published in February 2001 is a working draft – in other words, it isn't finished yet. In all, it encompasses over 90 per cent of the euchromatic region of the genome – some 2 950 000 000 base pairs – in which most genes are found and where the DNA adopts a relatively 'open' conformation. (Other portions of the genome, known as heterochromatic regions, which include the centres and ends of chromosomes, are considered to be in a 'closed' conformation.) Each base of the working draft has at least 4–5× coverage, and has an accuracy of at least 99.9 per cent. But some regions have been read many more times and, in fact, roughly one-third was already in 'finished' form, including two entire chromosomes (21 and 22), at the time of publication.

The working draft contains many gaps and ambiguities. It's like having a book with some pages missing, some words in the wrong order or on the wrong page, and lots of typographical errors. What causes gaps and misassemblies? Often they occur because the chemical properties of particular stretches of DNA make them harder to clone and sequence. Repetitive regions of DNA, of which the human genome has many, can be especially difficult to handle. They can be troublesome to clone because of their instability; they can be difficult to sequence because they confuse DNA polymerase; and they can complicate the assembly because identical-looking regions can be mistaken for one another.

The human genome sequence will be finished when all the gaps are closed and ambiguities resolved (as far as current technology allows). The goal of the HGP is to produce a finished version of the human genome sequence by the year 2003 – hopefully in time for the 50th anniversary of the discovery of the structure of DNA by Watson and Crick.

Making sense of the sequence

An assembled genome sequence can amount to little more than a long, featureless string of letters. To be of real value it must be 'annotated', meaning that information about genes, the proteins they encode, and other interesting features are ascribed to specific stretches of the sequence. Making sense of a DNA sequence the size of the human genome is a more formidable task than the sequencing itself.

Researchers enlist the power of computers to scrutinize the jumble of bases. These reveal patterns suggestive of genes or other features that help to deduce how the genome is wired. The development of computer-driven methods for extracting biological information from sequences is part of a rapidly expanding field of science known as bioinformatics, which straddles the boundaries of biology, computer science and mathematics. It involves the storage, retrieval, analysis and integration of biological data to identify genes, determine their structures and predict their functions. Analysis of DNA sequences is just one part of the field, which also encompasses analysis of protein structure, gene and protein functions, preclinical and clinical trial information, and studies of metabolic pathways in numerous species.

Finding human genes can be a difficult task. Not only do they make up a tiny percentage of the genome, but they tend to be split into a multitude of small coding 'exons' (the instructions for making proteins), separated by much longer stretches of non-coding sequence called 'introns'. There are several ways to identify genes in a long sequence. One is to look for the tell-tale signatures of gene structures. Another is to look for sequences that match known gene sequences from other organisms on the assumption that, if they have been conserved through evolution, such sequences are likely to be functional. A third approach is to search for evidence that a DNA sequence is copied into messenger RNA, suggesting that the sequence encodes a real gene. Although each of these approaches has limitations, they are powerful in combination.

The 'working draft' provides a good first view of the landscape of the human genome. It allows new genes to be identified and tentative predictions to be made about the functions of genes or families of genes. It can also be used to learn about the types and distribution of repeated elements in the human genome, to characterize duplications and to identify so-called 'pseudogenes', which are genes that no longer function but can provide insights into genetic evolution. At present, the speed at which sequence data are being acquired greatly exceeds our capacity to understand it and to put it into a proper biological context. Furthermore, as with the translation of a vast and complex script in an unknown language, there is potential for multiple interpretations. Researchers have designed annotation programs that can create significantly different genome schematics. The challenge of the future will be to resolve these differences and to refine our view of the genome.

Genome Gallery

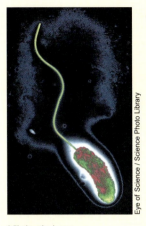

Vibrio cholerae
(2000) 4 030 000 bp
Cause of the disease cholera

Eye of Science / Science Photo Library

Arabidopsis thaliana
(2000) 120 000 000 bp
The first genome of a plant, the mustard weed

Images courtesy of J. Berger, T. Laux & E. Meyerowitz

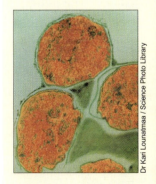

Mycobacterium leprae
(2001) 3 270 000 bp
Cause of the disease leprosy

Dr Kari Lounatmaa / Science Photo Library

Timeline of the human genome project

The completion of the draft human genome sequence by the HGP is the culmination of more than a decade of work, involving 20 sequencing centres in six countries. Here are some of the key moments.

Mid 1980s

Scientists begin to discuss plans for a mammoth project to sequence the complete human genome. Charles DeLisi, then director of health and environmental research at the Department of Energy (DOE), also explores the feasibility of such a project and DOE funding begins in 1987.

1988

The National Institutes of Health (NIH) establishes the Office of Human Genome Research in September 1988. Renamed the National Centre for Human Genome

Research (NCHGR) a year later, its director is James Watson, co-discoverer of the double helix structure of DNA. Watson's testimony to the US Congress, in which he pledged to devote a small fraction of the project's budget to 'ethical, legal and social' issues, had proved instrumental in garnering political support.

Early 1990s

With sequencing still slow and expensive, the genome project adopts a 'map-first, sequence-later' strategy. In the early 1990s, two Parisian laboratories, the Centre d'Etude du Polymorphisme Humain and Généthon, have an integral role in

mapping – underlining the project's international character. The labs' driving forces are Daniel Cohen and Jean Weissenbach. Later, the genome project constructs a higher-resolution map that is used to sequence and assemble the human genome.

1992

In April 1992, Watson resigns as head of NCHGR after clashing with then-NIH director Bernadine Healy over the patenting of gene fragments. Francis Collins (below)

of the University of Michigan is appointed director of NCHGR in April 1993.

In June 1992, Craig Venter leaves NIH to set up The Institute for Genomic Research (TIGR) in Rockville, Maryland. TIGR later sequences a host of bacterial genomes, starting with *Haemophilus influenzae*, the first free-living organism to be sequenced.

1996

In February 1996, at a meeting in Bermuda, international partners in the genome project agree to formalize the

SUPER MODELS

The complete genome sequences of model organisms are proving immensely valuable to biologist working on these species, and will also help interpret the human genome sequence. Published highlights to date include the yeast *Saccharomyces cerevisiae* (May 1997), the nematode *Caenorhabditis elegans* (December 1998), the fruit fly *Drosophila melanogaster* (March 2000, above), and the plant *Arabidopsis thaliana* (December 2000, right).

conditions of data access, including release of sequence data into public databases within 24 hours. These came to be known as the 'Bermuda principles'.

1998

In May 1998, Craig Venter (left) forms a company to sequence the human genome within three years. The company, later named Celera, will use an ambitious 'whole-genome shotgun' method, which involves assembling the genome without using maps. But its data release policy will not follow the Bermuda principles.

Volker Steger / Science Photo Library

1999

The public project responds to Venter's challenge. By early 1999, it is on track to produce a draft genome sequence by 2000. Increasingly, the bulk of the

sequencing takes place in five huge centres dubbed the 'G5': at the Whitehead Institute for Biomedical Research in Cambridge Massachusetts; the Sanger Centre near Cambridge, UK; Baylor College of Medicine in Houston; Washington University in St Louis; and the DOE's Joint Genome Institute (JGI) in

Walnut Creek, California. Here, Robert Waterson of Washington University in St Louis and John Sulston of the Sanger Centre are pictured in a rare moment of relaxation. Trevor Hawkins and Elbert Branscomb of JGI prepare samples (below).

1999–2000

The first sequence of a human chromosome – number 22 – is published in December 1999. Chromosome 21 follows in May 2000, a collaborative effort led by German and Japanese groups.

2000

On 26 June 2000, leaders of the public project and Celera announce completion of a working draft of the human genome sequence. Collins and Venter are seen

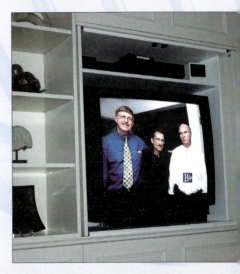

here on television with Ari Patrinos of the DOE, who cut through the animosity between the rival projects to broker the joint announcement at the White House in Washington.

2001

The assembly and analysis of the drafts of the human genome are published by the public project in *Nature* on 15 February and by Celera in *Science*'s 16 February issue.

We are grateful for contributions and input from Francis Collins, Richard Gibbs, Eric Green, Trevor Hawkins, Victor McKusick and John McPherson.

Wellcome Photo Library

TECHNICAL GURUS

Without advances in sequencing technology, we would still be waiting to unveil our genetic blueprint. Double Nobel laureate Fred Sanger (pictured) of the Laboratory of Molecular Biology in Cambridge invented the now commonly used technique of gene sequencing back in the 1970s. In 1986, Leroy Hood, then at the California Institute of Technology in Pasadena, introduced the first automated sequencing machine. During the 1990s, the development of high-throughput capillary sequencing machines enabled the rapid sequencing progress of the past two years. Assembling fragments of the genome into a complete sequence, meanwhile, depended heavily on computer programs developed by Philip Green of the University of Washington in Seattle.

Highlights of our genome

Twenty-four chromosomes, 3.2 billion bases and around 31 000 genes – some theories confirmed, a few surprises and a hatful of mysteries. Welcome to your genome. In this chapter, we highlight the key discoveries from the genome sequence so far, as reported in the seminal paper by the International Human Genome Sequencing Consortium, which is reproduced later in this book.

What do we have in hand? A map and a 'working draft' sequence for most of the 'euchromatic' region of the genome, comprising 2.95 billion bases of the 3.2-billion total. Most of the genes are located here and the DNA is in an 'open' conformation. The small remainder of the genome is composed of 'heterochromatin' – large blocks of densely packaged sequence that are commonly found at the centres and ends of chromosomes and consist almost entirely of repetitive DNA.

The term 'draft' may underplay the status of the work – in all it encompasses over 90 per cent of the genome, each base being sequenced with at least 99.9 per cent accuracy. At the time of publication, about one-third of the genome sequence was 'finished'; that is, sequenced at 99.99 per cent accuracy (less than 1 error in every 10 000 bases). The missing bits are mostly regions that have proved difficult to clone and/or sequence. So although there are acknowledged holes and ambiguities scattered across this vast four-letter scroll, the sequence provides an invaluable new resource.

Where are all the genes?

If a fly, a worm and a mustard weed can get by with 13 000, 18 000 and 26 000 genes respectively, how many genes are needed to make a human being? Previously, numbers up to and

Complexity is not in gene numbers. Humans seem to have only two or three times as many genes as worms and flies, and fewer than twice as many as mustard weed, indicating that complexity is not the sum of the genes. [Dominic Li / WellcomeTrust Medical Photographic Library]

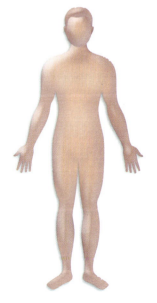

Human	Thale cress	Nematode worm	Fruit fly	Yeast	Tuberculosis microbe
31 000	**26 000**	**18 000**	**13 000**	**6 000**	**4 000**

beyond 150 000 have been bandied about. However, in something of a blow to our collective ego, it seems that there are only 30 000–40 000 human genes.

The proteins encoded by these genes can be grouped into families on the basis of their similarity to one another, and it turns out that we share most of the same protein families with worms, flies and plants – although the numbers of family members are greater in humans. This expansion is particularly evident in the genes that drive development by signalling between cells. Humans have thirty fibroblast growth-factor genes, whereas flies and worms have two each; similarly, humans have forty-two transforming growth-factor genes, of which flies and worms have nine and six respectively. Such differences are also apparent in the proteins of the immune system: humans have 765 genes that encode immunoglobulin subunits or 'domains' (such as those in antibodies), whereas the fly has 140, the worm has 64, and the mustard weed and yeast have none at all.

The remaining extra human genes are not primarily the result of the invention of new types of protein in vertebrates – only 7 per cent of identified protein and protein-domain families are truly unique to vertebrates. Rather, new proteins arise from reshuffling the number and order of protein domains, a process that is analogous to making different structures with the same Lego pieces.

If the increasing complexity of humans isn't due to a significant increase in gene numbers, then what might explain it? No single dominant property stands out. Instead, a mixed bag of features combines to enhance innovation greatly. One example is 'alternative splicing' of RNA; once RNA has been copied from a gene sequence, the non-coding intron sequences are spliced out to bring the coding sequences of the exons next to one another. By skipping an exon here and there, the splicing machinery can create new products. Around 60 per cent of human genes are predicted to have two or more alternatively spliced RNAs, compared, for example, with only 22 per cent in the worm. So instead of producing only one protein, the average human gene produces several.

Another factor is the lavish supply of proteins – called 'transcription factors' – that switch genes on or off. Some families of transcription factors, such as the 'zinc-finger' family, have expanded independently in humans, yeast, flies

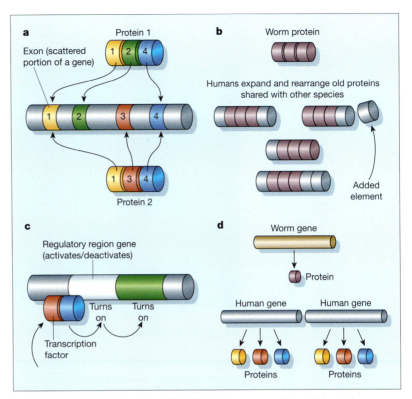

Manufacturing complexity

Although humans do not have many times more genes than simpler creatures, a variety of processes has created layers of complexity.

a) More than one protein can be made from the same gene by a process known as alternative splicing. The separate protein-encoding sections of genes, known as exons, can be skipped, giving rise to an assortment of proteins that either lack or contain the specific protein domains encoded by different exons.

b) Rather than creating entirely new protein domains, humans have rearranged the number and order of protein domains encoded within genes to come up with new combinations.

c) Humans have created unique patterns of gene expression by using different ways to modulate the times and places genes are turned on, including expanding the repertoire of transcription factors that control gene expression.

d) Simpler organisms, such as the worm, produce only one protein per gene. In contrast, an average human gene produces several different proteins, and sometimes as many as five. This is done by alternative splicing, as well as by modification of proteins after they are made, for example by addition of fats, sugars or chemical groups. [Majo Xeridat]

and worms – but humans still have twice as many zinc-finger proteins as flies and almost five times as many as worms. Meanwhile, proteins themselves can be modified, for example by enzymes snipping bits off, or by the addition of sugars or fats that alter their activity.

This builds into a picture of complex and exquisite control of genes and proteins, with genes being turned on and off, and up and down, with extraordinary subtlety. This finely tuned regulation of our gene activity drives our development from fertilized egg to adult, and maintains and repairs our bodies during the rigours of daily life.

Genome cartography

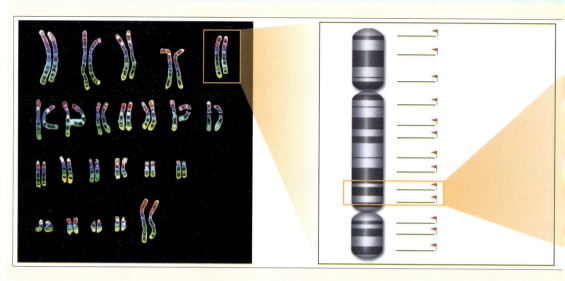

Above: Zooming in on the genome. [The Wellcome Trust]

Sequencers and mappers informed each other of their efforts throughout the Human Genome Project, thereby building up a contoured geography of the chromosomes. The sequence-analysis paper was accompanied by a series of different maps of the genome. What do the more panoramic views of the genome landscape tell us?

At the highest magnification, of course, there is the almost endless stream of As, Cs, Gs and Ts. From a distance, these can be resolved into gene structures, regulatory regions and other recognizable genetic features. One of the maps, showing single-nucleotide polymorphisms (SNPs), charts the naturally occurring genomic variation in the human population.

Zooming further out, the physical map provides a surveyor's chart of the whole genome, with almost every sequence captured in a cloned fragment. Finally, the bird's-eye view is of whole chromosomes decorated with tags at regular

Genome terrain

The general arrangement of the genome provides another startling jolt. In some ways it may resemble your garage/bedroom/refrigerator/life: highly individualistic and unkempt, with little evidence of organization and much accumulated clutter (referred to as 'junk' DNA). Virtually nothing is ever discarded, and the few patently valuable items are scattered indiscriminately, apparently carelessly.

These patently valuable items are the genes themselves. The actual sequence that codes for proteins, the exons, takes up only a few per cent of the genome, whereas the intervening sequences, the introns, account for 24 per cent. Genes are unevenly scattered across the genome and typically stick together, prompting geneticists to call the topography of our genome 'lumpy'. Some genes crowd together in particular regions like dense urban centres, whereas others are dotted across vast expanses of desolate 'deserts' of sequence. This exaggerated terrain of the human genome is distinct from that of other organisms, such as the fly, worm and mustard weed, in which the genes are much more regularly spaced.

Roughly half of the human genome consists of repetitive sequences, with the vast majority of

Genome cartography

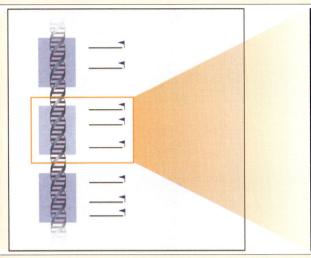

```
ATTAGAGGCTCACCGATTCATGTCGGAGATGGTCAGAAAAAC
CGTTTCAGAAGCAACCTTGGGCTTAGTCCCACCCTTTTTAGGC
GTGCCTAGAAAGATGACAACTCAAGCACCGACGTTTACGCAG
GTACTGGAGGGTAGTACCGCAACCTTTGAGGCTCACATTAGT
GTGAGCTGGTTTAGGGATGGCCAGGTGATTTCCACTTCCACTC
TCCTTTAGCGATGGCCGCGCTAAACTGACGATCCCCGCCGTGA
CGATATTCCCTGAAAGCCACCAATGGATCTGGACAAGCGACT
GTGAAAGCTGAGACAGCACCACCCAACTTCGTTCAACGACTG
CAAGGAAGCCAAGTGAGACTCCAAGTGAGAGTGACTGGAATC
TTCTACCGGGATGGAGCCGAAATCCAGAGCTCCCTTGATTTCC
GACCTCTACAGCTTACTGATTGCAGAAGCATACCCTGAGGACT
AATGCCACCAATAGCGTTGGAGAGCTACTTCGACTGCTGAAT
GAAGAGTACCTGCTAAAAAGACAAAGACAATTGTTTCGACT
AGACAAACCCGAATTGAAAAGAAGATTGAAGCCCACTTTGAT
GTTGAGATGGTCATAGATGGTGCCGCTGGGCAACAGCTGCCAC
ATTCCTCCGATCATAGA
```

intervals; this is the cytogenic map, the anchor at the whole-genome level upon which the sequence and clone maps depend.

During the formation of the sex cells, each of the twenty-three pairs of chromosomes briefly come together and exchange equivalent segments in the process of recombination. The advantage of this interchange is that it mixes the gene pool, allowing the creation of new combinations of traits. Analysis of the genome sequence reveals that the rate of recombination varies strikingly across the genome: short chromosome arms have a higher rate of recombination than long arms or the central regions of the chromosomes. Within this general pattern are recombination 'deserts', where recombination is sparse, and 'jungles', where exchange is frequent. Understanding these patterns of recombination frequency across the genome will be important for mapping the genes that cause disease.

these (around 45%) accounted for by repeats derived from 'parasitic' DNA sequences known as transposable elements, or transposons. These elements propagate by replicating and then inserting a new copy of themselves into another site in the genome. The sheer number of repeated elements is unprecedented in any other sequenced genome; repeats account for just 1.5 per cent of a typical bacterial genome and 3 per cent in the fly, 7 per cent in the worm and 11 per cent in the mustard weed.

Curiously, much of the repeat content of our genome represents ancient remnants of 'long-dead' transposons; in contrast, the fly and mouse genomes harbour large numbers of younger, more active elements. Only two mobile elements are known still to be active in the human genome: they are called long interspersed element 1 (LINE1) and Alu. Together, LINE1 and Alu account for more than 60 per cent of all repeated sequences in our genome. The sequence of LINE1 encodes the machinery

it needs to copy itself. Alu, however, cannot replicate by itself; it uses the machinery of LINE1 elements to reproduce and is therefore something of a freeloader. And a very successful freeloader it is: Alu is the most abundant transposable element, with a million copies littering the human genome.

A diverse genome landscape

Below: A look at the composition of our genome reveals a rich terrain. About 41 per cent is made up of either G or C bases; the rest is composed of A and T. Roughly half of our genome is repeat sequences. The majority of repeated sequences are transposable 'parasitic' elements, such as LINEs and SINEs, the most prolific member of the latter being Alu elements. About 5 per cent of the genome is made up of large duplicated regions. Genes occupy about a quarter of the genome, but only about 1.5 per cent encodes for proteins, the rest being non-coding stretches of DNA called introns that separate the protein-coding exons. [The Wellcome Trust]

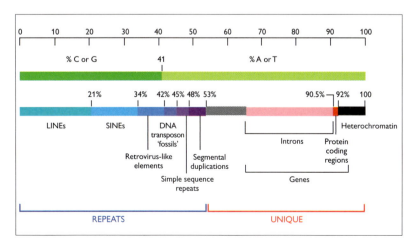

Most transposable elements entered our ancestral genome more than 100 million years ago, long before mammals with placentas (eutherian mammals) evolved. Some types of transposons flourished, such as LINE1 and Alu elements. Others seem to have found the environment unsavoury; for example, only faint traces of another group of transposable elements called LTR retrotransposons are detectable in the human genome – although they are alive and kicking in the mouse genome. DNA transposons, another type of repeat, have marked our genome with two bursts of activity: before and after the evolution of placental mammals.

Most repetitive elements are not under strict pressure to maintain the integrity of sequence as the genome is passed through generations,

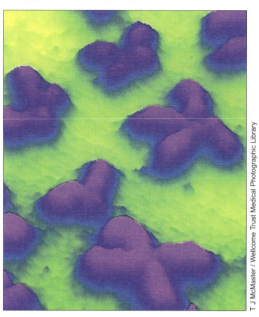

T J McMaster / Wellcome Trust Medical Photographic Library

Robert Boston / Washington University School of Medicine in St Louis

because they do not code for a functional product. So these repeats accumulate mutations and their sequences diverge over time. Tracking the migration of repetitive elements (by tracing the sequence changes) as they wandered through the genome during human evolution opens a

fascinating window on our past. Genetic archaeologists are sifting through the repeat sequences to figure out the pattern and timing of genetic changes in order to reconstruct the history of our genome and of the evolution of our species.

Why does the genome carry such a heavy load of parasitic DNA? Are we unusually sloppy at cleaning out the ancient debris of past invaders? Could we be considered simply as vehicles for the proliferation of these selfish elements? Or do we retain them because they serve some useful purpose? It is likely that there is some truth in each of these propositions. There is evidence that transposons shaped the evolution of the genome and mediated the creation of new genes. Fragments of transposons are found in the regulatory sequences that control the expression of several hundred genes. So it is not inconceivable that, at least in part, transposons are retained because they help to regulate the expression of certain genes.

Just as our genes are unevenly distributed, so too are the repeat elements. Most elements, including LINE1, are found in portions of the genome that are rich in A and T. This tends to separate them from genes, which usually congregate in areas of high GC content. Thus, the mobile elements are relegated to areas where they are less likely to disrupt gene sequences and damage their hosts. In contrast, Alu repeats are more common in GC-rich regions. The reason for this preferential distribution has puzzled geneticists, but it suggests that Alu may positively influence genes and contribute favourably towards their expression or evolution.

Seeing double

Duplications also seem to have played a significant role in genome evolution, as roughly 5 per cent of the sequence has arisen through duplication of large blocks (more than 10 000 base pairs) within and between chromosomes. This is a much more prevalent feature in the human genome than in the fly, worm and yeast genomes. Duplications enable one copy of a gene to relocate to a new site, where it may take on a distinct physiological function.

Duplicated regions are likely to have contributed much to the expansion of gene families in humans. The family of genes that encodes the olfactory receptors responsible for detecting smell provides an extreme example. In all, there are about a thousand olfactory genes scattered throughout the genome, which demonstrates the importance of smell to most mammals. Yet in humans about 60 per cent are non-functional pseudogenes, illustrating our reduced dependency on smell compared with other mammals.

Although duplication offers a means of increasing gene numbers, it can also cause problems. During the production of sperm or eggs, pairs of equivalent chromosomes swap DNA (a process called recombination), creating new combinations of traits in offspring. This process of DNA shuffling can be confused by chunks of almost identical sequence – leading to the deletion of large pieces of genome, which can result in human disease. Examples include DiGeorge and velocardiofacial syndromes, in which most patients harbour a deletion from chromosome 22 that is thought to have been mediated by duplicated sequences flanking the deleted region. As a result, several genes from this region are not present in patients with DiGeorge syndrome, causing a range of clinical symptoms including abnormal facial characteristics, heart defects, and immune and endocrine abnormalities. Similarly, duplications are thought to predispose to the rearrangement and deletion in the region of chromosome 7 associated with Williams–Beuren syndrome, in which patients have a characteristic facial appearance, heart and blood-vessel problems, dental and kidney abnormalities, and sensitive hearing.

Y under siege

The X and Y chromosomes were once a matching pair, like the other twenty-two pairs of chromosomes in the human genome. However, since the evolution of sexual differences in mammals, which is thought to have begun when one of the pair of chromosomes acquired a gene for 'maleness', the sex chromosomes have grown apart. As they diverged over time, the Y chromosome lost the ability to exchange segments with the X chromosome (except at the very tips) during recombination. As a consequence, many of the

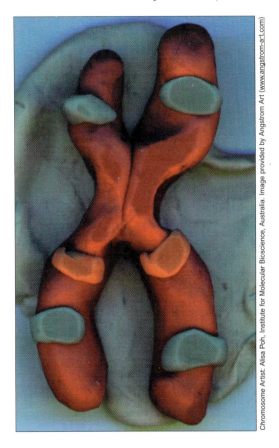

Chromosome Artist: Artist: Alisa Poh, Institute for Molecular Bioscience, Australia. Image provided by Angstrom Art (www.angstrom-art.com)

genes on the Y that are not essential have accumulated mutations and have either been shut down or lost. Now, the Y chromosome is reduced to a stump, clinging onto the genes that offer males a competitive advantage.

The Y chromosome has some other striking features. It has a large number of repeated chunks that are so similar that they are almost impossible to tell apart. Many of the male specific genes lie in these regions, such as those involved in testes development and sperm production. It is possible that these duplications are an attempt to ensure that the genes aren't lost (safety in numbers!).

Another surprise is a large stretch of sequence, around 3.5 million bases, which is in one orientation in some males, and in the opposite orientation in others. It is feasible, although merely a speculation, that this large inversion, as well as other variations in the number and orientation of the duplicated blocks, may underlie variations in male fertility and virility.

Mutations in the male genome are more common than in the female. Indeed, sperm cells are the main source of human genetic mutation, with mutation rates that are twice as high as in eggs. The reason for this is not known, but it may be because of the numerous cell divisions that occur during continuous sperm production throughout adult male life, whereas females undergo only a single round of egg production during development. Every time a cell divides, there is the potential for a mistake to occur when the DNA is copied. Sometimes these mutations can be a source of innovation that aids survival in changing environments; however, they can also disrupt gene function and cause disease.

Bacterial footprints

Have bacteria left their mark on our genome? Remarkably, a couple of hundred genes that are found in humans seem to be more similar to bacterial genes than to any in yeast, worms, flies or plants. This observation suggested that some genes might have been transferred from various bacterial genomes directly to vertebrate genomes.

Could it be a case of bacterial genes hitching an evolutionary ride, or is there something in it for us? Most of the bacterially derived genes encode enzymes and have been sequestered into specific pathways, such as stress responses and the metabolism of environmental chemicals, suggesting that they have been adapted to important physiological functions.

Since the publication of the human genome sequence, studies by others have provided evidence that at least some of these genes are unlikely to have been transferred from bacteria. Instead, they were probably present in an ancestral species and were lost in some lineages, such as those of worms and flies, but were retained in vertebrates. Both hypotheses are presently being debated, but the matter is unlikely to be settled until extensive sequence data from the genomes of other organisms have been acquired.

Hunting down disease genes

The draft genome sequence has already had a considerable impact on the identification of disease genes. Previously, the identification of a candidate gene in the absence of sequence and biochemical information was time consuming, costly and tedious. Now candidate genes can be readily detected using computers to interrogate public sequence databases, and confirmed by mutation screening of plausible candidates. The entire process can be completed in a matter of months.

With the aid of the draft genome sequence, more than thirty genes have already been identified, including genes involved in breast cancer, muscle disease, deafness and blindness. Meanwhile, the search for the many genes that underlie common chronic ailments, such as cardiovascular disease, arthritis, osteoporosis and cancer, will be greatly assisted by a knowledge of the sites in the genome that vary between individuals. The most common type of sequence variation is a change of an individual base, known as a single-nucleotide polymorphism (SNP). On average, there is likely to be one SNP in every 1000–2000 bases. The SNP Consortium – a collaboration between researchers, private companies and a charity — has gone a long way towards documenting human variation by constructing a map of 1.42 million SNPs, which was published in the same issue of *Nature* as the report of the human genome sequence.

Looking further ahead, a complete catalogue of genes opens up opportunities for drug development. Currently, our entire drug cabinet is derived from fewer than 500 targets. Even if a small proportion of genes and gene products prove to be valid targets, the genome sequence will provide enormous possibilities for new therapies. For example, when classic drug-target proteins were used to search the genome sequence database, more than a dozen new relatives were found, including neurotransmitter receptors and growth factors.

Collaboration, competition and conclusions

The Human Genome Project (HGP) is responsible for one of two draft sequences that were published contemporaneously. The other, from Celera Genomics, appeared in the journal *Science*. Was it, as some have argued, a waste of money to undertake two genome projects? Can a winner and a loser in the 'sequence race' be identified?

Our answer to both of these questions is 'no'. The two groups took different approaches to the sequencing of the genome, spurring developments in technology. Both were successful – the HGP independently and Celera in combining the (publicly accessible) HGP's data with its own major sequencing effort. The availability of two draft sequences inspires confidence in the conclusions drawn about the genome (which are, by and large, in close agreement). And the two groups provide a range of tools with which to analyse and exploit the sequence.

Beyond practical application, what does it mean to know the sequence of our genome? We have a description of our own inheritance, a glimpse into eons of gradual change. The genome is clearly not a well-ordered, tightly executed instruction manual, as some over-simplistic metaphors may have implied. Even with the genetic read-out in front of us, we have only tantalizing hints of how our genes work together, with extraordinary complexity, to build a human being. We are left with a sense of the quirkily human nature of the genome, with all its wonderful ingredients, that gets more interesting the deeper into it we go.

The genome in the news

There were two great waves of coverage of the human genome in the media. The first was in late June 2000, following press conferences at the White House and 10 Downing Street to announce that the genome had been sequenced. This was headline news and the coverage was celebratory in tone, epitomized by the use of colourful metaphors as political and scientific leaders sought to put the achievement into context. Remarkably, the publication of the research papers in February 2001 received still greater coverage in the media.

"President Clinton called the information 'the most important, most wondrous map ever produced'. He said that 'it outstripped in importance the map with which Lewis and Clarke opened America's way to the West'."

Other comparisons floated were to Galileo's discoveries in the solar system, the invention of the printing press, the race to develop the atomic bomb, and the space programme that put men on the moon.

The Boston Globe again:

"As a scientific enterprise, the race to map the human genome bears resemblance to the Manhattan Project, the furious World War II rush to develop an atomic bomb ..."

Richard Dawkins, quoted in the UK's *The Daily Telegraph*, went beyond the sciences for inspiration:

"Along with Bach's music, Shakespeare's sonnets and the Apollo Space Programme, the Human Genome Project is one of those achievements of the human spirit that makes me proud to be human."

This time the analysis was weightier, focusing on what was learned from the sequence, rather than on the achievement itself. Here's a selection of what was on offer.

Grappling with the concept

The sequencing of the genome is not an easy sell in news terms. How was it presented to bring out the significance?

The Boston Globe went for a winning combination of political clout and the appeal of history:

Others, such as Japanese Prime Minister Yoshiro Mori, quoted in *The Washington Post*, got straight to the point:

Groundbreaking 'map' to human genome due today

Genetic decoding holds surprise: Proteins seen playing major role

By Maggie Fox
REUTERS NEWS AGENCY

Our future may not lie in our genes, after all.

Two separate teams of researchers will report today that they have taken the first in-depth look at the human genetic code and found about half what they expected to find. Instead of 60,000 to 80,000 genes, humans have only 30,000 to 40,000.

Both teams agree this means that, in humans anyway, it is proteins that matter — much more so than genes.

The human body, it seems, is set up to adapt to its environment by cutting up and recombining the protein "products" of genes to make a protein suitable for the circumstance.

Each gene makes one protein — this is the basic function of any cell. Researchers had known

see ROLE, *page A10*

By Patricia Reaney
REUTERS NEWS AGENCY

Scientists will publish the initial sequence of the human genome today in a breakthrough that promises to revolutionize the understanding and treatment of diseases.

The sequencing of 3.1 billion letters of DNA shows humans are made up of about 30,000 to 40,000 genes, considerably fewer than earlier estimates of 60,000 to 100,000 genes, and only about twice as many as the earthworm and fruit fly.

Scientists say identifying all the genes and what they do will herald a new age in science and medicine, vastly expanding human knowledge and accelerating the diagno-

Holds key to aging, intractable illness

sis and treatment, as well as potential preventions and cures, for disease.

"It is going to revolutionize science and medicine," said Tim Hubbard of the Sanger Centre in Cambridge, England, who worked on the project.

"Everything about us is in the sequence."

The Human Genome Project, the publicly funded international collaboration of 20 groups of scientists from the United States, Britain, Japan, France, Germany and China, completed the working draft of the human genetic code in June.

All the information has now been arranged and appears in the scientific journal Nature with a dizzying array of reports, maps and analyses to explain what it all means.

Celera Genomics Inc. of Rockville, the privately owned company that raced to produce the first draft, reports its findings in the journal Science.

The sequence is just the beginning and will not be fully finished for several years, but it is already revealing its secrets — far fewer genes, where they come from, the complexity of proteins and what makes us different from other organisms.

see GENOME, *page A10*

"An immense step forward for humanity in deciphering the makeup of life itself."

Political leaders around the world concurred. French Research Minister Roger-Gerard Schwartzenberg said:

"The deciphering of the book of life, is a milestone in science."

Chinese President Jiang Zemin commented that it was:

"A great scientific project in the scientific history of human beings and one of vital importance to the development of life sciences, medicine and pharmaceutical study".

But for sheer bravado, the Director of the Wellcome Trust, quoted in *The Financial Times*, was unmatched:

"Mike Dexter ... said the breakthrough could have longer-lasting significance than the wheel. In fact, while the wheel might one day be obsolete, the genome would be useful so long as there were humans on this — or any other — planet."

A humbling gene number

If there was a stand-out finding, it was the dramatically lower estimate for gene number than had been anticipated. This was widely interpreted as a severe knock to the human ego. For example, *The Daily Telegraph* lamented:

"It took about only 12 000 more genes than a worm and around 17 000 more genes than a fruit fly to build Einstein ..."

Warming to a similar theme, *The New York Times* considered that:

"The human genome, besides being only just out of the worm league, seems to have almost too much in common with many other kinds of animal genomes."

Taking the theme of self-deprecation a stage further was Robert Waterston, one of the leaders of the Human Genome Project (HGP). Talking to *The Washington Post*, he said:

"It's a humbling perspective. You can't study the genome for very long before you start feeling that you're just a transient vehicle for making more DNA."

Not absolutely everyone, however, was angst-ridden over the revelation on gene numbers. Martin Bobrow, a professor of medical genetics, was sanguine in *The Daily Telegraph*:

"Knowing a fly only has slightly fewer genes than me doesn't make me feel degraded — they are pretty complex things; they have four wings and can fly — I can't do that."

Once the initial shock was overcome, the media regrouped to provide some sensible explanations for our biological complexity despite the dearth of genes. Australia's *Sydney Morning Herald* suggested that:

"What sets us apart from flies and worms is the complexity of our proteins. Our extra genes do not make lots of new kinds of proteins. Rather, they reshuffle the different bits of old protein in novel ways ...

What makes us human is our intricate mechanism for switching genes on and off at various stages of life."

SOURCES: Celera Genomics; Human Genome Project   AP/GLOBE GRAPHIC

First genome reading holds some surprises

Scientists find DNA of humans similar to that of 'lower' animals

By Richard Saltus
GLOBE STAFF

Humbled by their first look at man's complete set of genes, scientists yesterday said it will take years to understand how humankind evolved from a DNA blueprint that is remarkably similar to so-called "lower" animals.

That striking similarity was the biggest surprise to scientists from public and private teams who released the first readout of the complete set of DNA instructions in human cells — the "human genome."

At press conferences in Washington and in other countries, the scientists said the genome — likened to a recipe book or parts list for the human body — contains only about 30,000 genes, or separate messages — only about one-third as many as had been predicted and only about twice as many as fruit flies.

And it turns out there's nothing so special about human genes. The common mouse has genes that are similar to all but some 300 human genes, reported Craig Venter, leader of the private-sector genome effort. The two teams largely agreed on what the genome messages say.

"The abiding mystery of the genome is how we became so complex with such a relatively

Continued on next page

Startling finds in genome analysis
Refute long-held 'facts' of genetics carried in today's textbooks

By August Gribbin
THE WASHINGTON TIMES

The world's leading genetic scientists yesterday formally announced the first detailed analysis of the human genome — the so-called "blueprint of life" — they produced in June.

Their analysis held surprises.

Although it's difficult for the nonscientist to grasp the full significance of the startling finds the genome sequencing and its interpretation facilitate, most of the discoveries are immensely important.

They help researchers identify the lines of inquiry that ultimately are expected to revolutionize medicine and lead to new therapies and cures for devastating diseases like Alzheimer's, Parkinson's, diabetes and others.

There is no way of predicting when such cures might come. But Peter Bruns, a geneticist and vice president at Bethesda's Howard Hughes Medical Institute, said, "I think we'll see the scientific findings put to some practical uses very quickly. Not years. Days."

Even before the genome was sequenced, gene research had linked certain genes with specific illnesses like prostate cancer, glaucoma, muscular atrophy, hardening of the arteries of the heart and more.

And in Boston, physicians last summer successfully injected genes into the hearts of chronically ill heart-attack patients and provoked the growth of new blood vessels. Increasingly, medical pioneers are developing similar gene-based therapies.

Mr. Bruns says such work will be speeded because, "Now the findings are all on the Internet. It's like having a map of a city that's so detailed, you can see which homes have refrigerators and what kind.

"If there is spoiled food, you can look to see if it was caused by the

see GENOME, page A10

These views were backed up by Francis Collins, who was quoted in *The New York Times*:

"The main invention seems to have been cobbling things together to make a multitasked protein. Maybe evolution designed most of the basic folds that proteins could use a long time ago, and the major advances in the last 400 million years have been to figure out how to shuffle those in interesting ways. That gives another reason not to panic."

Far from panicking, another set of reports seized on gene number to make bold assertions about the human condition. These were spearheaded by the British Sunday newspaper, *The Observer*. A somewhat overstated lead story opined that:

"The discovery of our meagre gene numbers ... reveals that environmental influences are vastly more powerful in shaping the way humans act."

It speculated that:

"The discovery has critical implications for our understanding of the idea of free will".

Other papers, such as Italy's *La Repubblica*, acknowledged nature–nurture interaction in more balanced terms:

Tiny Gene Disparities Go a Long Way

■ **Science:** DNA of people of different races is unexpectedly alike, new genome findings show.

By ROSIE MESTEL
TIMES MEDICAL WRITER

The first detailed survey of the human genetic code is revealing many striking things about the blueprint for making a human being. Among them: how similar we all are to each other. And how different.

The findings, to be formally announced today and published later this week, reveal for instance that members of two different racial groups can be more alike than members of the same group.

The studies also reveal that two unrelated people are unexpectedly alike, differing on average at just 1 out of every 1,000 sites in our DNA.

Yet even that small difference adds up to roughly 3 million places in DNA where tiny disparities exist between two people's genetic codes. That's enough to create all the known genetic variety, from simple traits such as eye color to more complicated ones such as higher risks for depression or heart disease, according to the new findings published by two groups, the Human Genome Project and a privately funded effort by Celera Genomics Corp. The genome project was funded largely by the U.S. government and a British charity.

Today, because of the genome effort, places in our DNA where those differences occur have been cataloged in detail unimaginable just a few years ago. This new in-

Please see GENES, A5

"The French geneticist Jean-Michel Claverie writes in *Science* that such a low number of genes could represent a paradigm shift that could radically change our understanding of the complexity of organisms and of evolution. This paradigm shift would force us to re-examine the importance of genetic control on who we are and what we do, rescuing the role of the environment and its interaction with the genome."

Undermining racism

Another point picked up throughout the world was our extraordinary similarity at the genetic level. For instance, the UK's *The Independent* reported:

"The notion that skin colour is a useful method of predicting physical or mental variations between human beings has been refuted by the genome maps. Every person on earth shares 99.9% of the same genetic code and the differences within racial groups are often greater than those between people of different colours ..."

Hope was expressed that these findings could reduce racial prejudice. According to *The Daily Telegraph*:

"Prof Svante Paabo, of the Max Planck Institute of Evolutionary Anthropology in Leipzig, Germany, is optimistic that as we learn more about the genetic similarities and differences between individuals, this knowledge will encourage social tolerance and compassion.

It is already apparent that the gene pool in Africa contains more variation than elsewhere, and that the genetic variation found outside Africa contains only a subset of that found in Africa. 'From that perspective, all humans are therefore Africans, either residing in Africa or in recent exile,' he said.

He believes that studies of genetic variation in human populations may not be so easy to abuse, in terms of using the data as 'scientific support' for racism and other forms of bigotry. 'If anything, such studies will have the opposite effect because prejudice, oppression and racism feed on ignorance.'

The impact on health and society

Many newspapers described the possible impact that the genome sequence could have on health. They highlighted the potential for new medical approaches, as in the case of the UK's *The Guardian*:

"The human genome could open an era of a new kind of medicine – one tailored to a patient's unique genetic makeup".

And they covered target diseases, as reported in Spain's *El Pais*:

"For scientists, the publication of the sequence ... opens the door to a new era for medicine and biology, as well as for finding a cure to a large number of diseases that have a hereditary component such as cancer, Alzheimer or diabetes ..."

The *China People's Daily* commented:

"In addition to genetic disease, the knowledge of human biochemistry that is contained in the human genome could hold new insights into tackling infectious diseases such as AIDS and tuberculosis."

Bill Clinton drew up a rough timeframe for developments, as quoted in *The Financial Times*:

"It is conceivable that our children's children will know the term Cancer only as a constellation of stars."

Others were less sure of this timetable, with *The Washington Post* warning that:

"It will take the best part of the 21st century to fulfil the grand promises made at the announcement of the complete human blueprint."

Some unease about the practical consequences was also expressed. Displaying mixed emotions was *The Irish Times*:

"The effort has raised both immense hopes for curing diseases and stopping birth defects along with fears of genetic discrimination and selective breeding."

Discomfort registered a further notch in *The Boston Globe*:

"Hanging over the whole day, however, was the sense that the world had entered uncharted ethical waters. Will people be discriminated against based on their genes? Will babies be genetically engineered?"

The Hindustan Times was of like mind:

"However, many are concerned about the resulting risk of discrimination in insurance and employment, leading to 'nightmare' dilemmas about whether such tests should be permitted by employers, lenders and health insurers."

These thorny issues were grasped by UK Prime Minister Tony Blair, quoted in *The Financial Times*:

"The powerful information now at our disposal [must] be used to transform medicine, not abused to make man his own creator or invade individual privacy."

There was also concern from developing countries about access to advances for all peoples. *The Namibian Times*, for instance, said:

"Disabled people fear the information will be used to create perfect people and some scientists claim the benefits of the achievement will only be enjoyed by people living in wealthy countries."

The sequence and society

Like other great scientific advances, the sequence of the human genome can be harnessed for good or bad purposes. There are several factors that make human genomic information stand out. One is its personal nature – it deals with our fundamental make-up. Another is the power of its application – information about our genes potentially affects all of us in our daily lives, transforming medicine but also providing new knowledge that can be used outside the clinical context. Add the aspect of immediacy, namely that practical use of the sequence is already a reality, and it is clear that there is an urgent need for society to understand, debate and decide on the appropriate settings for the use of genetic information.

Fortunately, there has been a considerable amount of foresight and planning for these new developments in genetic science. Many countries are already grappling with complex ethical, social and legal issues in the light of growing genome knowledge. Expert committees have been set up, politicians have played an important role in fostering debate, and media interest has been invaluable in bringing concerns to the attention of the wider public. In this chapter, we consider some of the societal conundrums that the availability of the genome sequence raises.

Dealing with genetic risk

Genetic knowledge offers great hope for early detection, more accurate diagnosis and

Joe Heller, Wisconsin - *The Green Bay Press-Gazette*

The ELSI Program

The architects of the Human Genome Project (HGP) recognized that, in parallel with striving towards the scientific and technological goal of sequencing the human genome, it would be imperative to address the impact of this new science on individuals, families and society. To address these issues they established, in 1990, the Ethical, Legal and Social Implications (ELSI) Program, which receives 3–5 per cent of the annual HGP budgets of the US Department of Energy (DOE) and the National Human Genome Research Institute (NHGRI). This represents the world's largest bioethics programme. It is complemented by similar initiatives around the world.

The ELSI Program funds research in ethics, law, economics and other related fields to anticipate and address the issues arising from new genetic information and technology, and from their applications. In several instances, this has led to policy recommendations and proposed legislative solutions. Some of the specific topics include fair use of genetic information, maintaining the privacy and confidentiality of genetic information, issues of informed consent for genetic testing, commercialization of genetic research discoveries, and education of health professionals and policy-makers. Information about ELSI can be obtained at the NHGRI or DOE websites: www.nhgri.nih.gov; or www.ornl.gov/hgmis.

mine whether a person carries a gene mutation that could affect his or her children. To do a gene test, a DNA sample isolated from the blood or tissue of a patient is scanned for a specific mutation linked to a particular disorder.

Although there are several hundred genetic tests for different conditions, fewer than 100 tests are available commercially, and most are for rare disorders caused by a mutation in just a single gene. For instance, individuals with a family history of a genetic condition such as Huntington disease, an adult-onset neurodegenerative condition, may desire a genetic test because they want to know their chances of developing the disease later in life. Couples who

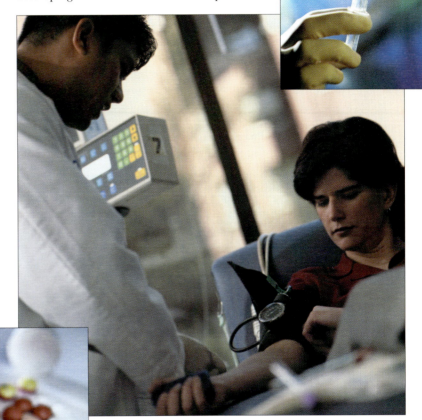

improved treatment of a variety of medical conditions. It is also likely to bring preventive medicine to the fore. Researchers will begin to understand how DNA variations underlie individual susceptibilities to certain diseases and responses to medical treatments. It may also be possible to design customized drugs that are more effective for particular groups of patients.

Genetic testing (or DNA-based testing) is one of the first commercial applications of genetic discoveries to medicine. It can be used, for example, to diagnose a condition, to estimate the likelihood of developing one, or to deter-

are asymptomatic but suspect that they may carry a disease-causing mutation may use genetic tests to find out the risk of passing on the condition to their children. Prenatal diagnosis, involving the testing of a foetus during pregnancy, can be used to diagnose a number of conditions, including Tay–Sachs disease, a fatal neurological disorder of early childhood.

Genetic tests present a number of personal dilemmas and ethical issues. Should a test be taken? What action should be taken in response to the results? How should the new information be handled, both individually and within the family? A further problem is that in some cases diagnosis has outpaced treatment. For an adult-onset disorder where a test for a genetic disorder is available but a treatment is not, is it preferable not to know and live in hope, or to be rid of the uncertainty?

Single-gene disorders are rare and account for only a very small percentage of all diseases. For the majority of common diseases – such as cancer, heart disease, asthma, obesity and diabetes – the causes are much more complex and are thought to involve a variety of gene mutations, perhaps acting in combination with each other and together with certain environmental factors. Untangling the genetic and environmental contributions to these complex diseases will ultimately offer opportunities to develop diagnostic markers for predisposition to certain conditions.

Knowing your genetic susceptibility to common diseases can offer some advantages. It might help you to make better choices about

lifestyle and behaviour. For example, people who are genetically predisposed to heart problems may be able to lower their chances of developing symptoms by not smoking, eating healthy foods, exercising regularly and taking certain medicines. For diseases such as colorectal cancer, where early detection can significantly improve prognosis, knowing one's genetic disposition can alert an individual to the need for close monitoring and regular check-ups.

There are concerns about how people will handle the uncertainty that surrounds genetic susceptibility to common diseases. Testing positive for a genetic indicator of disease predisposition only indicates a risk of developing the disorder; it does not mean that the condition will automatically develop. It may only be in the context of other gene mutations as well as the influence of specific environmental factors that the disease will occur. Conversely, a negative result for a genetic test does not necessarily mean an individual won't eventually develop the disease.

It is difficult to predict how different people will react to learning that they are genetically predisposed to a certain condition. Some may use this information to make better decisions about their lifestyle and habits. For others, however, it may provoke anxiety and stress. Learning that one is at risk, even a slight risk, of developing a particular disease could give way to 'genetic fatalism', whereby a person believes that genes alone determine future health, irrespective of changes to diet and behaviour.

All of us carry some variations that are potentially detrimental in particular contexts and environments. The degree of risk can vary for individuals in different populations and can be influenced by a host of factors, some of which are within a person's control and others outside any sort of control. Given that it will be possible to identify our individual genetic susceptibilities to disease in the not-too-distant future, it is essential that people are educated about the meaning of genetic risk, how best to handle this risk, and how to live with the uncertainty.

While genetic information is deeply personal, it is not just about an individual. It may reveal information about another family member. People contemplating genetic tests need to consider the consequences not only for them-

Genetic testing of children

- Genetic testing of children can confirm a medical diagnosis or make a predictive diagnosis.
- Often a child's health will greatly benefit from early detection and treatment of a genetic disorder.
- Genetic testing of children raises unique ethical concerns. For instance, 'informed consent' of the patient is required prior to genetic testing; obviously this presents difficulties in the case of a child.
- Telling a child about his or her genetic susceptibilities raises difficult issues, such as at what age and level of maturity to do so.
- While parents may believe that having a specific genetic test is in the best interests of the child, it is hard to predict how the child will deal with this information later in life.
- The testing of children for adult-onset disorders, such as Huntington disease, where there is no intervention that can decrease risk or prevent disease onset, is controversial and is typically not permitted.

NCT Publishing

selves but also for their relatives. Should other members of a family be informed of the test results, which may reveal their own risk of getting a particular disease?

As more genetic tests become available, mechanisms to assure the accuracy and reliability of such tests must be put in place. There is currently little regulation of tests but it is urgently needed. There is also a need to ensure equitable access to genetic tests and counselling. At present, genetic tests can range in cost from hundreds to thousands of dollars. There needs to be the infrastructure to enable everyone, regardless of their socioeconomic or other circumstances, to have the opportunity to take advantage of the beneficial information that genetic testing potentially offers.

Genetic discrimination and privacy

The potential abuse of genetic information, such as in the workplace or in the provision of health insurance, raises grave concern. There is a real danger that a genetic 'underclass' might develop.

Based on genetic information, employers might try to avoid hiring workers who they believe are likely to take sick leave, resign or retire early for health reasons. Cases of discrimination could occur even where an individual

The importance of genetic counselling

Because of the dilemmas it can present, genetic testing should always be contemplated and conducted with the support of genetic counselling. Professionally trained counsellors help people to understand the limitations of genetic tests, to anticipate how different test results might affect them, and to explain the choices available in the light of the results. The value of genetic counselling is not to make the decision for the patient, but rather to help them make informed decisions regarding all the possible implications.

Currently, medical geneticists in conjunction with trained genetic counsellors provide genetic testing. However, as it becomes cheaper, easier to perform and more readily available, genetic testing may be carried out by primary health-care providers. It is essential to establish practices that continue to ensure that people undergoing genetic testing are fully informed of the implications. All health-care providers will have to be equipped with the skills to provide appropriate counselling. And there needs to be knowledge on both sides – the patient and the health-care provider – so they can work together to enable the patient to reach the most informed decision and realize when additional, more specialized counselling should be sought.

may never show signs of disease, or where their genetic condition has no effect on their ability to perform work. Substantial gaps in the laws of many countries leave open the potential for employees to be discriminated against on the basis of genetic information. While existing anti-discrimination laws, such as those protecting people with disabilities, could be interpreted to include genetic discrimination, specific legislation is desirable.

There are concerns too about the health insurance industry using genetic information to discriminate against individuals. Cases have been reported where individuals with a genetic disorder or predisposition have been refused health insurance, or had their enrolment cancelled or premiums increased. There are fears that medical expenses for those suffering from genetic conditions will not be covered and children may be excluded from coverage because they are at risk of inheriting a genetic disease.

Information about genetic make-up is personal. Furthermore, it may not just reveal information about an individual but also about other family members. And yet it could be made available without a person's knowledge, or even against his or her wishes. For instance, an employer or insurance provider may require access to medical records which include the results of genetic tests. Disclosure of genetic information may be considered an invasion of privacy.

Many argue that holders of genetic information should be prohibited from releasing it without the individual's prior authorization and, to protect against information being used for purposes other than what it was originally collected for, an individual's consent should be required for each disclosure. While it may not be realistic to expect confidential genetic profiles of individuals never to be disclosed in an electronic information age, this underscores the need to have policies in place that serve not only to deter but also to penalize the inappropriate collection, disclosure and abuse of this information.

The DNA files

DNA technology has revolutionized forensic science and the criminal justice system. DNA has been a key 'witness' in numerous trials, helping police and the courts to identify the perpetrators of violent crimes with a very high degree of confidence. But, just as it can point the finger at the guilty, it can also exonerate the wrongly accused.

DNA fingerprinting, or DNA typing, involves comparing sections of DNA between samples – if all the sections match, mathematical formulae are used to estimate the odds that both samples come from the same person. DNA typing cannot absolutely prove a match, but it can come very close. Although DNA typing can't be used to prove guilt definitively – because there is always a chance, no matter how remote, that a suspect's DNA profile is a very close match to that of the real offender – it can nevertheless support corroborating evidence. DNA typing, however, can prove innocence because it can indisputably determine whether a person's genetic profile does not match the DNA evidence of the crime.

Hair, blood, saliva, semen, skin, teeth and nail clippings all contain cells with DNA, which can be used to provide useful information about a person, such as their sex or whether two people are related. Aside from its use in criminal cases, DNA typing can also help identify missing persons and murder or accident victims. It has also been used to prove or

---

DNA to the defence

"DNA aids the search for truth by exonerating the innocent. The criminal justice system is not infallible"

Janet Reno, US Attorney General during President Clinton's administration

Forensic DNA technology can help convict the guilty as well as exonerate the innocent. A number of factors can contribute to a wrongful conviction, such as mistaken eyewitnesses, incompetent counsel and erroneous laboratory results from other biological markers. But DNA testing provides scientific and irrefutable proof of innocence.

An example is the Innocence Project, founded in 1992 by attorneys Barry Scheck and Peter Neufeld – a programme operating out of the Benjamin N. Cardozo School of Law in New York. This programme provides free assistance to help wrongly convicted prison inmates prove their innocence through DNA tests.

Since the advent of forensic DNA testing in the late 1980s, more than 63 people in the United States have been exonerated and subsequently set free after DNA testing of the evidence. Some of these people had been convicted of capital crimes and were incarcerated or on death row. The Innocence Project represented or assisted in over 36 of these cases, and is presently handling over 200 cases, with a further 1000 or more pending evaluation.

Efforts such as this have been credited with expanding awareness amongst law-enforcement authorities, prosecutors, judges, politicians and the public that not only is DNA testing an important tool to fight crime, but it also provides a means for ensuring that the criminal justice system works fairly.

disprove paternity, that is whether a man is the father of a particular child.

Although DNA typing is a powerful tool for forensics, it does raise questions about the protection of privacy. DNA information collected by the police is stored in law-enforcement agencies' DNA databanks. While this information is typically only obtained for serious crimes, increasingly DNA is being collected and stored for anyone convicted of a crime, and some agencies are considering collecting DNA from all those arrested, in much the same way that fingerprints are currently taken. Some concerns have been raised about the confidentiality of this information and whether it could be used for purposes other than that for which it was collected.

Genes for cure

Genetic knowledge is being applied to diagnosis, treatment, cure and prevention of health problems. Genetic engineering of medicines and genetic testing provide two approaches – gene therapy is another. Gene therapy involves changing genes within a person's cells. It can be used, for example, to replace missing or defective proteins, or to introduce gene products that destroy cancer cells or render them more vulnerable to treatment.

The transfer of genes into cells of the body, other than sex cells (sperm or eggs), is called somatic gene therapy. It is still at an experimental stage and is usually only tried with patients who have diseases for which there is no cure. The technical challenges are enormous: a gene must be introduced into the appropriate cells of the body, and the cells must express the gene product in an active form, at the right time and in the right quantities. Hundreds of gene-therapy trials have been undertaken or are presently under way. Although there have been a number of promising studies, none has so far been able to achieve complete success.

There are also many ethical concerns with gene-therapy applications. The benefits of treating serious diseases may be very clear-cut, but for other conditions it may raise questions about what is a disorder or disability and what is normal. An example of this is growth disorders. Also, gene therapy is, and will probably remain, very expensive, raising questions about who will have access to these therapies and who will pay.

Even more contentious than somatic therapy is germline gene therapy. This involves changing genes in sex cells, that is the sperm and egg, so that the altered DNA is passed on to offspring. Germline gene therapy offers the potential for permanently ridding a family of an

Czar Nicholas II and his family
DNA analysis can also uncover history – and controversy. It has been at the centre of the debate over whether US President Thomas Jefferson fathered children with his slave Sally Hemings. And it has confirmed that the bodies in a mass grave were those of Czar Nicholas II and his family, who were murdered during the Russian Revolution. [Bettmann / CORBIS]

inherited genetic disorder. There are a number of concerns about this form of gene therapy, one of which is the potentially detrimental impact that new genes could have on foetal development. Another concern is the ethical implications of changing the genes of future generations without assurances of safety, not to mention the lack of consent of all the potentially affected persons. So controversial is the issue of germline gene therapy that it is not being actively researched. At the very least, its development should await a better understanding of all the implications.

Genetic enhancement

Eugenics left an indelible stain on the twentieth century. Eugenics (derived from the Greek word meaning 'wellborn') is the use of genetics to improve the quality of humankind. It became popular in North America and Europe during the early to mid-1900s. The eugenics movement led to the endorsement of the sterilization of thousands of individuals considered 'genetically unfit' (which included people deemed 'feebleminded'). This type of thinking reached a shameful peak in the desire for racial purity in Nazi Germany. But even today eugenic ideas are being applied. For example, in China a law forbids mentally disabled people from marrying unless they have been sterilized.

The ideas, and many of the experiments, of

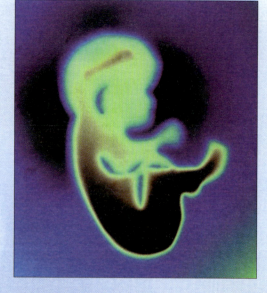

the original eugenicists have been largely discredited. However, there are fears that a new form of eugenics could emerge, whereby people might be tempted to tinker with genes on a wide scale in an attempt to select for 'desirable' traits for their children, such as physical attributes, IQ and personality. This raises the prospect of so-called 'designer babies'.

Fortunately, there are several barriers blocking such development. First, there are the technical hurdles. It is extremely difficult to select for several genes at once. Figuring out exactly which genes, in which combinations, create the complement of desired traits remains in the realm of science fiction, at least with current knowledge and technology. Also, environment and upbringing, and even experiences in the womb, play a big part in how a child develops.

While it may be technologically out of reach for some time, the issue of genetic enhancement of our future generations warrants ethical concern. Many would see the merits of using genetic knowledge and technology to help fix a faulty gene that would otherwise result in children being born with a painful, devastating disease that claims their lives early. However, we will inevitably reach the blurry boundaries between disease and normal variation.

Colour-enhanced scanning electron micrograph of a human egg. [Yorgos Nikas / Wellcome Trust Medical Photographic Library]

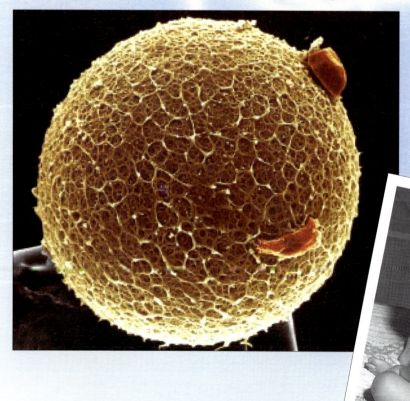

Who 'owns' our genes?

A patent is a legal agreement that gives an inventor rights over the invention for a period (typically 20 years), during which time others cannot make, use or sell the invention unless the inventor licenses it to them to do so. Patents were developed to make information about inventions publicly available, to encourage investment and to reward inventiveness.

Patenting of genes has, however, been controversial. Should a naturally occurring entity, such as a gene, be viewed as an invention? Should the genome be treated like property or stocks and shares, being divided up, bought, sold and traded?

Patenting offers an incentive for the research and development necessary to translate genetic discoveries into genetic medicines. However, there are concerns that excessive licensing charges could impede the development of genetic diagnostic and therapeutic products. This could be exacerbated by the fact that a novel diagnostic or therapeutic product might involve many distinct patents, each owned by a different individual or company. A single DNA sequence can be patented in several different ways, triggering fears that the requirement for payments to multiple patent owners or licensees might discourage the development of genetic tests and medicines. Others argue that cross-licensing agreements will be developed to overcome this potential obstacle.

The patenting of genetic tests can raise concerns for patients. Patients and their families often cooperate in the search for disease genes, providing blood samples and medical information about themselves and their families. However, when the gene discovery is commercialized, the very same families may be required to pay what they perceive to be unreasonable costs for tests and treatments derived from the gene that they helped identify.

In a recent example, a US patient advocacy group for Canavan disease has filed a lawsuit against the hospital and researcher who patented the gene that is mutated in the degenerative brain disorder. They claim that the gene was discovered using the genetic information and financial resources provided by Canavan families, and that the hospital charged royalties that limited the availability of testing for the disease. In contrast, the patient advocacy group for sufferers of pseudoxanthoma elasticum (a genetic disorder that causes connective tissue in the skin, eyes and arteries to calcify) played a significant role in organizing the research on

DNA crystals display conical and striated fan textures, and have a colour pattern that shifts through a range of different colours, such as from yellow to purple as illustrated in this photomicrograph. [Courtesy of Michael W. Davidson, Florida State University, Tallahassee]

Knowing our genetic selves

How does the human genome sequence affect the way we think about ourselves? The genome sequence will reveal a lot about each of us at an individual level, but it can also reveal a lot about our communities, our history and our species.

We shall learn about the 'sameness' and 'otherness' of people. Our genomes are about 99.9 per cent identical to each other. Identifying and deciphering that mere 0.1 per cent of genetic code that varies between humans will help us understand how genes influence our appearance, behaviour and susceptibility to disease.

The human genome sequence also potentially breaks down national and cultural boundaries; there are more genetic differences within a given population than between different populations. History lessons may require revision as scrutiny of the DNA sequence of different people resolves ambiguities in ancestral family trees, or records of the wanderings of earlier travellers. By extensive comparisons of human genomes, we shall be able to generate a higher-resolution picture of our cultural and geographical origins.

It also throws new light on how our species evolved and our relatedness to other species. We share about 99 per cent DNA sequence similarity with chimpanzees. Furthermore, at least 10 per cent of our genes have relatives in flies and worms. More refined comparison of the genetic differences between us and other animals will inevitably challenge notions of the uniqueness of the human species.

the disorder, and thereby garnered a place for their leader as an inventor on a patent application describing the gene discovery. This unique alliance provides a model for handling intellectual property emerging from collaboration between patients' groups and researchers.

A call for education

Genetics is a complex science, involving inferences from patterns of inheritance, some knowledge of statistics and sophisticated technologies. This presents challenges in understanding the enormous promise of the Human Genome Project as well as the allied concerns. It is therefore imperative to translate the genetic science into understandable terms and concepts for the general public.

As genetic information begins to influence mainstream medical practice, the need to educate both practitioners and the public on its implications for health care will escalate. The public's expectations are high and, as a result of the electronic information flow of the worldwide web, some patients may be better informed than their physicians. The basic principles of genetics, as well as an awareness of current trends and recent developments, will need to be integrated into fundamental levels of training for all health-care practitioners, and also into continuing medical education.

Aside from the clinical setting, genomic knowledge is being used in other contexts, for example in the courtrooms and in evaluations of health and life insurance. It is incumbent on those who can act on the basis of this information to have a solid grounding in genetics.

It is imperative to guard against extreme reactions to the sequencing of the genome, both the unreasonably negative and the unrealistically positive. The issues relating to the use (or abuse) of genetic knowledge are complex and concerns from all sides must be discussed openly. Many of the worries relate to application of the science rather than to the science itself. Despite the pessimism of some, many of these concerns appear to have potential remedies. Nonetheless, a major feature in resolving doubts and fears will be ensuring that the public is educated about the science and has a say in how it is utilized.

At the other extreme, it is essential to keep expectations at a realistic level. Deciphering the human genome sequence is just a start; the hard part now begins as we try to understand what genes do, how they interact and what goes awry during disease. It will be many years before the full benefits are realized. If this message fails to get through, then widespread disappointment may result and, worse still, create a backlash against genetic science.

There are numerous benefits in the genomic era. But it will require an educated, prepared and receptive society to realize this potential fully.

The genome and beyond

"... the more we learn about the human genome, the more there is to explore."

*International Human Genome Sequencing Consortium; penultimate
sentence of the paper describing the human genome sequence*

The Human Genome Project is continuing apace. The plan is to have all gaps closed and all ambiguities resolved by the year 2003. Using this completely 'finished' sequence, a comprehensive inventory of genes and sequence diversity should be well within reach. But identifying all human genes is only the first step; equipped with this new information, scientists aim to discover how genes and their protein products work and how they influence human health and disease.

Traditional methods of deciphering gene function typically work on a 'one gene in one experiment' basis. Now, thanks to the development of 'high-throughput' technologies, the functions of thousands of genes and the proteins they encode can be studied simultaneously. This large-scale analysis of the genome and its products is known as 'functional genomics'; it includes the study of where and when genes are expressed, and the process of determining the structure and function of proteins.

Learning from others

Genetically speaking, we are very much like other organisms. The structure of DNA, and the code that translates proteins from genes, are virtually identical across all life forms. So we can learn a lot about ourselves by studying non-human organisms.

Comparing DNA sequences of different species – an approach called 'comparative genomics' – can identify genes and provide clues as to their function. The concept is simple: DNA segments that have a function are more likely to retain their sequence during evolution than non-functional segments. Laboratory experiments using less complex organisms test these predictions. Yeast, for example, has been studied extensively. Most of the genes that control yeast cell division have been identified. By looking for their relatives in the human genome, a role in cell division can tentatively be assigned to those human genes and then be tested through specific experiments. A wide range of basic biological processes – such as DNA replication, repair, protein manufacturing, and general cell housekeeping – are shared by virtually all life forms, including single-celled organisms like bacteria and yeast.

Humans, of course, are far more complex than single-cell organisms. But insight into how our trillions of cells communicate and work together to form a human being can be gained from simpler animals. The worm *C. elegans*, for example, has one of the least complex nervous systems – it has only 302 neurons compared with about a hundred billion neurons in a human brain. Nonetheless, the development and function of the worm's nervous system can shed light on the workings of more complex brains.

Finding function

Knowing when and where genes are expressed can provide clues about function. This information can be gathered using microarrays, also known as 'DNA chips', thumbnail-sized glass or silicon wafers to which thousands of unique

Protein structures can give important clues about function, and can aid in the design of new drugs. [Wellcome Medical Photographic Library]

fragments of genes are attached. Microarrays enable scientists to monitor the expression levels of thousands of genes at a time.

To look at gene expression in a particular tissue at a particular time, scientists extract messenger RNA (mRNA) sequences, which are the transcribed intermediaries between DNA and proteins. The solution containing mRNAs is then washed over the microarray. The mRNA complementary to the fragments on the DNA chip 'sticks' to the microarrays and indicates which genes are expressed in the cells. By labelling mRNA samples with coloured dyes, it is possible to compare gene expression in different cells and tissues directly in one experiment.

An example of where microarrays have revealed useful information is in the study of cancer. By comparing expression profiles of tumour versus normal cells, it is possible to identify genes that may play a role in the development or progression of cancer. Tumours, even of the same cancer type, can vary in terms of how quickly they grow, how they respond to treatment and their likelihood of metastasis. While they can be indistinguishable by microscopy, specific patterns of gene expression can serve as markers to classify tumours and predict prognosis and response to treatment.

Profiling proteins

With the human genome nearly complete, scientists are increasingly turning their attention to proteins. The emerging field of 'proteomics' encompasses the study of the location, interaction, structure and function of proteins. Proteins are more challenging to work with than DNA. They often have complicated three-dimensional structures that determine their function. A suite of tools is being developed to characterize the full complement of proteins within a given cell, known as the 'proteome'.

'Protein chips', equivalent to DNA chips, are currently under development. These can measure changes in the quantities of proteins in a cell in response to drug treatment, or identify proteins with particular enzymatic properties.

The location of a protein within a cell provides clues to its function. A protein found on the cell surface, for example, may function as a receptor, while one sequestered in the nucleus may modulate gene expression. Using tagging and tracing techniques, the cellular locations of proteins are rapidly being discovered, building up a picture of the cellular landscape.

The binding partners of proteins provide another piece of the functional puzzle. If the function of a protein is known, it becomes possible to infer the function of proteins to which it binds. Techniques exist in which proteins are used as 'bait' to 'fish' for interacting partners in the cellular pond. Cataloguing these interactions provides an overview of protein networks and biological processes in a cell. Once again, model organisms such as yeast, in which the physical interactions between every protein have been mapped, provide invaluable templates to delineate similar networks in human cells.

One of the most powerful ways to analyse a protein is to determine its three-dimensional structure. This can provide insight into how proteins work at the molecular level and can aid rational drug design. Determining the structures of all human proteins presents an even greater challenge than sequencing the genome and a similar industrial-scale approach is under way.

Integrating information

The effort to unravel the functions of the genome is multidisciplinary, drawing in scientists from many fields – including biology, medicine, chemistry, physics, mathematics, engineering and computer science. The greatest promise lies in the integration of all this information to create a detailed picture of biological processes, and find out how these processes go awry in disease. Integral to this is 'bioinformatics', a new field of science that manages and analyses data to extract biological meaning. Databases and software tools will need to become much more sophisticated to help scientists make sense of the staggering amount of emerging computational and laboratory data.

While the potential is enormous, we are still a long way from a full understanding of biological processes. And it will be considerably longer before genome-based information delivers the anticipated practical solutions to medical problems. In the meantime, preparation and debate on the responsible use of genetic information are required. So, while pushing ahead with the research and technology, parallel steps are needed to ensure that public awareness and understanding – as well as attention to ethical, legal and societal concerns – keep pace with the science. The sequencing of the human genome is only a beginning.

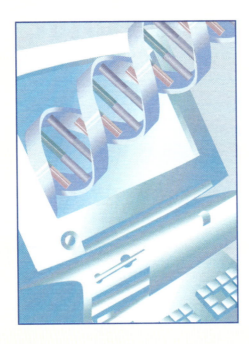

The research papers

The following section is reprinted directly from the pages of the human genome issue of *Nature* published on 15 February 2001. The centrepiece is the main research paper from the International Human Genome Sequencing Consortium, which describes the assembly and analysis of the human genome. It is the first comprehensive survey of the human genome sequence and summarizes more than a decade of work.

There is tradition, rigour and protocol in how scientists publish their discoveries. Research papers are submitted to scientific journals such as *Nature* for review and evaluation by editors and scientific peers; papers are accepted for publication only when the reviewers' queries have been satisfied. While the readership of *Nature* is primarily scientists, the authors went to some lengths to present the paper in a language and style that are as accessible as possible, within the constraints of a scientific journal. It is the longest scientific report ever published in the 132 years of the journal's existence. We are delighted to reproduce the work here, not only as a historic document but also as a fine example of scientific communication.

The engaging series of News and Views articles that introduced the scientific work is also reprinted here. *Nature*'s News and Views articles, introduced into the journal in 1927, provide context and accessibility for key findings. They are contributed by prominent independent researchers who have the expertise to comment on the relevance and impact of the research.

Our genome unveiled

David Baltimore

The draft sequences of the human genome are remarkable achievements. They provide an outline of the information needed to create a human being and show, for the first time, the overall organization of a vertebrate's DNA.

I've seen a lot of exciting biology emerge over the past 40 years. But chills still ran down my spine when I first read the paper that describes the outline of our genome and now appears on page 860 of this issue[1]. Not that many questions are definitively answered — for conceptual impact, it does not hold a candle to Watson and Crick's 1953 paper[2] describing the structure of DNA. Nonetheless, it is a seminal paper, launching the era of post-genomic science.

This milestone of biology's megaproject is the long-promised draft DNA sequence from the International Human Genome Sequencing Consortium (the public project). The sequence itself is available to all those connected to the Internet[3]. In the paper in this issue, we are presented with a description of the strategy used to decipher the structures of the huge DNA molecules that constitute the genome, and with analyses of the content encoded in the genome. It is the achievement of a coordinated effort involving 20 laboratories and hundreds of people around the world. It reflects the scientific community at its best: working collaboratively, pooling its resources and skills, keeping its focus on the goal, and making its results available to all as they were acquired.

Simultaneously, another draft sequence is being published[4]. It is less freely available because it was generated by a company, Celera Genomics, that hopes to sell the information. This week's *Science* contains an account of the history of that project and the analyses of its data, while another of the papers in this issue contains a comparison of the quality of the two sequences[5]. To those who saw this as a competitive sport, the papers make it appear to be roughly a tie. However, it is important to remember that Celera had the advantage of all of the public project's data. Nevertheless, Celera's achievement of producing a draft sequence in only a year of data-gathering is a testament to what can be realized today with the new capillary sequencers, sufficient computing power and the faith of investors.

Answers

What have we learned from all of these AGCTs? The best way to answer the question is to read the analytical sections of the papers. I will only make some general comments. It is important to remember that no statements can be made with high precision because the draft sequences have holes and imperfections, and the tools for analysis remain limited (as described in a further paper[6] in this issue, page 828). However, the answers provided by the draft will be of interest to many investigators, and the value of having the draft published in its imperfect form is unquestionable.

The sequences are about 90% complete for the euchromatic (weakly staining, gene-rich) regions of the human chromosomes. The estimated total size of the genome is 3.2 Gb (that is gigabases, the latest escalation of units needed to contain the fruits of modern technology). Of that, about 2.95 Gb is euchromatic. Only 1.1% to 1.4% is sequence that

actually encodes protein; that is just 5% of the 28% of the sequence that is transcribed into RNA. Over half of the DNA consists of repeated sequences of various types: 45% in four classes of parasitic DNA elements, 3% in repeats of just a few bases, and about 5% in recent duplications of large segments of DNA. The amounts in the first and third classes will certainly grow as our ability to characterize them increases in effectiveness and we examine the darkly staining, heterochromatin regions of chromosomes. As the co-discoverer of reverse transcriptase (the enzyme that reverses the common mode of information transfer from DNA to RNA), I find it striking that most of the parasitic DNA came about by reverse transcription from RNA. In places, the genome looks like a sea of reverse-transcribed DNA with a small admixture of genes.

Repeats

By contrast, the puffer fish — another vertebrate — has a genome that contains very few repeats. But it encodes a perfectly functional creature, so it seems likely that most of the repeats are simply parasitic, selfish DNA elements that use the genome as a convenient host. People call this 'junk DNA', but from the DNA's point of view it deserves more respect. In most places in the human genome the selfish elements are tolerated, and in some places —near the ends of chromosomes, for instance, or near the chromosome constrictions called centromeres — it builds up to form huge segments. However, the repeated DNA may have both negative and positive effects. For instance, the paucity of repeats in certain highly regulated regions of the genome suggests that insertions there can disrupt gene regulation and are deleterious. Conversely, the enrichment of the so-called Alu class of repeated sequences in the gene-rich, high-GC regions of the genome implies that they have a positive function. The repeats can also be fodder for evolving new functions and act as loci for gene rearrangements.

In humans, virtually all of the parasitic DNA repeats seem old and enfeebled, with little evidence of continuing reinsertions. However, there has been very little evolutionary scouring of these repeats from the human genome, making it a rich record of evolutionary history. The mouse genome, by contrast, has many actively reinserting parasitic sequences and is scoured more intensely, making it a much younger and more dynamic genome. This difference might reflect the shorter generation time of mice or something about their physiology, but I find it an intriguingly enigmatic observation.

Much of what we learn about the global organization of the genome is an elaboration of previous notions. For instance, we knew that the genome had regions with a relatively high content of GC bases and regions high in AT, but now we have a very complete appreciation of this architecture. What maintains the patchiness of the GC/AT ratio in the genome remains an unanswered question. As was expected, most genes are located outside the heterochromatic regions; interestingly, however, in regions of the genome rich in GC bases, the gene density is greater and the average intron size is lower. These introns — made up of largely meaningless sequence that breaks up the protein-coding sequences (exons) of genes — are much longer in human DNA than in the genomes previously sequenced. Their dilution of the coding sequence is one element that makes finding genes by computer so difficult in human DNA.

A major interest of the genome sequence to many biologists will be the opportunity it provides to discover new genes in their favourite systems — for instance, cell biologists will search for new genes for signalling proteins, and neurobiologists will look for new ion channels. This data-mining exercise

was carried out by various groups which report their initial findings in papers that appear on pages 824–859 of this issue. They found some new and interesting genes, but surprisingly few, and occasionally could not find the full extent of genes that they knew were there. The paucity of discoveries reflects their concentration on systems that were previously heavily studied.

Gene-regulatory sequences are now there for all to see, but initial attempts to find them were also disappointing. This is where the genomic sequences of other species — in which the regulatory sequences, but not the functionally insignificant DNA, are likely to be much the same — will open up a cornucopia. Basically, the human sequence at its present level of analysis allows us to answer many global questions fairly well, but the

detailed questions remain open for the future.

What interested me most about the genome? The number of genes is high on the list. The public project estimates that there are 31,000 protein-encoding genes in the human genome, of which they can now provide a list of 22,000. Celera finds about 26,000. There are also about 740 identified genes that make the non-protein-coding RNAs involved in various cell housekeeping duties, with many more to be found. The number of coding genes in the human sequence compares with 6,000 for a yeast cell, 13,000 for a fly, 18,000 for a worm and 26,000 for a plant. None of the numbers for the multicellular organisms is highly accurate because of the limitations of gene-finding programs. But unless the human genome contains a lot of genes that are

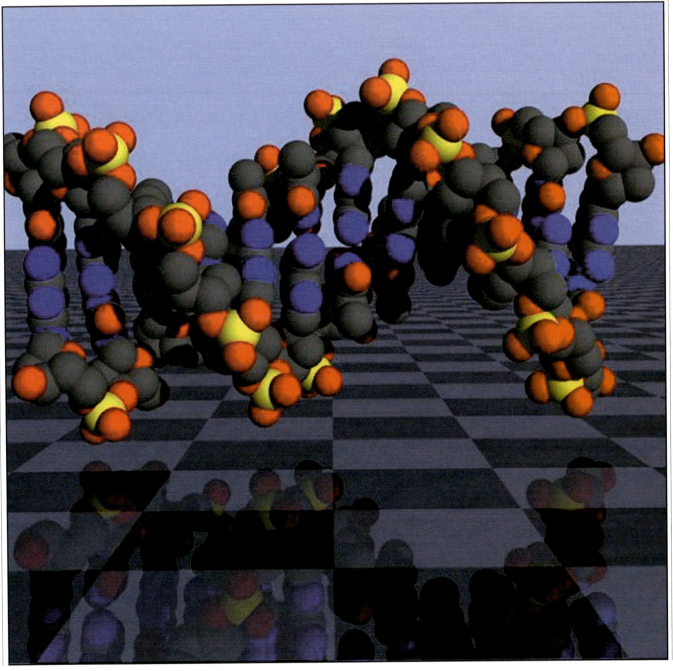

opaque to our computers, it is clear that we do not gain our undoubted complexity over worms and plants by using many more genes. Understanding what does give us our complexity — our enormous behavioural repertoire, ability to produce conscious action, remarkable physical coordination (shared with other vertebrates), precisely tuned alterations in response to external variations of the environment, learning, memory… need I go on? — remains a challenge for the future.

Complexity

Where do our genes come from? Mostly from the distant evolutionary past. In fact, only 94 of 1,278 protein families in our genome appear to be specific to vertebrates. The most elementary of cellular functions — basic metabolism, transcription of DNA into RNA, translation of RNA into protein, DNA replication and the like — evolved just once and have stayed pretty well fixed since the evolution of single-celled yeast and bacteria. The biggest difference between humans and worms or flies is the complexity of our proteins: more domains (modules) per protein and novel combinations of domains. The history is one of new architectures being built from old pieces. A few of our genes seem to have come directly from bacteria, rather than by evolution from bacteria — apparently bacterial genomes can be direct donors of genes to vertebrates. So DNA chimaeras consisting of the genes from several organisms can arise naturally as well as artificially (opponents of 'genetically modified foods' take note).

The most exciting new vista to come from the human genome is not tackling the question "What makes us human?", but addressing a different one: "What differentiates one organism from another?". The first question, imprecise as it is, cannot be answered by staring at a genome. The second, however, can be answered this way because our differences from plants, worms and flies are mainly a consequence of our genetic endowments. The Celera team[4] presents the more detailed analysis of the numbers of different protein motifs and protein types, in extensive tables. From them, it is easy to see what types of proteins and motifs have been amplified for specific types of organisms. In vertebrates, not surprisingly, we see elaboration and the *de novo* appearance of two types of genes: those for specific vertebrate abilities (such as neuronal complexity, blood-clotting and the acquired immune response), and those that provide increased general capabilities (such as genes for intra- and intercellular signalling, development, programmed cell death, and control of gene transcription). Someday soon we will have the mouse genome, and then those of fish and dogs, and probably the kangaroo genome from the Australians. Each of these will fill in a piece of the evolutionary puzzle and will provide exciting comparisons.

We wait with bated breath to see the chimpanzee genome. But knowing now how few genes humans have, I wonder if we will learn much about the origins of speech, the elaboration of the frontal lobes and the opposable thumb, the advent of upright posture, or the sources of abstract reasoning ability, from a simple genomic comparison of human and chimp. It seems likely that these features and abilities have mainly come from subtle changes — for example, in gene regulation, in the efficiency with which introns are spliced out of RNA, and in protein–protein interactions — that are not now easily visible to our computers and will require much more experimental study to tease out. Another half-century of work by armies of biologists may be needed before this key step of evolution is fully elucidated.

What is next? Lots of hard work, but with new tools and new aims. First, we have to stay the course and get the most precise representation of the genome that we can: this is a matter of filling the cracks, cleaning up the errors, and getting rid of the uncertainties that plague each of the analytical methods. Second, we need to see more genomes, with each one giving us a deeper insight into our own. Third, we need to learn how to take advantage of this book of life. Tools for scanning the activity levels of genes in different cells, tissues and settings are becoming available and are already revolutionizing how we do biological investigation. But we will have to move back from the general to the particular, because each gene is a story in itself and its full significance can be learned only from concentrating on its particular properties.

Fourth, we need to turn our new genomic information into an engine of pharmaceutical discovery. Individual humans differ from one another by about one base pair per thousand. These 'single nucleotide polymor-

phisms' (SNPs) are markers that can allow epidemiologists to uncover the genetic basis of many diseases. They can also provide information about our personal responses to medicines — in this way, the pharmaceutical industry will get new targets and new tools to sharpen drug specificity. Moreover, the analysis of SNPs will provide us with the power to uncover the genetic basis of our individual capabilities such as mathematical ability, memory, physical coordination, and even, perhaps, creativity.

Biology today enters a new era, mainly with a new methodology for answering old questions. Those questions are some of the deepest and simplest: "Daddy, where did I come from?"; "Mommy, why am I different from Sally?". As these and other questions get robust answers, biology will become an engine of transformation of our society. Instead of guessing about how we differ one from another, we will understand and be able to tailor our life experiences to our inheritance. We will also be able, to some extent, to control that inheritance. We are creating a world in which it will be imperative for each individual person to have sufficient scientific literacy to understand the new riches of knowledge, so that we can apply them wisely.

David Baltimore is at the California Institute of Technology, 1200 East California Boulevard, Mail Code 204-31, Pasadena, California 91125, USA.
e-mail: baltimo@caltech.edu

1. International Human Genome Sequencing Consortium *Nature* **409,** 860–921 (2001).
2. Watson, J. D. & Crick, F. H. C. *Nature* **171,** 737–738 (1953).
3. http://genome.cse.ucsc.edu/
4. Venter, J. C. *et al. Science* **291,** 1304–1351 (2001).
5. Aach, J. *et al. Nature* **409,** 856–859 (2001).
6. Birney, E., Bateman, A., Clamp, M. E. & Hubbard, T. J. *Nature* **409,** 827–828 (2001).

The maps

Clone by clone by clone

Maynard V. Olson

The public project's sequencing strategy involved producing a map of the human genome, and then pinning sequence to it. This helps to avoid errors in the sequence, especially in repetitive regions.

This issue of *Nature* celebrates a halfway point in the implementation of the 'map first, sequence later' strategy adopted by the Human Genome Project in the mid-1980s[1]. The results suggest that the strategy was basically sound. It led, as hoped, to a project that could be distributed internationally across many genome-sequencing centres, and that would allow sequenced fragments of the human genome to be anchored to mapped genomic landmarks long before the complete sequence coalesced

into one long string of Gs, As, Ts and Cs.

The centrepiece of the suite of mapping papers in this issue is on page 934, where the International Human Genome Mapping Consortium describes a 'clone-based' physical map of the human genome[2]. A map like this not only charts the genome, giving a structure on which to hang sequence data, but also provides a starting point for sequencing. Figure 1 shows the basics of the approach. I drew this figure in 1981, using India ink and a Leroy lettering set. Both

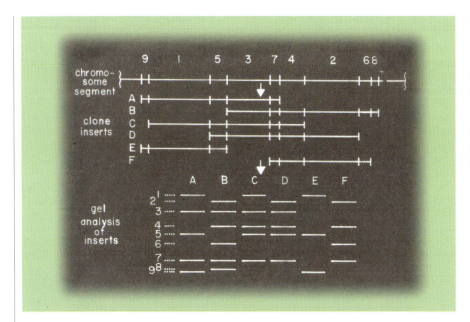

Figure 1 **Clone-based physical mapping. The top line shows the location of 'restriction' sites (vertical bars) in a particular region of the genome. Restriction sites are places at which a site-specific restriction endonuclease cleaves DNA. The fragments produced by cleavage at every possible point in this region are numbered 1 to 9. Below the line are several clones with random end points, labelled A to F. Clones are produced by first partially digesting many copies of the genome with different restriction endonucleases; the resulting large segments are then inserted into bacteria and replicated (cloned). Each clone is digested with a restriction endonuclease, and the resulting fragments are separated, by size, on an electrophoretic gel ('gel analysis of inserts'). This process yields a distinctive pattern ('fingerprint') for each clone. The map-assembly problem requires working backwards (upwards in this figure) from the fingerprints to a clone-overlap map and restriction-site map of the chromosome segment. To finish the analysis of this region of the genome, the natural choice of clones to sequence would be A and B.**

graphical and mapping technologies have come a long way since then, but the principles behind clone-based physical mapping have not changed.

The clone-based approach works as follows. Many copies of the genome are cut up into segments of about 150,000 base pairs by partial digestion with site-specific restriction endonucleases — enzymes that cleave DNA in specific places. ('Partial' digestion means that the reaction is not carried out for long enough to allow every possible cleavage to be made.) The large DNA segments are plugged into bacterial 'artificial chromosomes' (BACs) and inserted into bacteria, where they are copied exactly each time the bacteria divide. The process produces 'clones' of identical DNA molecules that can be purified for further analysis. Next, each clone is completely digested with a restriction endonuclease, chosen to produce a characteristic pattern of small fragments, or a 'fingerprint', for each clone. Comparison of the patterns reveals overlap between the clones, allowing them to be lined up in order, while the sites in the genome at which the restriction endonuclease cleaves are charted. The result is a physical map.

Individual BAC clones are then sheared into smaller fragments and cloned; the resulting 'small-insert' subclones are sequenced. The sequence of an individual

BAC clone is assembled from the sequences of an 'oversampled' set of subclones (in other words, enough subclones are sequenced to ensure that each part of the original clone is analysed several times). Finally, the whole genome sequence is assembled by melding together the sequences of a set of BACs that spans the genome.

This approach is similar to that used in the 1980s and early 1990s to map and sequence the genomes of the nematode *Caenorhabditis elegans* and the yeast *Saccharomyces cerevisiae*[3,4]. What is new in the human project is its staggering scale, and the speed with which it has been completed. By way of comparison, although the nematode and yeast genomes are, respectively, only 3% and 0.5% the size of the human genome, these early mapping projects spanned the better part of a decade, as opposed to two years for the much larger human project.

One weakness of clone-based physical mapping is that the maps often have poor continuity. For example, there is not always a BAC clone to cover every part of the genome; and overlaps between clones can be obscured by data errors or the presence of large-scale repeats in the genome. The current map[2] has more than 1,000 discontinuities. These will cause some difficulties as the Human Genome Project moves to its next phase, which will involve ensuring accuracy and

filling in any gaps in the sequence. Nonetheless, the current map typically maintains continuity for several million base pairs at a stretch. These continuous segments are big enough to allow the clone-based map to be overlaid on various lower-resolution maps. In this way, the mapped segments can be ordered and orientated, much as a discontinuous patchwork of high-resolution maps of the Earth's surface can be orientated by overlaying them on a satellite photograph of the whole Earth.

Two particularly interesting low-resolution maps are the genetic and cytogenetic maps, on pages 951 and 953 of this issue[5,6]. The genetic map[5] is based on the probability of the occurrence of recombination — the swapping of corresponding, nearly identical segments of DNA between maternally and paternally derived chromosomes as the genome is passed from one generation to the next. The cytogenetic map[6] is based on subtle variations in the staining properties of different regions of the genome, as viewed by light microscopy. Yet more papers describe different approaches to clone-based mapping[7–9]. These methods were applied to particular chromosomes simply because different sequencing centres chose to rely on the whole-genome map to different degrees.

But was all this cartography even necessary? Another draft of the human genome sequence is described in this week's *Science* by Celera Genomics[10]. This group adopted a different approach, which involved preparing small-insert clones directly from genomic DNA rather than from mapped BACs. The major rationale for the BAC-by-BAC approach[2] was to make easier the finishing phase of the Human Genome Project, which lies ahead. The consortium now plans to upgrade the 30,000 BAC sequences by sequencing more subclones from each BAC (the 'topping-up' phase) and then resolving internal gaps and discrepancies (the 'finishing' phase). Segmenting the finishing phase into BAC-sized portions provides an enormous advantage in dealing with blocks of sequence that are repeated at many different places within the genome. The power of this strategy is nicely illustrated by the mapping of the Y chromosome, whose repetitive structure is unusually complex (page 943 of this issue[11]).

Nature readers should not expect any real answer to the question of which of these two approaches is the better one. But it is likely that the only players still on the field when the toughest finishing issues are confronted will be the public consortium's BAC brigade. In the future, as genome sequencing moves on to other mammals, the context will have changed; the human sequence will provide an invaluable guide to assembling long stretches of sequence that are shared among all mammalian genomes. So the sequencing of the human genome is likely to be the only large sequencing project carried to completion by the methods described in this issue. Genome sequencing will get easier from here.

Looking ahead, there are two threats to producing a quality finished product. One is simple exhaustion on the part of the consortium's members: each new round of press conferences announcing that the human genome has been sequenced saps the morale of those who must come to work each day actually to do what they read in the newspapers has already been done.

We may also expect to hear the argument that the current sequence is good enough for most purposes, and that remaining problems should be resolved by users as the need for accurate sequence in specific regions arises. What we have now is certainly a lot better than what we had yesterday. But biologists in the future will be comparing vast data sets to the reference sequence of the human genome. They must be able to do so with confidence that the discrepancies they encounter are due to the limitations of their

own data or, more interestingly, to biology. They should not need to expend time, energy and imagination compensating for a failure now to pursue the Human Genome Project to a grand conclusion. We must move on and finish the job, even as the bright lights of media attention shift elsewhere.

Maynard V. Olson is in the Departments of Medicine and Genetics, Fluke Hall, Mason Road, University of Washington, Seattle, Washington 98195-2145, USA.
e-mail: mvo@u.washington.edu

1. National Research Council *Mapping and Sequencing the Human Genome* (National Academy Press, Washington DC, 1988).
2. The International Human Genome Mapping Consortium *Nature* **409,** 934–941 (2001).
3. Coulson, A., Sulston, J., Brenner, S. & Karn, J. *Proc. Natl Acad. Sci. USA* **83,** 7821–7825 (1986).
4. Olson, M. V. *et al. Proc. Natl Acad. Sci. USA* **83,** 7826–7830 (1986).
5. Yu, A. *et al. Nature* **409,** 951–953 (2001).
6. The BAC Resource Consortium *Nature* **409,** 953–958 (2001).
7. Montgomery, K. T. *et al. Nature* **409,** 945–946 (2001).
8. Brüls, T. *et al. Nature* **409,** 947–948 (2001).
9. Bentley, D. R. *et al. Nature* **409,** 942–943 (2001).
10. Venter, J. C. *et al. Science* **291,** 1304–1351 (2001).
11. Tilford, C. A. *et al. Nature* **409,** 943–945 (2001).

The draft sequences

Filling in the gaps

Peer Bork and Richard Copley

Two rough drafts of the human genome sequence are now published. Completion of the sequences lies ahead, but the implications for studying human diseases and for biotechnology are already profound.

With the publication of the human genome sequence — described and analysed on page 860 of this issue[1] and in this week's *Science*[2] — we cross a border on the route to a better understanding of our biological selves. But unlike the previously published sequences of human chromosomes 21 and 22 (refs 3,4), the present sequences of the whole human genome are not considered complete. The bulk of the data make up what is called a 'rough draft'. So what is all the fuss about? What exactly does 'rough draft' mean, and what can we learn from sequences such as this?

In the draft from the publicly funded International Human Genome Sequencing Consortium[1], around 90% of the gene-rich — euchromatic — portion of the genome has been sequenced and 'assembled', the term used to describe the process of using a computer to join up bits of sequence into a larger whole. Each base pair of this 90% was sequenced four times on average, ensuring reasonable precision. Only about a quarter of the whole genome is considered 'finished' — another bit of genomics jargon, which basically means that each base pair has been sequenced eight to ten times on average, with gaps in the sequence existing only because of the limitations of present technology. Nonetheless, the sequence of base pairs in

the draft is very accurate, and is unlikely to change much; 91% of the euchromatin sequenced has an error rate of less than one base in 10,000 (ref. 1).

For the other draft, that produced by Celera Genomics[2], a variety of methods suggest that between 88% and 93% of the euchromatin has been sequenced and assembled. But direct comparison of these numbers with the public consortium's draft is almost impossible — different procedures and measures were used to process the data and to estimate accuracy. Both projects also have sequence data that were not used in the assembly process, raising the real level of coverage by a few percentage points.

These numbers might seem rather arbitrary, but even when the first genome of an animal species was published[5], it was clear that simple, practical finish lines do not exist (Box 1, Fig. 1). The present level of coverage of the human genome reflects the point where a shift of focus occurs, from sequencing the genome many times over to producing a high-quality, continuous sequence[6]. There is some way to go yet.

Essentially, 'rough draft' refers to the fact that the sequences are not continuous — there are gaps (Box 1). If there are too many gaps, it can be impossible to order and orientate the many small strings of bases that are the raw products of genome sequencing. This might, for example, hamper projects that seek to identify genes involved in inherited diseases. A first step to finding such genes is to work out which region of which chromosome they are on. The complete genome sequence should be immensely useful for the next step — identifying the relevant gene at that region. But gaps and errors in ordering and placing the strings of sequence will make this difficult.

Another problem of incompleteness is that it is difficult to make definitive

---

Box 1 What makes a completely sequenced genome?

When is sequencing work on a genome complete? No genome for a eukaryotic organism — roughly, those organisms whose cells contain a nucleus — has been sequenced to 100%. There are regions, often highly repetitive, that are difficult or impossible to clone (one of the initial steps in a sequencing project) or sequence with current technology. Fortunately, such regions are expected to contain relatively few protein-coding genes[4,10].

The extent of these regions varies widely in different species. So, rather than applying a universal gold standard, each sequencing project has made pragmatic decisions as to what constitutes a sufficient level of coverage for a particular genome. For example, as much as one-third of the sequence of the fruitfly *Drosophila melanogaster* was not stable in the cloning systems used, and so was not sequenced. But 97% of the so-called euchromatic portion — where most genes are thought to reside — was sequenced[11] (Fig. 1).

For the human genome, one definition of 'finished' is that fewer than one base in 10,000 is incorrectly assigned[6]; more than 95% of the euchromatic regions are sequenced; and each gap is smaller than 150 kilobases[12]. Such standards represent realistic goals given current technology. By this standard, over a quarter of the public consortium's sequence[1] is considered finished at present, including the previously published long arms of chromosomes 21 and 22 (refs 3,4; Fig. 1). The Celera sequences of chromosomes 21 and 22 are slightly more gappy than those from the public consortium, but the converse seems to be true for the other chromosomes[2]. But again, as different protocols were used, it is not easy to compare the overall status of the two assemblies. In the longer term, as much of the heterochromatin — which is harder to sequence, and contains few genes — as possible must be sequenced, because we might otherwise miss important features.

P.B. & R.C.

Organism	Year	Millions of bases sequenced	Total coverage (%)	Coverage of euchromatin (%)	Predicted number of genes	Number of genes per million bases sequenced
Saccharomyces cerevisiae	1996	12	93	100	5,800	483
Caenorhabditis elegans	1998	97	99	100	19,099	197
Drosophila melanogaster	2000	116	64	97	13,601	117
Arabidopsis thaliana	2000	115	92	100	25,498	221
Human chromosome 21	2000	34	75	100	225	7
Human chromosome 22	1999	34	70	97	545	16
Human genome rough draft (public sequence)	2001	2,693	84	90	31,780	12
Human genome rough draft (Celera sequence)	2001	2,654	83	88—93	39,114	15

Figure 1 **Sequenced eukaryotic genomes. Total coverage uses an estimate of the total genome size and includes heterochromatin (condensed genomic areas that were originally characterized by staining techniques, and are thought to be highly repetitive and gene-poor). The gene-rich areas make up euchromatin. Gene numbers are taken from the original sequence publications**[1–6,14,15]**; most numbers have since changed slightly and different sources give different estimates depending on protocols. The data for the public consortium's rough draft of the human genome are taken from ref. 1, Table 8, page 872. The estimate of total coverage for the Celera data is based on the public consortium's estimate of the full genome size (3,200 million base pairs); the percentage of euchromatin covered is taken from ref. 2. The predicted numbers of human genes are discussed further in the text.**

Box 2 When is a predicted gene a gene?

How many genes are encoded in the human genome? This is a simple question without — as yet — a straightforward answer[13]. The density of genes in the human genome is much lower than for any other genome sequenced so far (Fig. 1), making it particularly difficult to predict where genes are.

Both Celera and the public sequencing consortium used computational algorithms to model genes and make predictions, but such methods are far from perfect. Not only can the start and end positions of a predicted gene be wrong, but exons (the coding parts of a gene) can be missed entirely or wrongly predicted to exist. To reduce this latter effect, the public sequencing consortium required the exons of predicted genes to be 'confirmed', by showing significant similarity to a known sequence (DNA or protein) in a database. But this requirement might be too conservative, making it difficult to predict the presence of new gene families. Celera has required similar confirmation of predictions, but its mouse-genome sequencing project may have provided evidence for further vertebrate-specific genes.

Spurious prediction is also a problem. All genes are expressed by being copied (transcribed) into messenger RNA; most messenger RNAs are then translated into proteins. But even evidence that a stretch of DNA is transcribed does not definitively show that stretch to be a gene. We do not know how efficiently cells control transcription; indeed, it seems likely that non-gene DNA sequences are transcribed relatively frequently[12]. Nor do we know how well the cell identifies transcripts that cannot be translated into a functioning protein. Moreover, proteins that cannot serve any useful function (for example, because they cannot fold correctly) could be made, but rapidly removed. To arrive at a true set of protein-encoding genes, we cannot rely on computational techniques alone, but must continue to characterize proteins and their functions.

These problems provide scope for estimates of human gene number to vary widely. Although recent estimates are converging in the 30,000–40,000 range (as opposed to earlier estimates of 100,000 or so), it could be many years before we have the final answer. **P.B & R.C.**

statements about which genes are unique to other species and do not have relatives in the human genome. So it might be prudent not to place too much emphasis on such 'missing' genes at this stage. Even so, they are running out of places to hide, particularly because the level of coverage of the human genome is probably higher than reported here[1,2] — there are other chunks of unassembled genome sequence in public databases, such as in independent collections of so-called expressed sequence tags.

But ensuring high quality and high coverage are only two aspects of producing a finished genome. For most biologists, the real interest is in the genes themselves. Here, the picture is less rosy, although the problems are caused not so much by the draft nature of the sequence as by the difficulty in finding genes among the other genomic DNA (Box 2).

Even coming up with a rough count of the number of genes is not straightforward. The public consortium's initial set contains about 32,000 genes, made up of around 15,000 known genes and 17,000 predictions. But these 32,000 genes are estimated to come from around 24,500 actual genes — some predicted genes could be 'pseudogenes', or just fragments of real genes. On the other hand, the sensitivity of prediction tends to be only about 60%, so it is reasonable to assume that another 6,800 or so genes (40% of 17,000) have been overlooked. This is how the present estimate of about 31,000 genes (6,800 plus 24,500) was reached[1]. Celera predicts that there are around 39,000 genes, but warns that the evidence for some 12,000 of these is weak[2]. The two groups use different gene-identification techniques, so these numbers are not directly comparable. Minor changes in procedures or data could alter either figure considerably. For example, such changes led to a recent estimate being lowered[7,8] from 120,000 to fewer than 81,000 — and both now seem untenable. Much is a matter of interpretation.

Fortunately, there is every reason to believe that the quality of gene prediction will rapidly improve, and an experimental technique for doing so is discussed on page 922 (ref. 9). With the sequencing of the genomes of other vertebrates, our ability to detect genes by their similarity to known sequences will get better. This is because, thanks to natural selection, gene sequences tend to be altered less during evolution than the DNA surrounding them. In a couple of years we should have at least a more complete list of testable gene candidates.

Despite all this, the information now available has profound implications. For example, there are already many heavily hunted disease-associated genes that have been identified using the public draft (ref. 1, Table 26, page 912). Together with studies of single nucleotide polymorphisms — the base differences from human to human — the draft also provides a framework for understanding the genetic basis and evolution of many human characteristics.

With the draft in hand, researchers have a new tool for studying the regulatory regions and networks of genes. Comparisons with other genomes should reveal common regulatory elements, and the environments of genes shared with other species may offer insight into function and regulation beyond the level of individual genes. The draft is also a starting point for studies of the three-dimensional packing of the genome into a cell's nucleus. Such packing is likely to influence gene regulation.

On a more applied note, the information can be used to exploit technologies such as chips made using DNA or proteins, complementing more traditional approaches. Such chips could now, for instance, contain all the members of a protein family, making it possible to find out which are active in particular diseased tissues. A new world of biotechnology will provide tools and information by exploiting genome data.

Sequencing the tough leftovers of the human genome will be essential. Without a finished sequence, we will not know what we are missing. Each missed gene is potentially a missed drug target, and even gene-poor areas might be critical for gene regulation. Nevertheless, we must now confront the fact that the era of rapid growth in human genomic information is over. The challenge we face is nothing less than understanding

how this comparatively small set of genes creates the diversity of phenomena and characteristics that we see in human life. The human genome lies before us, ready for interpretation.

Peer Bork and Richard Copley are at EMBL, Meyerhofstrasse 1, 69012 Heidelberg, Germany. Peer Bork is at the Max-Delbrück Center for Molecular Medicine, Robert-Rössle-Strasse 10, 13125 Berlin-Buch, Germany.
e-mails: Peer.Bork@EMBL-Heidelberg.de Richard.Copley@EMBL-Heidelberg.de

1. International Human Genome Sequencing Consortium *Nature* **409**, 860–921 (2001).
2. Venter, J. C. *et al. Science* **291**, 1304–1351 (2001).
3. Dunham, I. *et al. Nature* **402**, 489–495 (1999).
4. The Chromosome 21 Mapping and Sequencing Consortium *Nature* **405**, 311–319 (2000).
5. The *C. elegans* Sequencing Consortium *Science* **282**, 2012–2018 (1998).
6. Collins, F. S. *et al. Science* **282**, 682–689 (1998).
7. Liang, F. *et al. Nature Genet.* **25**, 239–240 (2000).
8. Liang, F. *et al. Nature Genet.* **26**, 501 (2000).
9. Shoemaker, D. D. *et al. Nature* **409**, 922–927 (2001).
10. The Arabidopsis Sequencing Consortium *Cell* **100**, 377–386 (2000).
11. Adams, M. D. *et al. Science* **287**, 2185–2195 (2000).
12. Normile, D. & Pennisi, E. *Science* **285**, 2038–2039 (1999).
13. Aparicio, S. *Nature Genet.* **25**, 129–130 (2000).
14. Goffeau, A. *et al. Nature* **387** (suppl.), 1–105 (1997).
15. The Arabidopsis Genome Initiative *Nature* **408**, 796–815 (2000).

are predicted to exist, 60% have some sequence similarity to proteins from other species whose genomes have been sequenced. Just over 40% of the predicted human proteins share similarity with fruitfly or worm proteins. And 61% of fruitfly proteins, 43% of worm proteins and 46% of yeast proteins have sequence similarities to predicted human proteins.

But what about the proteins whose sequences show no strong similarity to known proteins from other species? Over a third of the yeast, fruitfly, worm and human proteins fall into this class. These proteins might retain similar functions, even though their sequences have diverged. Or they might have acquired species-specific functions.

Alternatively, we may need to entertain the possibility that the open reading frames that encode these proteins are maintained in a new way, one that is independent of the precise amino-acid sequence and thus is free to evolve rapidly. (An open reading frame is the part of a gene encoding the amino-acid sequence of its protein product.) After all, we know that cells have at least one mechanism, called nonsense-mediated decay of mRNA, for detecting imperfect open reading frames irrespective of the amino-acid sequence that they encode[8].

It will be interesting to see the extent to which the number of human proteins in this rapidly evolving class decreases as the genomes of other vertebrates, such as mice, are sequenced. This will give us an indication of just how fast these proteins are changing. Indeed, there is already evidence from studies of flies[9] and worms[10] that these rapidly evolving proteins are less likely to have essential functions, consistent with their being less likely to be conserved during evolution.

Such comparisons of distantly related genomes are fascinating from an evolutionary point of view. But comparison of closely related genomes will be much more important in addressing the key problem now facing genomics — determining the function of individual DNA segments. The concept is simple: segments that have a function are more likely to retain their sequence during evolution than non-functional segments. So DNA segments that are conserved between species are likely to have important functions. The ideal species for comparison are those whose form, physiology and behaviour are as similar as possible, but whose genomes have evolved sufficiently that non-functional sequences have had time to diverge. In practice, there may be no one ideal species, because different genes and regulatory sites evolve at different rates. Nevertheless, this approach has a long history of success, and becomes progressively more efficient as the cost of DNA sequencing declines.

One use of such sequence comparisons is

The draft sequences

Comparing species

Gerald M. Rubin

Comparing the human genome sequences with those of other species will not only reveal what makes us genetically different. It may also help us understand what our genes do.

How are the differences between humans and other organisms reflected in our genomes? How similar are the numbers and types of proteins in humans, fruitflies, worms, plants and yeast? And what does all of this tell us about what makes a species unique? With the publication of the draft human genome sequences, on page 860 of this issue[1] and in this week's *Science*[2], we can start to compare the sequences of vertebrate, invertebrate and plant genomes in an attempt to answer these questions.

An obvious place to start our comparison is the total number of genes in each species. Here is a real surprise: the human genome probably contains between 25,000 and 40,000 genes, only about twice the number needed to make a fruitfly[3], worm[4] or plant[5]. We know that there is a higher degree of 'alternative splicing' in humans than in other species. In other words, there are often many more ways in which a gene's protein-coding sections (exons) can be joined together to create a functional messenger RNA molecule, ready to be translated into protein. So more proteins are encoded per gene in humans than in other species.

Even so, we cannot escape the conclusion

— drawn previously from comparisons of simpler genomes[6] — that physical and behavioural differences between species are not related in any simple way to gene number. Many researchers, struck by the fact that there are four times as many genes in some gene families in the human genome compared with fruitflies[7], extrapolated from these cases and suggested that the human genome might be the product of two doublings of the whole of a simpler genome found in the common ancestor of fruitflies and humans. But, as the analyses of the human genome show[1,2], if such doublings did occur, the evidence for them has since been obscured by massive gene loss and amplification of particular gene families in the human genome.

Individual proteins often feature discrete structural units, called domains, that are conserved in evolution. More than 90% of the domains that can be identified in human proteins are also present in fruitfly and worm proteins, although they have been shuffled to create nearly twice as many different arrangements in humans[1,2]. Thus, vertebrate evolution has required the invention of few new domains. Of the human proteins that

to determine the structure of genes — which parts (the exons) make their way into a functional mRNA molecule and which do not (the introns). The high degree of alternative splicing in vertebrates makes this comparative approach particularly important. Gene-finding computational algorithms cannot easily predict the existence of alternative forms of an mRNA without experimental information, but this information is difficult to come by in the case of rare mRNAs. For example, an exon that is used in only a few cells of the human brain might never be experimentally detected in an mRNA. But that exon's sequence would probably be conserved in the mouse genome.

Comparing the genomes of closely related species can also help in identifying gene-control regions. This approach has been used for over two decades[11], and has been validated by showing that the conserved sequences indeed correspond to functional control elements in individual genes[12]. But this computational problem is more difficult than identifying exons, and it will be challenging to scale up to a genome-wide level. The proteins that control gene expression by recognizing regulatory regions often detect sequence features that elude the best computer algorithms, and may use information from contacts with other proteins that is difficult to model. Proteins are simply cleverer than computers.

That said, our knowledge of the DNA-binding properties of individual proteins, as well as the structural features of the DNA sites to which they bind, continues to increase. Moreover, we can use experimental evidence; for example, genes that are expressed together might be expected to share control elements. And, as methods for comparing sequences continue to improve, we can expect to learn more about elusive features of the genome, such as genes encoding RNAs that do not encode proteins[13], start points of DNA replication, and genetic elements that control chromosome structure.

Gerald M. Rubin is in the Department of Molecular and Cell Biology, University of California at Berkeley, Berkeley, California 94708-3200, and the Howard Hughes Medical Institute, 4000 Jones Bridge Road, Chevy Chase, Maryland 20815-6789, USA.
e-mail: gerry@fruitfly.berkeley.edu

1. International Human Genome Sequencing Consortium *Nature* **409**, 860–921 (2001).
2. Venter, J. C. *et al. Science* **291**, 1304–1351 (2001).
3. Adams, M. D. *et al. Science* **287**, 2185–2195 (2000).
4. The *C. elegans* Sequencing Consortium *Science* **282**, 2012–2018 (1998).
5. The Arabidopsis Genome Initiative *Nature* **408**, 796–815 (2000).
6. Rubin, G. M. *et al. Science* **287**, 2204–2215 (2000).
7. Spring, J. *FEBS Lett.* **400**, 2–8 (1997).
8. Hentze, M. W. & Kulozik, A. E. *Cell* **96**, 307–310 (1999).
9. Ashburner, M. *et al. Genetics* **153**, 179–219 (1999).
10. Fraser, A. G. *et al. Nature* **408**, 325–330 (2000).
11. Ravetch, J. V., Kirsch, I. R. & Leder, P. *Proc. Natl Acad. Sci. USA* **77**, 6734–6738 (1980).
12. Fortini, M. E. & Rubin, G. M. *Genes Dev.* **4**, 444–463 (1990).
13. Lee, R. C., Feinbaum, R. L. & Ambros, V. *Cell* **75**, 843–854 (1993).

Single nucleotide polymorphisms

From the evolutionary past...

Mark Stoneking

Single nucleotide polymorphisms are the bread-and-butter of DNA sequence variation. They provide a rich source of information about the evolutionary history of human populations.

Studies of genetic variation in human populations began inauspiciously[1]. The first such study — of ABO blood-group frequencies — was carried out by two Polish immunologists, Ludwik and Hanka Hirszfeld, at the end of the First World War. This work was notable for its broad coverage of the world's populations, large sample sizes and scrupulous attention to anthropological details. Yet the Hirszfelds still ran into difficulties in publishing in *The Lancet*, the premier medical journal of the time. The editor could not see the relevance of their work, and so this seminal study of human genetic variation first appeared in an obscure anthropological journal[2]. The relevance became abundantly clear when Felix Bernstein subsequently used the Hirszfelds' data to demonstrate that the ABO blood-group frequencies were better explained by a single gene with three variants (alleles), and not — as prevailing wisdom then held — two genes each with two alleles[3].

Happily, times have changed, diversity is now all the rage[4,5], and editors have become more appreciative of the importance of human genetic variation. The latest evidence of that is the paper on page 928 of this issue[6], which reports the identification and mapping of 1.4 million single nucleotide polymorphisms (SNPs, pronounced 'snips') in the human genome. The paper is the result of the labours of a large collaboration, The International SNP Map Working Group.

So, what are SNPs? Quite simply, they are the bread-and-butter of DNA sequence variation — polymorphism, to those in the business. A DNA sequence is a linear combination of four nucleotides; compare two sequences, position by position, and wherever you come across different nucleotides at the same position, that's a SNP (see Fig. 1 on page 823). So SNPs reflect past mutations that were mostly (but not exclusively) unique events, and two individuals sharing a variant allele are thereby marked with a common evolutionary heritage. In other words, our genes have ancestors, and analysing shared patterns of SNP variation can identify them.

However, the real importance of SNPs is that there are so many of them. One estimate[7] is that comparing two human DNA sequences results in a SNP every 1,000–2,000 nucleotides. That may not sound like much until you realize that there are 3.2 billion nucleotides in the human genome, which translates into 1.6 million–3.2 million SNPs. And that's just from comparing two sequences — the total number of SNPs in humans is obviously much more. Most human variation that is influenced by genes can be traced to SNPs, especially in such medically (and commercially) important traits as how likely you are to become afflicted with a particular disease, or how you might respond to a particular pharmaceutical treatment, as discussed by Chakravarti[8] on the following page. And even when a SNP is not directly responsible, the sheer number of SNPs means they can also be used to locate genes that influence such traits.

The deluge of SNPs reported by the SNP working group[6] also promises great things for those of us who analyse patterns of molecular genetic variation to reconstruct the evolutionary history of human populations. Our genes contain the signature of an expansion from Africa within the past 150,000 years or so[9]. But there is still debate as to whether the modern humans from Africa completely replaced archaic non-African populations with no interbreeding, or whether we perhaps carry the vestiges of Neanderthal or other archaic non-African genes.

Demonstrating a recent African origin for every single one of our 3.2 billion nucleotides goes beyond the bounds of reason or necessity, but there is still much to be learned. For a start, most of our insights into molecular anthropology arise from DNA in mitochondria and (more recently) polymorphisms of the Y chromosome. This is because these DNA sequences are haploid — that is, represented just once in each cell, in contrast to the other chromosomes, which are represented twice — and they are inherited from just one parent, so they do not undergo the usual sequence shuffling (recombination) during egg and sperm production. This makes them easier to analyse and extremely informative. But both suffer from the drawback that, in the absence of recombination, they behave as single genes, and the history of any single gene can differ from that of a population or species because of natural selection or chance events involving that gene.

Accurate inferences concerning popula-

69

tion history demand the analysis of several genes, with the most promising approach involving haplotypes[10], which consist of several closely spaced (linked) polymorphisms. The advantage of haplotypes over simply analysing polymorphisms at random is that there is valuable information in the associations between linked polymorphisms — the whole is greater than the sum of the parts. So the 1.4 million SNPs are a welcome resource that will greatly help in identifying haplotypes for tracing human evolutionary history, especially those that might reveal archaic non-African ancestry.

However, answering all of our questions about human evolutionary history will not be as simple as mining the SNP database and determining haplotypes in a representative sample of worldwide populations. There are four main reasons for that.

First, to be really useful, the SNPs in the database should really be SNPs, and not errors or artefacts, and they should be polymorphic in other samples, not just the sample of individuals used to find the SNPs. An important aspect of the SNP working group's data is that 1,585 SNPs were chosen for further verification, of which about 95% turned out to be true SNPs, which is good news indeed. Moreover, 1,276 SNPs were tested on additional population samples and at least 82% were polymorphic, which is reassuring.

Second, one might ask why only 0.1% of the 1.4 million SNPs were verified and tested. The answer is that our ability to determine allele frequencies efficiently and inexpensively for large numbers of SNPs lags behind our ability to simply identify them. This situation is reminiscent of the beginnings of the Human Genome Project, when developing technology was a primary concern and it was not at all clear how the 3.2 billion nucleotides were going to be determined. But human ingenuity won out then, and given the number of bright and capable minds now wrestling with the SNP-typing problem, one or more solutions should soon be at hand (especially with the motivation of lucrative commercial applications).

Third, a problem known as ascertainment bias can complicate the interpretation of results based on SNPs. For example, SNPs that were found to be polymorphic in European populations will overestimate genetic diversity in European as opposed to non-European populations. Moreover, the probability of finding a SNP, and the frequency of polymorphism at a SNP, depends on how many times a particular DNA segment was sequenced, and from how many individuals. The SNP working group report some intriguing preliminary findings regarding how SNP diversity is apportioned among chromosomes. But further work is required to see if these are truly biological differences, or if they instead reflect

ascertainment biases. Ascertainment bias is not an insurmountable problem — statistical geneticists love this sort of challenge and are already coming up with creative solutions[11]. Even so, SNP-finders must keep careful track of how their SNPs were ascertained.

Fourth, the emphasis in the SNP database is on SNPs where both of the alleles occur at high frequency, because these will be most useful for disease-association studies. In general, the higher the frequency of a SNP allele, the older the mutation that produced it, so high-frequency SNPs largely predate human population diversification. But many questions in human evolution involve specific migrations (such as the colonization of Polynesia or the Americas) for which population-specific alleles are most informative — indeed, this is one of the attractions of mitochondrial-DNA and Y-chromosome analyses for such questions, because population-specific alleles can be readily found. It is unlikely that Polynesian-specific SNPs are present in the database, so more work will be required to find such informative, population-specific SNPs.

Still, one can imagine that in the not-too-distant future the details of human population history will have been fleshed out, at least to the extent possible by analysing genetic variation in extant populations. What then? One area that is receiving increasing attention is the detection of the effects of natural selection in human populations[12]. Using SNPs to find chromosomal regions with abnormally low levels of varia-

tion is a particularly promising way of detecting the genomic signature of selection for favourable mutations[13].

Another area of increasing interest is identifying the molecular genetic basis of 'normal' phenotypic variation[4] — that is, variation of the old-fashioned, morphological kind, which is a traditional concern of anthropology. Molecular anthropology has for the most part concentrated on the molecules and what their diversity tells us about human evolution. With the advent of the human genome sequence and the SNP database, the ultimate in molecular tools, we are ironically now poised to focus on phenotypes and what their diversity tells us about human evolution — thereby bringing the anthropology back into molecular anthropology.

Mark Stoneking is at the Max Planck Institute for Evolutionary Anthropology, Inselstrasse 22, D-04103 Leipzig, Germany.
e-mail: stoneking@eva.mpg.de

1. Mourant, A. E. *Blood Relations* p.13 (Oxford Univ. Press, 1983).
2. Hirszfeld, L. & Hirszfeld, H. *Anthropologie* **29**, 505–537 (1919).
3. Crow, J. F. *Genetics* **133**, 4–7 (1993).
4. Weiss, K. M. *Genome Res.* **8**, 691–697 (1998).
5. Collins, F. S., Brooks, L. D. & Chakravarti, A. *Genome Res.* **8**, 1229–1231 (1998).
6. The International SNP Map Working Group *Nature* **409**, 928–933 (2001).
7. Li, W. H. & Sadler, L. A. *Genetics* **129**, 513–523 (1991).
8. Chakravarti, A. *Nature* **409**, 822–823 (2001).
9. Stoneking, M. *Evol. Anthropol.* **2**, 60–73 (1993).
10. Tishkoff, S. A. *et al. Science* **271**, 1380–1387 (1996).
11. Kuhner, M. K., Beerli, P., Yamato, J. & Felsenstein, J. *Genetics* **156**, 439–447 (2000).
12. Przeworski, M., Hudson, R. R. & Di Rienzo, A. *Trends Genet.* **16**, 296–302 (2000).
13. Nurminsky, D., De Aguiar, D., Bustamante, C. D. & Hartl, D. L. *Science* **291**, 128–130 (2001).

Single nucleotide polymorphisms

...to a future of genetic medicine

Aravinda Chakravarti

Single base differences between human genomes underlie differences in susceptibility to, or protection from, a host of diseases. Hence the great potential of such information in medicine.

The beginning of the Human Genome Project, over a decade ago, was accompanied by a cantankerous debate over whose genome was to be sequenced. Would it be a single individual? A celebrity, perhaps (widely rumoured to be Jim Watson, co-discoverer of the structure of DNA)? Or would several genomes, from many individuals, be studied? The discussion struck at the very heart of genetics. As the study of inherited variation between individuals, genetics might not immediately benefit from the sequence of a single genome. But even one genome would be immensely revealing to the science of deciphering the molecular blueprint of a species. Fortunately, geneticists were not forced to make this choice. Papers in this issue describe not only a single,

history-making human genome sequence, composed of little bits from many humans[1] (page 860), but also some 1.4 million sites of variation mapped along that reference sequence[2] (page 928).

But why this preoccupation with sequence variation, with the fact that no two humans (except identical twins) are genetically the same? The answer is that such variations, or 'polymorphisms', are markers of genes and genomes with which researchers perform genetic analysis in an outbred species where matings cannot be controlled. The fields of human and medical genetics simply cannot exist without understanding this variation.

It has become clear that the two 'genomes' that each of us carry, inherited

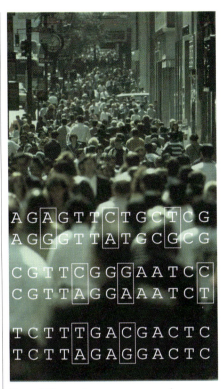

Figure 1 **The most common sources of variation between humans are single nucleotide polymorphisms (SNPs) — single base differences between genome sequences. Fragments of two sequences, with eight SNPs, are shown.**

from our parents, most often differ — from each other, and from the genomes of other humans — in terms of single base changes[1] (Fig. 1). The twentieth century saw the identification of only a few thousand of these so-called single nucleotide polymorphisms (SNPs, or 'snips' to the streetwise). In just the first year of the new century, this number has been increased one-thousand-fold[2]. Beyond the numbers, the excitement today comes from precise knowledge of where these sites of variation are in the genome[2]. The 1.42 million known SNPs are found at a density of one SNP per 1.91 kilobases. This means that more than 90% of any stretches of sequence 20 kilobases long will contain one or more SNPs. The density is even higher in regions containing genes. The International SNP Map Working Group[2] estimates that they have identified 60,000 SNPs within genes ('coding' SNPs), or one coding SNP per 1.08 kilobases of gene sequence. Moreover, 93% of genes contain a SNP, and 98% are within 5 kilobases of a SNP. For the first time, nearly every human gene and genomic region is marked by a sequence variation.

These data provide interesting first glimpses into the pattern of variation across the genome. Variation is commonly assessed by nucleotide diversity — the number of base differences between two genomes, divided by the number of base pairs

compared. Nucleotide diversity is a sensitive indicator of biological and historical factors that have affected the human genome[3]. The nucleotide diversity in gene-containing regions has been estimated to be 8 differences per 10 kilobases[4,5]; we now know that the genome-wide average is similar, 7.51 differences per 10 kilobases (ref. 2). The variation between individual non-sex chromosomes is small, and lies in the range 5.19 (for chromosome 21) to 8.79 (for chromosome 15) differences per 10 kilobases (ref. 2).

Strikingly, humans vary least in their sex chromosomes. The variation between different X chromosomes is about 4.69 differences per 10 kilobases, and it is very much lower for the Y chromosome (1.51 differences per 10 kilobases). This is because the sex chromosomes have patterns of mutation and recombination (the swapping of similar DNA segments during the generation of eggs and sperm) that differ both from each other and from the non-sex chromosomes. Moreover, fewer ancestors have contributed to the sex chromosomes, which are therefore less variable than the non-sex chromosomes.

Perhaps not surprisingly, some genomic regions have significantly lower or higher diversity than the average. For example, the HLA locus, which encodes proteins that present antigens to the immune system, shows the greatest diversity. Such comparisons within genomes will be essential to our understanding of how variation shapes biochemical and cellular functions, and in illuminating past human evolution, as discussed in ref. 3, and by Stoneking in the preceding article (page 821; ref. 6).

But the main use of the human SNP map will be in dissecting the contributions of individual genes to diseases that have a complex, multigene basis. Knowledge of genetic variation already affects patient care to some degree. For example, gene variants lead to tissue and organ incompatibility, affecting the success of transplants. And the mainstay of medical genetics has been the study of the rare gene variants that lie behind inherited diseases such as cystic fibrosis.

But variations in genome sequences underlie differences in our susceptibility to, or protection from, all kinds of diseases; in the age of onset and severity of illness; and in the way our bodies respond to treatment. For example, we already know that single base differences in the *APOE* gene are associated with Alzheimer's disease, and that a simple deletion within the chemokine-receptor gene *CCR5* leads to resistance to HIV and AIDS. The benefit of the SNP map is that it covers the entire genome. So, by comparing patterns and frequencies of SNPs in patients and controls, researchers can identify which SNPs are associated with which diseases[7–9]. Such research will bring about 'genetic medicine', in which knowledge of our uniqueness

will alter all aspects of medicine, perceptibly and forever.

Studies of SNPs and diseases will become more efficient when a few more problems are solved[3]. First, although 82% of SNP variants are found at a frequency of more than 10% in the global human population, the 'microdistribution' of SNPs in individual populations is not known. Second, not all SNPs are created equal, and it will be essential to know as much as possible about their effects from computational analyses before studying their involvement in disease. For example, each SNP can be classified by whether it is coding or not. Coding SNPs can be classified by whether they alter the sequence of the protein encoded by the altered gene. Changes that alter protein sequences can be classified by their effects on protein structure. And non-coding SNPs can be classified according to whether they are found in gene-regulating segments of the genome[10] — many complex diseases may arise from quantitative, rather than qualitative, differences in gene products. Third, the technology for assaying thousands of SNPs, in thousands of patients and controls[7], is not yet fully developed, although there are some creative ideas around.

In the twentieth century, humans were not the geneticists' species of choice. The emphasis then was on understanding gene structure and function. Now, geneticists will concentrate increasingly on understanding physical and behavioural characteristics. Here, our species, with its obsession with self-examination, will make a superior subject. We will also see more studies of how natural variation leads to each one of our qualities. To some, there is a danger of genomania, with all differences (or similarities, for that matter) being laid at the altar of genetics[11]. But I hope this does not happen. Genes and genomes do not act in a vacuum, and the environment is equally important in human biology. By identifying variation across the whole genome, the SNP map[2] may be our best route yet to a better understanding of the roles of nature and (not versus) nurture. ■

Aravinda Chakravarti is at the McKusick–Nathans Institute of Genetic Medicine, Johns Hopkins University School of Medicine, 600 North Wolfe Street, Jefferson Street Building 2-109, Baltimore, Maryland 21287, USA.
e-mail: aravinda@jhmi.edu

1. International Human Genome Sequencing Consortium *Nature* **409**, 860–921 (2001).
2. The International SNP Map Working Group *Nature* **409**, 928–933 (2001).
3. Chakravarti, A. *Nature Genet.* **21** (suppl.), 56–60 (1999).
4. Halushka, M. K. *et al. Nature Genet.* **22**, 239–247 (1999).
5. Cargill, M. *et al. Nature Genet.* **22**, 231–238 (1999).
6. Stoneking, M. *Nature* **409**, 821–822 (2001).
7. Risch, N. & Merikangas, K. *Science* **273**, 1516–1517 (1996).
8. Lander, E. S. *Science* **274**, 536–539 (1996).
9. Collins, F. S., Guyer, M. S. & Chakravarti, A. *Science* **278**, 1580–1581 (1997).
10. Loots, G. G. *et al. Science* **288**, 136–140 (1999).
11. Lewontin, R. *It Ain't Necessarily So: The Dream of the Human Genome and Other Illusions* (New York Review of Books, 2000).

Initial sequencing and analysis of the human genome

International Human Genome Sequencing Consortium*

A partial list of authors appears on the opposite page. Affiliations are listed at the end of the paper.

The human genome holds an extraordinary trove of information about human development, physiology, medicine and evolution. Here we report the results of an international collaboration to produce and make freely available a draft sequence of the human genome. We also present an initial analysis of the data, describing some of the insights that can be gleaned from the sequence.

The rediscovery of Mendel's laws of heredity in the opening weeks of the 20th century[1-3] sparked a scientific quest to understand the nature and content of genetic information that has propelled biology for the last hundred years. The scientific progress made falls naturally into four main phases, corresponding roughly to the four quarters of the century. The first established the cellular basis of heredity: the chromosomes. The second defined the molecular basis of heredity: the DNA double helix. The third unlocked the informational basis of heredity, with the discovery of the biological mechanism by which cells read the information contained in genes and with the invention of the recombinant DNA technologies of cloning and sequencing by which scientists can do the same.

The last quarter of a century has been marked by a relentless drive to decipher first genes and then entire genomes, spawning the field of genomics. The fruits of this work already include the genome sequences of 599 viruses and viroids, 205 naturally occurring plasmids, 185 organelles, 31 eubacteria, seven archaea, one fungus, two animals and one plant.

Here we report the results of a collaboration involving 20 groups from the United States, the United Kingdom, Japan, France, Germany and China to produce a draft sequence of the human genome. The draft genome sequence was generated from a physical map covering more than 96% of the euchromatic part of the human genome and, together with additional sequence in public databases, it covers about 94% of the human genome. The sequence was produced over a relatively short period, with coverage rising from about 10% to more than 90% over roughly fifteen months. The sequence data have been made available without restriction and updated daily throughout the project. The task ahead is to produce a finished sequence, by closing all gaps and resolving all ambiguities. Already about one billion bases are in final form and the task of bringing the vast majority of the sequence to this standard is now straightforward and should proceed rapidly.

The sequence of the human genome is of interest in several respects. It is the largest genome to be extensively sequenced so far, being 25 times as large as any previously sequenced genome and eight times as large as the sum of all such genomes. It is the first vertebrate genome to be extensively sequenced. And, uniquely, it is the genome of our own species.

Much work remains to be done to produce a complete finished sequence, but the vast trove of information that has become available through this collaborative effort allows a global perspective on the human genome. Although the details will change as the sequence is finished, many points are already clear.

● The genomic landscape shows marked variation in the distribution of a number of features, including genes, transposable elements, GC content, CpG islands and recombination rate. This gives us important clues about function. For example, the developmentally important HOX gene clusters are the most repeat-poor regions of the human genome, probably reflecting the very complex coordinate regulation of the genes in the clusters.

● There appear to be about 30,000–40,000 protein-coding genes in the human genome—only about twice as many as in worm or fly. However, the genes are more complex, with more alternative splicing generating a larger number of protein products.

● The full set of proteins (the 'proteome') encoded by the human genome is more complex than those of invertebrates. This is due in part to the presence of vertebrate-specific protein domains and motifs (an estimated 7% of the total), but more to the fact that vertebrates appear to have arranged pre-existing components into a richer collection of domain architectures.

● Hundreds of human genes appear likely to have resulted from horizontal transfer from bacteria at some point in the vertebrate lineage. Dozens of genes appear to have been derived from transposable elements.

● Although about half of the human genome derives from transposable elements, there has been a marked decline in the overall activity of such elements in the hominid lineage. DNA transposons appear to have become completely inactive and long-terminal repeat (LTR) retroposons may also have done so.

● The pericentromeric and subtelomeric regions of chromosomes are filled with large recent segmental duplications of sequence from elsewhere in the genome. Segmental duplication is much more frequent in humans than in yeast, fly or worm.

● Analysis of the organization of Alu elements explains the long-standing mystery of their surprising genomic distribution, and suggests that there may be strong selection in favour of preferential retention of Alu elements in GC-rich regions and that these 'selfish' elements may benefit their human hosts.

● The mutation rate is about twice as high in male as in female meiosis, showing that most mutation occurs in males.

● Cytogenetic analysis of the sequenced clones confirms suggestions that large GC-poor regions are strongly correlated with 'dark G-bands' in karyotypes.

● Recombination rates tend to be much higher in distal regions (around 20 megabases (Mb)) of chromosomes and on shorter chromosome arms in general, in a pattern that promotes the occurrence of at least one crossover per chromosome arm in each meiosis.

● More than 1.4 million single nucleotide polymorphisms (SNPs) in the human genome have been identified. This collection should allow the initiation of genome-wide linkage disequilibrium mapping of the genes in the human population.

In this paper, we start by presenting background information on the project and describing the generation, assembly and evaluation of the draft genome sequence. We then focus on an initial analysis of the sequence itself: the broad chromosomal landscape; the repeat elements and the rich palaeontological record of evolutionary and biological processes that they provide; the human genes and proteins and their differences and similarities with those of other

Genome Sequencing Centres (Listed in order of total genomic sequence contributed, with a partial list of personnel. A full list of contributors at each centre is available as Supplementary Information.)

Whitehead Institute for Biomedical Research, Center for Genome Research: Eric S. Lander[1]*, Lauren M. Linton[1], Bruce Birren[1]*, Chad Nusbaum[1]*, Michael C. Zody[1], Jennifer Baldwin[1], Keri Devon[1], Ken Dewar[1], Michael Doyle[1], William FitzHugh[1]*, Roel Funke[1], Diane Gage[1], Katrina Harris[1], Andrew Heaford[1], John Howland[1], Lisa Kann[1], Jessica Lehoczky[1], Rosie LeVine[1], Paul McEwan[1], Kevin McKernan[1], James Meldrim[1], Jill P. Mesirov[1]*, Cher Miranda[1], William Morris[1], Jerome Naylor[1], Christina Raymond[1], Mark Rosetti[1], Ralph Santos[1], Andrew Sheridan[1], Carrie Sougnez[1], Nicole Stange-Thomann[1], Nikola Stojanovic[1], Aravind Subramanian[1] & Dudley Wyman[1]

The Sanger Centre: Jane Rogers[2], John Sulston[2]*, Rachael Ainscough[2], Stephan Beck[2], David Bentley[2], John Burton[2], Christopher Clee[2], Nigel Carter[2], Alan Coulson[2], Rebecca Deadman[2], Panos Deloukas[2], Andrew Dunham[2], Ian Dunham[2], Richard Durbin[2]*, Lisa French[2], Darren Grafham[2], Simon Gregory[2], Tim Hubbard[2]*, Sean Humphray[2], Adrienne Hunt[2], Matthew Jones[2], Christine Lloyd[2], Amanda McMurray[2], Lucy Matthews[2], Simon Mercer[2], Sarah Milne[2], James C. Mullikin[2]*, Andrew Mungall[2], Robert Plumb[2], Mark Ross[2], Ratna Shownkeen[2] & Sarah Sims[2]

Washington University Genome Sequencing Center: Robert H. Waterston[3]*, Richard K. Wilson[3], LaDeana W. Hillier[3]*, John D. McPherson[3], Marco A. Marra[3], Elaine R. Mardis[3], Lucinda A. Fulton[3], Asif T. Chinwalla[3]*, Kymberlie H. Pepin[3], Warren R. Gish[3], Stephanie L. Chissoe[3], Michael C. Wendl[3], Kim D. Delehaunty[3], Tracie L. Miner[3], Andrew Delehaunty[3], Jason B. Kramer[3], Lisa L. Cook[3], Robert S. Fulton[3], Douglas L. Johnson[3], Patrick J. Minx[3] & Sandra W. Clifton[3]

US DOE Joint Genome Institute: Trevor Hawkins[4], Elbert Branscomb[4], Paul Predki[4], Paul Richardson[4], Sarah Wenning[4], Tom Slezak[4], Norman Doggett[4], Jan-Fang Cheng[4], Anne Olsen[4], Susan Lucas[4], Christopher Elkin[4], Edward Uberbacher[4] & Marvin Frazier[4]

Baylor College of Medicine Human Genome Sequencing Center: Richard A. Gibbs[5]*, Donna M. Muzny[5], Steven E. Scherer[5], John B. Bouck[5]*, Erica J. Sodergren[5], Kim C. Worley[5]*, Catherine M. Rives[5], James H. Gorrell[5], Michael L. Metzker[5], Susan L. Naylor[6], Raju S. Kucherlapati[7], David L. Nelson, & George M. Weinstock[8]

RIKEN Genomic Sciences Center: Yoshiyuki Sakaki[9], Asao Fujiyama[9], Masahira Hattori[9], Tetsushi Yada[9], Atsushi Toyoda[9], Takehiko Itoh[9], Chiharu Kawagoe[9], Hidemi Watanabe[9], Yasushi Totoki[9] & Todd Taylor[9]

Genoscope and CNRS UMR-8030: Jean Weissenbach[10], Roland Heilig[10], William Saurin[10], Francois Artiguenave[10], Philippe Brottier[10], Thomas Bruls[10], Eric Pelletier[10], Catherine Robert[10] & Patrick Wincker[10]

GTC Sequencing Center: Douglas R. Smith[11], Lynn Doucette-Stamm[11], Marc Rubenfield[11], Keith Weinstock[11], Hong Mei Lee[11] & JoAnn Dubois[11]

Department of Genome Analysis, Institute of Molecular

Biotechnology: André Rosenthal[12], Matthias Platzer[12], Gerald Nyakatura[12], Stefan Taudien[12] & Andreas Rump[12]

Beijing Genomics Institute/Human Genome Center: Huanming Yang[13], Jun Yu[13], Jian Wang[13], Guyang Huang[14] & Jun Gu[15]

Multimegabase Sequencing Center, The Institute for Systems Biology: Leroy Hood[16], Lee Rowen[16], Anup Madan[16] & Shizen Qin[16]

Stanford Genome Technology Center: Ronald W. Davis[17], Nancy A. Federspiel[17], A. Pia Abola[17] & Michael J. Proctor[17]

Stanford Human Genome Center: Richard M. Myers[18], Jeremy Schmutz[18], Mark Dickson[18], Jane Grimwood[18] & David R. Cox[18]

University of Washington Genome Center: Maynard V. Olson[19], Rajinder Kaul[19] & Christopher Raymond[19]

Department of Molecular Biology, Keio University School of Medicine: Nobuyoshi Shimizu[20], Kazuhiko Kawasaki[20] & Shinsei Minoshima[20]

University of Texas Southwestern Medical Center at Dallas: Glen A. Evans[21]†, Maria Athanasiou[21] & Roger Schultz[21]

University of Oklahoma's Advanced Center for Genome Technology: Bruce A. Roe[22], Feng Chen[22] & Huaqin Pan[22]

Max Planck Institute for Molecular Genetics: Juliane Ramser[23], Hans Lehrach[23] & Richard Reinhardt[23]

Cold Spring Harbor Laboratory, Lita Annenberg Hazen Genome Center: W. Richard McCombie[24], Melissa de la Bastide[24] & Neilay Dedhia[24]

GBF—German Research Centre for Biotechnology: Helmut Blöcker[25], Klaus Hornischer[25] & Gabriele Nordsiek[25]

* Genome Analysis Group (listed in alphabetical order, also includes individuals listed under other headings): Richa Agarwala[26], L. Aravind[26], Jeffrey A. Bailey[27], Alex Bateman[2], Serafim Batzoglou[1], Ewan Birney[28], Peer Bork[29,30], Daniel G. Brown[1], Christopher B. Burge[31], Lorenzo Cerutti[28], Hsiu-Chuan Chen[26], Deanna Church[26], Michele Clamp[2], Richard R. Copley[30], Tobias Doerks[29,30], Sean R. Eddy[32], Evan E. Eichler[27], Terrence S. Furey[33], James Galagan[1], James G. R. Gilbert[2], Cyrus Harmon[34], Yoshihide Hayashizaki[35], David Haussler[36], Henning Hermjakob[28], Karsten Hokamp[37], Wonhee Jang[26], L. Steven Johnson[32], Thomas A. Jones[32], Simon Kasif[38], Arek Kaspryzk[28], Scot Kennedy[39], W. James Kent[40], Paul Kitts[26], Eugene V. Koonin[26], Ian Korf[3], David Kulp[34], Doron Lancet[41], Todd M. Lowe[42], Aoife McLysaght[37], Tarjei Mikkelsen[38], John V. Moran[43], Nicola Mulder[28], Victor J. Pollara[1], Chris P. Ponting[44], Greg Schuler[26], Jörg Schultz[30], Guy Slater[28], Arian F. A. Smit[45], Elia Stupka[28], Joseph Szustakowki[38], Danielle Thierry-Mieg[26], Jean Thierry-Mieg[26], Lukas Wagner[26], John Wallis[3], Raymond Wheeler[34], Alan Williams[34], Yuri I. Wolf[26], Kenneth H. Wolfe[37], Shiaw-Pyng Yang[3] & Ru-Fang Yeh[31]

Scientific management: National Human Genome Research Institute, US National Institutes of Health: Francis Collins[46]*, Mark S. Guyer[46], Jane Peterson[46], Adam Felsenfeld[46]* & Kris A. Wetterstrand[46]; **Office of Science, US Department of Energy:** Aristides Patrinos[47]; **The Wellcome Trust:** Michael J. Morgan[48]

organisms; and the history of genomic segments. (Comparisons are drawn throughout with the genomes of the budding yeast *Saccharomyces cerevisiae*, the nematode worm *Caenorhabditis elegans*, the fruitfly *Drosophila melanogaster* and the mustard weed *Arabidopsis thaliana*; we refer to these for convenience simply as yeast, worm, fly and mustard weed.) Finally, we discuss applications of the sequence to biology and medicine and describe next steps in the project. A full description of the methods is provided as Supplementary Information on *Nature*'s web site (http://www.nature.com).

We recognize that it is impossible to provide a comprehensive analysis of this vast dataset, and thus our goal is to illustrate the range of insights that can be gleaned from the human genome and thereby to sketch a research agenda for the future.

Background to the Human Genome Project

The Human Genome Project arose from two key insights that emerged in the early 1980s: that the ability to take global views of genomes could greatly accelerate biomedical research, by allowing researchers to attack problems in a comprehensive and unbiased fashion; and that the creation of such global views would require a communal effort in infrastructure building, unlike anything previously attempted in biomedical research. Several key projects helped to crystallize these insights, including:

(1) The sequencing of the bacterial viruses ΦX174[4,5] and lambda[6], the animal virus SV40[7] and the human mitochondrion[8] between 1977 and 1982. These projects proved the feasibility of assembling small sequence fragments into complete genomes, and showed the value of complete catalogues of genes and other functional elements.

(2) The programme to create a human genetic map to make it possible to locate disease genes of unknown function based solely on their inheritance patterns, launched by Botstein and colleagues in 1980 (ref. 9).

(3) The programmes to create physical maps of clones covering the yeast[10] and worm[11] genomes to allow isolation of genes and regions based solely on their chromosomal position, launched by Olson and Sulston in the mid-1980s.

(4) The development of random shotgun sequencing of complementary DNA fragments for high-throughput gene discovery by Schimmel[12] and Schimmel and Sutcliffe[13], later dubbed expressed sequence tags (ESTs) and pursued with automated sequencing by Venter and others[14–20].

The idea of sequencing the entire human genome was first proposed in discussions at scientific meetings organized by the US Department of Energy and others from 1984 to 1986 (refs 21, 22). A committee appointed by the US National Research Council endorsed the concept in its 1988 report[23], but recommended a broader programme, to include: the creation of genetic, physical and sequence maps of the human genome; parallel efforts in key model organisms such as bacteria, yeast, worms, flies and mice; the development of technology in support of these objectives; and research into the ethical, legal and social issues raised by human genome research. The programme was launched in the US as a joint effort of the Department of Energy and the National Institutes of Health. In other countries, the UK Medical Research Council and the Wellcome Trust supported genomic research in Britain; the Centre d'Etude du Polymorphisme Humain and the French Muscular Dystrophy Association launched mapping efforts in France; government agencies, including the Science and Technology Agency and the Ministry of Education, Science, Sports and Culture supported genomic research efforts in Japan; and the European Community helped to launch several international efforts, notably the programme to sequence the yeast genome. By late 1990, the Human Genome Project had been launched, with the creation of genome centres in these countries. Additional participants subsequently joined the effort, notably in Germany and China. In addition, the Human Genome Organization (HUGO) was founded to provide a forum for international coordination of genomic research. Several books[24–26] provide a more comprehensive discussion of the genesis of the Human Genome Project.

Through 1995, work progressed rapidly on two fronts (Fig. 1). The first was construction of genetic and physical maps of the human and mouse genomes[27–31], providing key tools for identification of disease genes and anchoring points for genomic sequence. The second was sequencing of the yeast[32] and worm[33] genomes, as

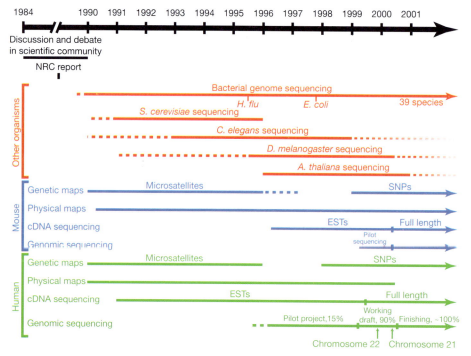

Figure 1 Timeline of large-scale genomic analyses. Shown are selected components of work on several non-vertebrate model organisms (red), the mouse (blue) and the human (green) from 1990; earlier projects are described in the text. SNPs, single nucleotide polymorphisms; ESTs, expressed sequence tags.

well as targeted regions of mammalian genomes[34–37]. These projects showed that large-scale sequencing was feasible and developed the two-phase paradigm for genome sequencing. In the first, 'shotgun', phase, the genome is divided into appropriately sized segments and each segment is covered to a high degree of redundancy (typically, eight- to tenfold) through the sequencing of randomly selected subfragments. The second is a 'finishing' phase, in which sequence gaps are closed and remaining ambiguities are resolved through directed analysis. The results also showed that complete genomic sequence provided information about genes, regulatory regions and chromosome structure that was not readily obtainable from cDNA studies alone.

In 1995, genome scientists considered a proposal[38] that would have involved producing a draft genome sequence of the human genome in a first phase and then returning to finish the sequence in a second phase. After vigorous debate, it was decided that such a plan was premature for several reasons. These included the need first to prove that high-quality, long-range finished sequence could be produced from most parts of the complex, repeat-rich human genome; the sense that many aspects of the sequencing process were still rapidly evolving; and the desirability of further decreasing costs.

Instead, pilot projects were launched to demonstrate the feasibility of cost-effective, large-scale sequencing, with a target completion date of March 1999. The projects successfully produced finished sequence with 99.99% accuracy and no gaps[39]. They also introduced bacterial artificial chromosomes (BACs)[40], a new large-insert cloning system that proved to be more stable than the cosmids and yeast artificial chromosomes (YACs)[41] that had been used previously. The pilot projects drove the maturation and convergence of sequencing strategies, while producing 15% of the human genome sequence. With successful completion of this phase, the human genome sequencing effort moved into full-scale production in March 1999.

The idea of first producing a draft genome sequence was revived at this time, both because the ability to finish such a sequence was no longer in doubt and because there was great hunger in the scientific community for human sequence data. In addition, some scientists favoured prioritizing the production of a draft genome sequence over regional finished sequence because of concerns about commercial plans to generate proprietary databases of human sequence that might be subject to undesirable restrictions on use[42–44].

The consortium focused on an initial goal of producing, in a first production phase lasting until June 2000, a draft genome sequence covering most of the genome. Such a draft genome sequence, although not completely finished, would rapidly allow investigators to begin to extract most of the information in the human sequence. Experiments showed that sequencing clones covering about 90% of the human genome to a redundancy of about four- to fivefold ('half-shotgun' coverage; see Box 1) would accomplish this[45,46]. The draft genome sequence goal has been achieved, as described below.

The second sequence production phase is now under way. Its aims are to achieve full-shotgun coverage of the existing clones during 2001, to obtain clones to fill the remaining gaps in the physical map, and to produce a finished sequence (apart from regions that cannot be cloned or sequenced with currently available techniques) no later than 2003.

Strategic issues

Hierarchical shotgun sequencing

Soon after the invention of DNA sequencing methods[47,48], the shotgun sequencing strategy was introduced[49–51]; it has remained the fundamental method for large-scale genome sequencing[52–54] for the past 20 years. The approach has been refined and extended to make it more efficient. For example, improved protocols for fragmenting and cloning DNA allowed construction of shotgun

libraries with more uniform representation. The practice of sequencing from both ends of double-stranded clones ('double-barrelled' shotgun sequencing) was introduced by Ansorge and others[37] in 1990, allowing the use of 'linking information' between sequence fragments.

The application of shotgun sequencing was also extended by applying it to larger and larger DNA molecules—from plasmids (~ 4 kilobases (kb)) to cosmid clones[37] (40 kb), to artificial chromosomes cloned in bacteria and yeast[55] (100–500 kb) and bacterial genomes[56] (1–2 megabases (Mb)). In principle, a genome of arbitrary size may be directly sequenced by the shotgun method, provided that it contains no repeated sequence and can be uniformly sampled at random. The genome can then be assembled using the simple computer science technique of 'hashing' (in which one detects overlaps by consulting an alphabetized look-up table of all k-letter words in the data). Mathematical analysis of the expected number of gaps as a function of coverage is similarly straightforward[57].

Practical difficulties arise because of repeated sequences and cloning bias. Small amounts of repeated sequence pose little problem for shotgun sequencing. For example, one can readily assemble typical bacterial genomes (about 1.5% repeat) or the euchromatic portion of the fly genome (about 3% repeat). By contrast, the human genome is filled (> 50%) with repeated sequences, including interspersed repeats derived from transposable elements, and long genomic regions that have been duplicated in tandem, palindromic or dispersed fashion (see below). These include large duplicated segments (50–500 kb) with high sequence identity (98–99.9%), at which mispairing during recombination creates deletions responsible for genetic syndromes. Such features complicate the assembly of a correct and finished genome sequence.

There are two approaches for sequencing large repeat-rich genomes. The first is a whole-genome shotgun sequencing approach, as has been used for the repeat-poor genomes of viruses, bacteria and flies, using linking information and computational

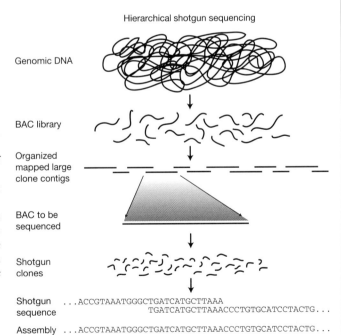

Hierarchical shotgun sequencing

Genomic DNA

BAC library

Organized mapped large clone contigs

BAC to be sequenced

Shotgun clones

Shotgun sequence ...ACCGTAAATGGGCTGATCATGCTTAAA
　　　　　　　　　　TGATCATGCTTAAACCCTGTGCATCCTACTG...

Assembly ...ACCGTAAATGGGCTGATCATGCTTAAACCCTGTGCATCCTACTG...

Figure 2 Idealized representation of the hierarchical shotgun sequencing strategy. A library is constructed by fragmenting the target genome and cloning it into a large-fragment cloning vector; here, BAC vectors are shown. The genomic DNA fragments represented in the library are then organized into a physical map and individual BAC clones are selected and sequenced by the random shotgun strategy. Finally, the clone sequences are assembled to reconstruct the sequence of the genome.

analysis to attempt to avoid misassemblies. The second is the 'hierarchical shotgun sequencing' approach (Fig. 2), also referred to as 'map-based', 'BAC-based' or 'clone-by-clone'. This approach involves generating and organizing a set of large-insert clones (typically 100–200 kb each) covering the genome and separately performing shotgun sequencing on appropriately chosen clones. Because the sequence information is local, the issue of long-range misassembly is eliminated and the risk of short-range misassembly is reduced. One caveat is that some large-insert clones may suffer rearrangement, although this risk can be reduced by appropriate quality-control measures involving clone fingerprints (see below).

The two methods are likely to entail similar costs for producing finished sequence of a mammalian genome. The hierarchical approach has a higher initial cost than the whole-genome approach, owing to the need to create a map of clones (about 1% of the total cost of sequencing) and to sequence overlaps between clones. On the other hand, the whole-genome approach is likely to require much greater work and expense in the final stage of producing a finished sequence, because of the challenge of resolving misassemblies. Both methods must also deal with cloning biases, resulting in under-representation of some regions in either large-insert or small-insert clone libraries.

There was lively scientific debate over whether the human genome sequencing effort should employ whole-genome or hierarchical shotgun sequencing. Weber and Myers[58] stimulated these discussions with a specific proposal for a whole-genome shotgun approach, together with an analysis suggesting that the method could work and be more efficient. Green[59] challenged these conclusions and argued that the potential benefits did not outweigh the likely risks.

In the end, we concluded that the human genome sequencing effort should employ the hierarchical approach for several reasons. First, it was prudent to use the approach for the first project to sequence a repeat-rich genome. With the hierarchical approach, the ultimate frequency of misassembly in the finished product would probably be lower than with the whole-genome approach, in which it would be more difficult to identify regions in which the assembly was incorrect.

Second, it was prudent to use the approach in dealing with an outbred organism, such as the human. In the whole-genome shotgun method, sequence would necessarily come from two different copies of the human genome. Accurate sequence assembly could be complicated by sequence variation between these two copies—both SNPs (which occur at a rate of 1 per 1,300 bases) and larger-scale structural heterozygosity (which has been documented in human chromosomes). In the hierarchical shotgun method, each large-insert clone is derived from a single haplotype.

Third, the hierarchical method would be better able to deal with inevitable cloning biases, because it would more readily allow targeting of additional sequencing to under-represented regions. And fourth, it was better suited to a project shared among members of a diverse international consortium, because it allowed work and responsibility to be easily distributed. As the ultimate goal has always been to create a high-quality, finished sequence to serve as a foundation for biomedical research, we reasoned that the advantages of this more conservative approach outweighed the additional cost, if any.

A biotechnology company, Celera Genomics, has chosen to incorporate the whole-genome shotgun approach into its own efforts to sequence the human genome. Their plan[60,61] uses a mixed strategy, involving combining some coverage with whole-genome shotgun data generated by the company together with the publicly available hierarchical shotgun data generated by the International Human Genome Sequencing Consortium. If the raw sequence reads from the whole-genome shotgun component are made available, it may be possible to evaluate the extent to which the sequence of the human genome can be assembled without the need

for clone-based information. Such analysis may help to refine sequencing strategies for other large genomes.

Technology for large-scale sequencing

Sequencing the human genome depended on many technological improvements in the production and analysis of sequence data. Key innovations were developed both within and outside the Human Genome Project. Laboratory innovations included four-colour fluorescence-based sequence detection[62], improved fluorescent dyes[63–66], dye-labelled terminators[67], polymerases specifically designed for sequencing[68–70], cycle sequencing[71] and capillary gel electrophoresis[72–74]. These studies contributed to substantial improvements in the automation, quality and throughput of collecting raw DNA sequence[75,76]. There were also important advances in the development of software packages for the analysis of sequence data. The PHRED software package[77,78] introduced the concept of assigning a 'base-quality score' to each base, on the basis of the probability of an erroneous call. These quality scores make it possible to monitor raw data quality and also assist in determining whether two similar sequences truly overlap. The PHRAP computer package (http://bozeman.mbt.washington.edu/phrap.docs/phrap.html) then systematically assembles the sequence data using the base-quality scores. The program assigns 'assembly-quality scores' to each base in the assembled sequence, providing an objective criterion to guide sequence finishing. The quality scores were based on and validated by extensive experimental data.

Another key innovation for scaling up sequencing was the development by several centres of automated methods for sample preparation. This typically involved creating new biochemical protocols suitable for automation, followed by construction of appropriate robotic systems.

Coordination and public data sharing

The Human Genome Project adopted two important principles with regard to human sequencing. The first was that the collaboration would be open to centres from any nation. Although potentially less efficient, in a narrow economic sense, than a centralized approach involving a few large factories, the inclusive approach was strongly favoured because we felt that the human genome sequence is the common heritage of all humanity and the work should transcend national boundaries, and we believed that scientific progress was best assured by a diversity of approaches. The collaboration was coordinated through periodic international meetings (referred to as 'Bermuda meetings' after the venue of the first three gatherings) and regular telephone conferences. Work was shared flexibly among the centres, with some groups focusing on particular chromosomes and others contributing in a genome-wide fashion.

The second principle was rapid and unrestricted data release. The centres adopted a policy that all genomic sequence data should be made publicly available without restriction within 24 hours of assembly[79,80]. Pre-publication data releases had been pioneered in mapping projects in the worm[11] and mouse genomes[30,81] and were prominently adopted in the sequencing of the worm, providing a direct model for the human sequencing efforts. We believed that scientific progress would be most rapidly advanced by immediate and free availability of the human genome sequence. The explosion of scientific work based on the publicly available sequence data in both academia and industry has confirmed this judgement.

Generating the draft genome sequence

Generating a draft sequence of the human genome involved three steps: selecting the BAC clones to be sequenced, sequencing them and assembling the individual sequenced clones into an overall draft genome sequence. A glossary of terms related to genome sequencing and assembly is provided in Box 1.

The draft genome sequence is a dynamic product, which is regularly updated as additional data accumulate en route to the ultimate goal of a completely finished sequence. The results below are based on the map and sequence data available on 7 October 2000, except as otherwise noted. At the end of this section, we provide a brief update of key data.

Clone selection

The hierarchical shotgun method involves the sequencing of overlapping large-insert clones spanning the genome. For the Human Genome Project, clones were largely chosen from eight large-insert libraries containing BAC or P1-derived artificial chromosome (PAC) clones (Table 1; refs 82–88). The libraries were made by

partial digestion of genomic DNA with restriction enzymes. Together, they represent around 65-fold coverage (redundant sampling) of the genome. Libraries based on other vectors, such as cosmids, were also used in early stages of the project.

The libraries (Table 1) were prepared from DNA obtained from anonymous human donors in accordance with US Federal Regulations for the Protection of Human Subjects in Research (45CFR46) and following full review by an Institutional Review Board. Briefly, the opportunity to donate DNA for this purpose was broadly advertised near the two laboratories engaged in library

---

Box 1

Genome glossary

Sequence

Raw sequence Individual unassembled sequence reads, produced by sequencing of clones containing DNA inserts.

Paired-end sequence Raw sequence obtained from both ends of a cloned insert in any vector, such as a plasmid or bacterial artificial chromosome.

Finished sequence Complete sequence of a clone or genome, with an accuracy of at least 99.99% and no gaps.

Coverage (or depth) The average number of times a nucleotide is represented by a high-quality base in a collection of random raw sequence. Operationally, a 'high-quality base' is defined as one with an accuracy of at least 99% (corresponding to a PHRED score of at least 20).

Full shotgun coverage The coverage in random raw sequence needed from a large-insert clone to ensure that it is ready for finishing; this varies among centres but is typically 8–10-fold. Clones with full shotgun coverage can usually be assembled with only a handful of gaps per 100 kb.

Half shotgun coverage Half the amount of full shotgun coverage (typically, 4–5-fold random coverage).

Clones

BAC clone Bacterial artificial chromosome vector carrying a genomic DNA insert, typically 100–200 kb. Most of the large-insert clones sequenced in the project were BAC clones.

Finished clone A large-insert clone that is entirely represented by finished sequence.

Full shotgun clone A large-insert clone for which full shotgun sequence has been produced.

Draft clone A large-insert clone for which roughly half-shotgun sequence has been produced. Operationally, the collection of draft clones produced by each centre was required to have an average coverage of fourfold for the entire set and a minimum coverage of threefold for each clone.

Predraft clone A large-insert clone for which some shotgun sequence is available, but which does not meet the standards for inclusion in the collection of draft clones.

Contigs and scaffolds

Contig The result of joining an overlapping collection of sequences or clones.

Scaffold The result of connecting contigs by linking information from paired-end reads from plasmids, paired-end reads from BACs, known messenger RNAs or other sources. The contigs in a scaffold are ordered and oriented with respect to one another.

Fingerprint clone contigs Contigs produced by joining clones inferred to overlap on the basis of their restriction digest fingerprints.

Sequenced-clone layout Assignment of sequenced clones to the physical map of fingerprint clone contigs.

Initial sequence contigs Contigs produced by merging overlapping sequence reads obtained from a single clone, in a process called sequence assembly.

Merged sequence contigs Contigs produced by taking the initial sequence contigs contained in overlapping clones and merging those found to overlap. These are also referred to simply as 'sequence contigs' where no confusion will result.

Sequence-contig scaffolds Scaffolds produced by connecting sequence contigs on the basis of linking information.

Sequenced-clone contigs Contigs produced by merging overlapping sequenced clones.

Sequenced-clone-contig scaffolds Scaffolds produced by joining sequenced-clone contigs on the basis of linking information.

Draft genome sequence The sequence produced by combining the information from the individual sequenced clones (by creating merged sequence contigs and then employing linking information to create scaffolds) and positioning the sequence along the physical map of the chromosomes.

N50 length A measure of the contig length (or scaffold length) containing a 'typical' nucleotide. Specifically, it is the maximum length L such that 50% of all nucleotides lie in contigs (or scaffolds) of size at least L.

Computer programs and databases

PHRED A widely used computer program that analyses raw sequence to produce a 'base call' with an associated 'quality score' for each position in the sequence. A PHRED quality score of X corresponds to an error probability of approximately $10^{-X/10}$. Thus, a PHRED quality score of 30 corresponds to 99.9% accuracy for the base call in the raw read.

PHRAP A widely used computer program that assembles raw sequence into sequence contigs and assigns to each position in the sequence an associated 'quality score', on the basis of the PHRED scores of the raw sequence reads. A PHRAP quality score of X corresponds to an error probability of approximately $10^{-X/10}$. Thus, a PHRAP quality score of 30 corresponds to 99.9% accuracy for a base in the assembled sequence.

GigAssembler A computer program developed during this project for merging the information from individual sequenced clones into a draft genome sequence.

Public sequence databases The three coordinated international sequence databases: GenBank, the EMBL data library and DDBJ.

Map features

STS Sequence tagged site, corresponding to a short (typically less than 500 bp) unique genomic locus for which a polymerase chain reaction assay has been developed.

EST Expressed sequence tag, obtained by performing a single raw sequence read from a random complementary DNA clone.

SSR Simple sequence repeat, a sequence consisting largely of a tandem repeat of a specific k-mer (such as $(CA)_{15}$). Many SSRs are polymorphic and have been widely used in genetic mapping.

SNP Single nucleotide polymorphism, or a single nucleotide position in the genome sequence for which two or more alternative alleles are present at appreciable frequency (traditionally, at least 1%) in the human population.

Genetic map A genome map in which polymorphic loci are positioned relative to one another on the basis of the frequency with which they recombine during meiosis. The unit of distance is centimorgans (cM), denoting a 1% chance of recombination.

Radiation hybrid (RH) map A genome map in which STSs are positioned relative to one another on the basis of the frequency with which they are separated by radiation-induced breaks. The frequency is assayed by analysing a panel of human–hamster hybrid cell lines, each produced by lethally irradiating human cells and fusing them with recipient hamster cells such that each carries a collection of human chromosomal fragments. The unit of distance is centirays (cR), denoting a 1% chance of a break occuring between two loci.

---

construction. Volunteers of diverse backgrounds were accepted on a first-come, first-taken basis. Samples were obtained after discussion with a genetic counsellor and written informed consent. The samples were made anonymous as follows: the sampling laboratory stripped all identifiers from the samples, applied random numeric labels, and transferred them to the processing laboratory, which then removed all labels and relabelled the samples. All records of the labelling were destroyed. The processing laboratory chose samples at random from which to prepare DNA and immortalized cell lines. Around 5–10 samples were collected for every one that was eventually used. Because no link was retained between donor and DNA sample, the identity of the donors for the libraries is not known, even by the donors themselves. A more complete description can be found at http://www.nhgri.nih.gov/Grant_info/Funding/Statements/RFA/human_subjects.html.

During the pilot phase, centres showed that sequence-tagged sites (STSs) from previously constructed genetic and physical maps could be used to recover BACs from specific regions. As sequencing expanded, some centres continued this approach, augmented with additional probes from flow sorting of chromosomes to obtain long-range coverage of specific chromosomes or chromosomal regions[89–94].

For the large-scale sequence production phase, a genome-wide physical map of overlapping clones was also constructed by systematic analysis of BAC clones representing 20-fold coverage of the human genome[86]. Most clones came from the first three sections of the RPCI-11 library, supplemented with clones from sections of the RPCI-13 and CalTech D libraries (Table 1). DNA from each BAC clone was digested with the restriction enzyme HindIII, and the sizes of the resulting fragments were measured by agarose gel electrophoresis. The pattern of restriction fragments provides a 'fingerprint' for each BAC, which allows different BACs to be distinguished and the degree of overlaps to be assessed. We used these restriction-fragment fingerprints to determine clone overlaps, and thereby assembled the BACs into fingerprint clone contigs.

The fingerprint clone contigs were positioned along the chromosomes by anchoring them with STS markers from existing genetic and physical maps. Fingerprint clone contigs were tied to specific STSs initially by probe hybridization and later by direct search of the sequenced clones. To localize fingerprint clone contigs that did not contain known markers, new STSs were generated and placed onto chromosomes[95]. Representative clones were also positioned by fluorescence in situ hybridization (FISH) (ref. 86 and C. McPherson, unpublished).

We selected clones from the fingerprint clone contigs for sequencing according to various criteria. Fingerprint data were reviewed[86,90] to evaluate overlaps and to assess clone fidelity (to bias against rearranged clones[83,96]). STS content information and BAC end sequence information were also used[91,92]. Where possible, we tried to select a minimally overlapping set spanning a region. However, because the genome-wide physical map was constructed concurrently with the sequencing, continuity in many regions was low in early stages. These small fingerprint clone contigs were nonetheless useful in identifying validated, nonredundant clones

Table 1 Key large-insert genome-wide libraries

Library name*	GenBank abbreviation	Vector type	Source DNA	Library segment or plate numbers	Enzyme digest	Average insert size (kb)	Total number of clones in library	Number of fingerprinted clones†	BAC-end sequence (ends/clones/clones with both ends sequenced)‡	Number of clones in genome layout§	Sequenced clones used in construction of the draft genome sequence		
											Number‖	Total bases (Mb)¶	Fraction of total from library
Caltech B	CTB	BAC	987SK cells	All	HindIII	120	74,496	16	2/1/1	528	518	66.7	0.016
Caltech C	CTC	BAC	Human sperm	All	HindIII	125	263,040	144	21,956/ 14,445/ 7,255	621	606	88.4	0.021
Caltech D1 (CITB-H1)	CTD	BAC	Human sperm	All	HindIII	129	162,432	49,833	403,589/ 226,068/ 156,631	1,381	1,367	185.6	0.043
Caltech D2 (CITB-E1)		BAC	Human sperm	All									
				2,501–2,565	EcoRI	202	24,960						
				2,566–2,671	EcoRI	182	46,326						
				3,000–3,253	EcoRI	142	97,536						
RPCI-1	RP1	PAC	Male, blood	All	MboI	110	115,200	3,388		1,070	1,053	117.7	0.028
RPCI-3	RP3	PAC	Male, blood	All	MboI	115	75,513			644	638	68.5	0.016
RPCI-4	RP4	PAC	Male, blood	All	MboI	116	105,251			889	881	95.5	0.022
RPCI-5	RP5	PAC	Male, blood	All	MboI	115	142,773			1,042	1,033	116.5	0.027
RPCI-11	RP11	BAC	Male, blood	All		178	543,797	267,931	379,773/ 243,764/ 134,110	19,405	19,145	3,165.0	0.743
				1	EcoRI	164	108,499						
				2	EcoRI	168	109,496						
				3	EcoRI	181	109,657						
				4	EcoRI	183	109,382						
				5	MboI	196	106,763						
Total of top eight libraries							1,482,502	321,312	805,320/ 484,278/ 297,907	25,580	25,241	3,903.9	0.916
Total all libraries								354,510	812,594/ 488,017/ 100,775	30,445	29,298	4,260.5	1

* For the CalTech libraries[82], see http://www.tree.caltech.edu/lib_status.html; for RPCI libraries[83], see http://www.chori.org/bacpac/home.htm.
† For the FPC map and fingerprinting[84–86], see http://genome.wustl.edu/gsc/human/human_database.shtml.
‡ The number of raw BAC end sequences (clones/ends/clones with both ends sequenced) available for use in human genome sequencing. Typically, for clones in which sequence was obtained from both ends, more than 95% of both end sequences contained at least 100 bp of nonrepetitive sequence. BAC-end sequencing of RPCI-11 and of the CalTech libraries was done at The Institute for Genomic Research, the California Institute of Technology and the University of Washington High Throughput Sequencing Center. The sources for the Table were http://www.ncbi.nlm.nih.gov/genome/clone/BESstat.shtml and refs 87, 88.
§ These are the clones in the sequenced-clone layout map (http://genome.wustl.edu/gsc/human/Mapping/index.shtml) that were pre-draft, draft or finished.
‖ The number of sequenced clones used in the assembly. This number is less than that in the previous column owing to removal of a small number of obviously contaminated, combined or duplicated projects; in addition, not all of the clones from completed chromosomes 21 and 22 were included here because only the available finished sequence from those chromosomes was used in the assembly.
¶ The number reported is the total sequence from the clones indicated in the previous column. Potential overlap between clones was not removed here, but Ns were excluded.

that were used to 'seed' the sequencing of new regions. The small fingerprint clone contigs were extended or merged with others as the map matured.

The clones that make up the draft genome sequence therefore do not constitute a minimally overlapping set—there is overlap and redundancy in places. The cost of using suboptimal overlaps was justified by the benefit of earlier availability of the draft genome sequence data. Minimizing the overlap between adjacent clones would have required completing the physical map before undertaking large-scale sequencing. In addition, the overlaps between BAC clones provide a rich collection of SNPs. More than 1.4 million SNPs have already been identified from clone overlaps and other sequence comparisons[97].

Because the sequencing project was shared among twenty centres in six countries, it was important to coordinate selection of clones across the centres. Most centres focused on particular chromosomes or, in some cases, larger regions of the genome. We also maintained a clone registry to track selected clones and their progress. In later phases, the global map provided an integrated view of the data from all centres, facilitating the distribution of effort to maximize coverage of the genome. Before performing extensive sequencing on a

clone, several centres routinely examined an initial sample of 96 raw sequence reads from each subclone library to evaluate possible overlap with previously sequenced clones.

Sequencing

The selected clones were subjected to shotgun sequencing. Although the basic approach of shotgun sequencing is well established, the details of implementation varied among the centres. For example, there were differences in the average insert size of the shotgun libraries, in the use of single-stranded or double-stranded cloning vectors, and in sequencing from one end or both ends of each insert. Centres differed in the fluorescent labels employed and in the degree to which they used dye-primers or dye-terminators. The sequence detectors included both slab gel- and capillary-based devices. Detailed protocols are available on the web sites of many of the individual centres (URLs can be found at www.nhgri.nih.gov/genome_hub). The extent of automation also varied greatly among the centres, with the most aggressive automation efforts resulting in factory-style systems able to process more than 100,000 sequencing reactions in 12 hours (Fig. 3). In addition, centres differed in the amount of raw sequence data typically obtained for each clone (so-called half-shotgun, full shotgun and finished sequence). Sequence information from the different centres could be directly integrated despite this diversity, because the data were analysed by a common computational procedure. Raw sequence traces were processed and assembled with the PHRED and PHRAP software packages[77,78] (P. Green, unpublished). All assembled contigs of more than 2 kb were deposited in public databases within 24 hours of assembly.

The overall sequencing output rose sharply during production (Fig. 4). Following installation of new sequence detectors beginning in June 1999, sequencing capacity and output rose approximately eightfold in eight months to nearly 7 million samples processed per month, with little or no drop in success rate (ratio of useable reads to attempted reads). By June 2000, the centres were producing raw sequence at a rate equivalent to onefold coverage of the entire human genome in less than six weeks. This corresponded to a continuous throughput exceeding 1,000 nucleotides per second, 24 hours per day, seven days per week. This scale-up resulted in a concomitant increase in the sequence available in the public databases (Fig. 4).

A version of the draft genome sequence was prepared on the basis of the map and sequence data available on 7 October 2000. For this version, the mapping effort had assembled the fingerprinted BACs into 1,246 fingerprint clone contigs. The sequencing effort had sequenced and assembled 29,298 overlapping BACs and other large-insert clones (Table 2), comprising a total length of 4.26 gigabases (Gb). This resulted from around 23 Gb of underlying raw shotgun sequence data, or about 7.5-fold coverage averaged across the genome (including both draft and finished sequence). The various contributions to the total amount of sequence deposited in the HTGS division of GenBank are given in Table 3.

Figure 3 The automated production line for sample preparation at the Whitehead Institute, Center for Genome Research. The system consists of custom-designed factory-style conveyor belt robots that perform all functions from purifying DNA from bacterial cultures through setting up and purifying sequencing reactions.

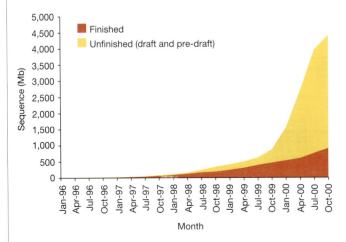

Figure 4 Total amount of human sequence in the High Throughput Genome Sequence (HTGS) division of GenBank. The total is the sum of finished sequence (red) and unfinished (draft plus predraft) sequence (yellow).

Table 2 Total genome sequence from the collection of sequenced clones, by sequence status

Sequence status	Number of clones	Total clone length (Mb)	Average number of sequence reads per kb*	Average sequence depth†	Total amount of raw sequence (Mb)
Finished	8,277	897	20–25	8–12	9,085
Draft	18,969	3,097	12	4.5	13,395
Predraft	2,052	267	6	2.5	667
Total					23,147

*The average number of reads per kb was estimated based on information provided by each sequencing centre. This number differed among sequencing centres, based on the actual protocols used.

†The average depth in high quality bases (≥99% accuracy) was estimated from information provided by each sequencing centre. The average varies among the centres, and the number may vary considerably for clones with the same sequencing status. For draft clones in the public databases (keyword: HTGS_draft), the number can be computed from the quality scores listed in the database entry.

By agreement among the centres, the collection of draft clones produced by each centre was required to have fourfold average sequence coverage, with no clone below threefold. (For this purpose, sequence coverage was defined as the average number of times that each base was independently read with a base-quality score corresponding to at least 99% accuracy.) We attained an overall average of 4.5-fold coverage across the genome for draft clones. A few of the sequenced clones fell below the minimum of threefold sequence coverage or have not been formally designated by centres as meeting draft standards; these are referred to as predraft (Table 2). Some of these are clones that span remaining gaps in the draft genome sequence and were in the process of being sequenced on 7 October 2000; a few are old submissions from centres that are no longer active.

The lengths of the initial sequence contigs in the draft clones vary as a function of coverage, but half of all nucleotides reside in initial sequence contigs of at least 21.7 kb (see below). Various properties of the draft clones can be assessed from instances in which there was substantial overlap between a draft clone and a finished (or nearly finished) clone. By examining the sequence alignments in the overlap regions, we estimated that the initial sequence contigs in a draft sequence clone cover an average of about 96% of the clone and are separated by gaps with an average size of about 500 bp.

Although the main emphasis was on producing a draft genome sequence, the centres also maintained sequence finishing activities during this period, leading to a twofold increase in finished sequence from June 1999 to June 2000 (Fig. 4). The total amount of human sequence in this final form stood at more than 835 Mb on 7 October 2000, or more than 25% of the human genome. This includes the finished sequences of chromosomes 21 and 22 (refs 93, 94). As centres have begun to shift from draft to finished sequencing in the last quarter of 2000, the production of finished sequence has increased to an annualized rate of 1 Gb per year and is continuing to rise.

In addition to sequencing large-insert clones, three centres generated a large collection of random raw sequence reads from whole-genome shotgun libraries (Table 4; ref. 98). These 5.77 million successful sequences contained 2.4 Gb of high-quality bases; this corresponds to about 0.75-fold coverage and would be statistically expected to include about 50% of the nucleotides in the human genome (data available at http://snp.cshl.org/data). The primary objective of this work was to discover SNPs, by comparing these random raw sequences (which came from different individuals) with the draft genome sequence. However, many of these raw sequences were obtained from both ends of plasmid clones and thereby also provided valuable 'linking' information that was used in sequence assembly. In addition, the random raw sequences provide sequence coverage of about half of the nucleotides not yet represented in the sequenced large-insert clones; these can be used as probes for portions of the genome not yet recovered.

Assembly of the draft genome sequence

We then set out to assemble the sequences from the individual large-insert clones into an integrated draft sequence of the human genome. The assembly process had to resolve problems arising from the draft nature of much of the sequence, from the variety of clone sources, and from the high fraction of repeated sequences in the human genome. This process involved three steps: filtering, layout and merging.

The entire data set was filtered uniformly to eliminate contamination from nonhuman sequences and other artefacts that had not already been removed by the individual centres. (Information about contamination was also sent back to the centres, which are updating the individual entries in the public databases.) We also identified instances in which the sequence data from one BAC clone was substantially contaminated with sequence data from another (human or nonhuman) clone. The problems were resolved in most instances; 231 clones remained unresolved, and these were eliminated from the assembly reported here. Instances of lower levels of cross-contamination (for example, a single 96-well microplate misassigned to the wrong BAC) are more difficult to detect; some undoubtedly remain and may give rise to small spurious sequence contigs in the draft genome sequence. Such issues are readily resolved as the clones progress towards finished sequence, but they necessitate some caution in certain applications of the current data.

The sequenced clones were then associated with specific clones on the physical map to produce a 'layout'. In principle, sequenced clones that correspond to fingerprinted BACs could be directly assigned by name to fingerprint clone contigs on the fingerprint-based physical map. In practice, however, laboratory mixups occasionally resulted in incorrect assignments. To eliminate such problems, sequenced clones were associated with the fingerprint clone contigs in the physical map by using the sequence data to calculate a

---

Table 3 Total human sequence deposited in the HTGS division of GenBank

Sequencing centre	Total human sequence (kb)	Finished human sequence (kb)
Whitehead Institute, Center for Genome Research*	1,196,888	46,560
The Sanger Centre*	970,789	284,353
Washington University Genome Sequencing Center*	765,898	175,279
US DOE Joint Genome Institute	377,998	78,486
Baylor College of Medicine Human Genome Sequencing Center	345,125	53,418
RIKEN Genomic Sciences Center	203,166	16,971
Genoscope	85,995	48,808
GTC Sequencing Center	71,357	7,014
Department of Genome Analysis, Institute of Molecular Biotechnology	49,865	17,788
Beijing Genomics Institute/Human Genome Center	42,865	6,297
Multimegabase Sequencing Center; Institute for Systems Biology	31,241	9,676
Stanford Genome Technology Center	29,728	3,530
The Stanford Human Genome Center and Department of Genetics	28,162	9,121
University of Washington Genome Center	24,115	14,692
Keio University	17,364	13,058
University of Texas Southwestern Medical Center at Dallas	11,670	7,028
University of Oklahoma Advanced Center for Genome Technology	10,071	9,155
Max Planck Institute for Molecular Genetics	7,650	2,940
GBF – German Research Centre for Biotechnology	4,639	2,338
Cold Spring Harbor Laboratory Lita Annenberg Hazen Genome Center	4,338	2,104
Other	59,574	35,911
Total	4,338,224	842,027

Total human sequence deposited in GenBank by members of the International Human Genome Sequencing Consortium, as of 8 October 2000. The amount of total sequence (finished plus draft plus predraft) is shown in the second column and the amount of finished sequence is shown in the third column. Total sequence differs from totals in Tables 1 and 2 because of inclusion of padding characters and of some clones not used in assembly. HTGS, high throughput genome sequence.
*These three centres produced an additional 2.4 Gb of raw plasmid paired-end reads (see Table 4), consisting of 0.99 Gb from Whitehead Institute, 0.66 Gb from The Sanger Centre and 0.75 Gb from Washington University.

Table 4 Plasmid paired-end reads

	Total reads deposited*	Read pairs†	Size range of inserts (kb)
Random-sheared	3,227,685	1,155,284	1.8–6
Enzyme digest	2,539,222	761,010	0.8–4.7
Total	5,766,907	1,916,294	

The plasmid paired-end reads used a mixture of DNA from a set of 24 samples from the DNA Polymorphism Discovery Resource (http://locus.umdnj.edu/nigms/pdr.html). This set of 24 anonymous US residents contains samples from European-Americans, African-Americans, Mexican-Americans, Native Americans and Asian-Americans, although the ethnicities of the individual samples are not identified. Informed consent to contribute samples to the DNA Polymorphism Discovery Resource was obtained from all 450 individuals who contributed samples. Samples from the European-American, African-American and Mexican-American individuals came from NHANES (http://www.cdc.gov/nchs/nhanes.htm); individuals were recontacted to obtain their consent for the Resource project. New samples were obtained from Asian-Americans whose ancestry was from a variety of East and South Asian countries. New samples were also obtained for the Native Americans; tribal permission was obtained first, and then individual consents. See http://www.nhgri.nih.gov/Grant_info/Funding/RFA/discover_polymorphisms.html and ref. 98.
*Reflects data deposited with and released by The SNP Consortium (see http://snp.cshl.org/data).
† Read pairs represents the number of cases in which sequence from both ends of a genomic cloned fragment was determined and used in this study as linking information.

partial list of restriction fragments *in silico* and comparing that list with the experimental database of BAC fingerprints. The comparison was feasible because the experimental sizing of restriction fragments was highly accurate (to within 0.5–1.5% of the true size, for 95% of fragments from 600 to 12,000 base pairs (bp))[84,85]. Reliable matching scores could be obtained for 16,193 of the clones. The remaining sequenced clones could not be placed on the map by this method because they were too short, or they contained too many small initial sequence contigs to yield enough restriction fragments, or possibly because their sequences were not represented in the fingerprint database.

An independent approach to placing sequenced clones on the physical map used the database of end sequences from fingerprinted BACs (Table 1). Sequenced clones could typically be reliably mapped if they contained multiple matches to BAC ends, with all corresponding to clones from a single genomic region (multiple matches were required as a safeguard against errors known to exist in the BAC end database and against repeated sequences). This approach provided useful placement information for 22,566 sequenced clones.

Altogether, we could assign 25,403 sequenced clones to fingerprint clone contigs by combining *in silico* digestion and BAC end sequence match data. To place most of the remaining sequenced clones, we exploited information about sequence overlap or BAC-end paired links of these clones with already positioned clones. This left only a few, mostly small, sequenced clones that could not be placed (152 sequenced clones containing 5.5 Mb of sequence out of 29,298 sequenced clones containing more than 4,260 Mb of sequence); these are being localized by radiation hybrid mapping of STSs derived from their sequences.

The fingerprint clone contigs were then mapped to chromosomal locations, using sequence matches to mapped STSs from four human radiation hybrid maps[95,99,100], one YAC and radiation hybrid map[29], and two genetic maps[101,102], together with data from FISH[86,90,103]. The mapping was iteratively refined by comparing the order and orientation of the STSs in the fingerprint clone contigs and the various STS-based maps, to identify and refine discrepancies (Fig. 5). Small fingerprint clone contigs (< 1 Mb) were difficult to orient and, sometimes, to order using these methods. In all, 942 fingerprint clone contigs contained sequenced clones. (An additional 304 of the 1,246 fingerprint clone contigs did not contain sequenced clones, but these tended to be extremely small and together contain less than 1% of the mapped clones. About one-third have been targeted for sequencing. A few derive from the Y chromosome, for which the map was constructed separately[89]. Most of the remainder are fragments of other larger contigs or represent other artefacts. These are being eliminated in subsequent versions of the database.) Of these 942 contigs with sequenced clones, 852 (90%, containing 99.2% of the total sequence) were localized to specific chromosome locations in this way. An additional 51 fingerprint clone contigs, containing 0.5% of the sequence, could be assigned to a specific chromosome but not to a precise position.

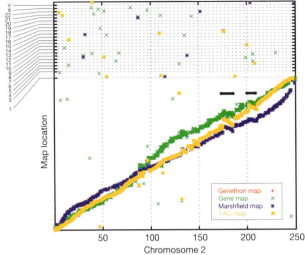

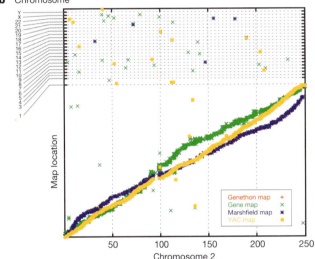

Figure 5 Positions of markers on previous maps of the genome (the Genethon[101] genetic map and Marshfield genetic map (http://research.marshfieldclinic.org/genetics/genotyping_service/mgsver2.htm), the GeneMap99 radiation hybrid map[100], and the Whitehead YAC and radiation hybrid map[29]) plotted against their derived position on the draft sequence for chromosome 2. The horizontal units are Mb but the vertical units of each map vary (cM, cR and so on) and thus all were scaled so that the entire map spans the full vertical range. Markers that map to other chromosomes are shown in the chromosome lines at the top. The data sets generally follow the diagonal, indicating that order and orientation of the marker sets on the different maps largely agree (note that the two genetic maps are completely superimposed). In **a**, there are two segments (bars) that are inverted in an earlier version draft sequence relative to all the other maps. **b**, The same chromosome after the information was used to reorient those two segments.

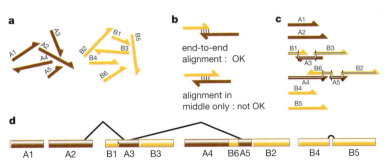

Figure 6 The key steps (**a–d**) in assembling individual sequenced clones into the draft genome sequence. A1–A5 represent initial sequence contigs derived from shotgun sequencing of clone A, and B1–B6 are from clone B.

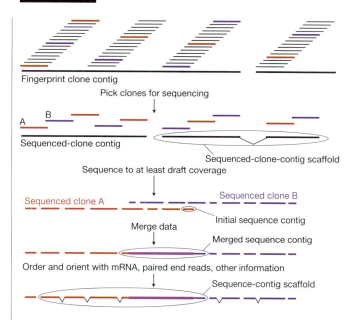

Figure 7 Levels of clone and sequence coverage. A 'fingerprint clone contig' is assembled by using the computer program FPC[84,451] to analyse the restriction enzyme digestion patterns of many large-insert clones. Clones are then selected for sequencing to minimize overlap between adjacent clones. For a clone to be selected, all of its restriction enzyme fragments (except the two vector-insert junction fragments) must be shared with at least one of its neighbours on each side in the contig. Once these overlapping clones have been sequenced, the set is a 'sequenced-clone contig'. When all selected clones from a fingerprint clone contig have been sequenced, the sequenced-clone contig will be the same as the fingerprint clone contig. Until then, a fingerprint clone contig may contain several sequenced-clone contigs. After individual clones (for example, A and B) have been sequenced to draft coverage and the clones have been mapped, the data are analysed by GigAssembler (Fig. 6), producing merged sequence contigs from initial sequence contigs, and linking these to form sequence-contig scaffolds (see Box 1).

The remaining 39 contigs containing 0.3% of the sequence were not positioned at all.

We then merged the sequences from overlapping sequenced clones (Fig. 6), using the computer program GigAssembler[104]. The program considers nearby sequenced clones, detects overlaps between the initial sequence contigs in these clones, merges the overlapping sequences and attempts to order and orient the sequence contigs. It begins by aligning the initial sequence contigs from one clone with those from other clones in the same fingerprint clone contig on the basis of length of alignment, per cent identity of the alignment, position in the sequenced clone layout and other factors. Alignments are limited to one end of each initial sequence contig for partially overlapping contigs or to both ends of an initial sequence contig contained entirely within another; this eliminates internal alignments that may reflect repeated sequence or possible misassembly (Fig. 6b). Beginning with the highest scoring pairs, initial sequence contigs are then integrated to produce 'merged sequence contigs' (usually referred to simply as 'sequence contigs'). The program refines the arrangement of the clones within the fingerprint clone contig on the basis of the extent of sequence overlap between them and then rebuilds the sequence contigs. Next, the program selects a sequence path through the sequence contigs (Fig. 6c). It tries to use the highest quality data by preferring longer initial sequence contigs and avoiding the first and last 250 bases of initial sequence contigs where possible. Finally, it attempts to order and orient the sequence contigs by using additional information, including sequence data from paired-end plasmid and BAC reads, known messenger RNAs and ESTs, as well as additional linking information provided by centres. The sequence contigs are thereby linked together to create 'sequence-contig scaffolds' (Fig. 6d). The process also joins overlapping sequenced clones into sequenced-clone contigs and links sequenced-clone contigs to form sequenced-clone-contig scaffolds. A fingerprint clone contig may contain several sequenced-clone contigs, because bridging clones remain to be sequenced. The assembly contained 4,884 sequenced-clone

Table 5 The draft genome sequence

Chromosome	Sequence from clones (kb)			Sequence from contigs (kb)		
	Finished clones	Draft clones	Pre-draft clones	Contigs containing finished clones	Deep coverage sequence contigs	Draft/predraft sequence contigs
All	826,441	1,734,995	131,476	958,922	840,815	893,175
1	50,851	149,027	12,356	61,001	78,773	72,461
2	46,909	167,439	7,210	53,775	81,569	86,214
3	22,350	152,840	11,057	26,959	79,649	79,638
4	15,914	134,973	17,261	19,096	66,165	82,887
5	37,973	129,581	2,160	48,895	61,387	59,431
6	75,312	76,082	6,696	93,458	28,204	36,428
7	94,845	47,328	4,047	103,188	14,434	28,597
8	14,538	102,484	7,236	16,659	47,198	60,400
9	18,401	77,648	10,864	24,030	42,653	40,230
10	16,889	99,181	11,066	21,421	54,054	51,662
11	13,162	111,092	4,352	16,145	65,147	47,314
12	32,156	84,653	7,651	37,519	43,995	42,946
13	16,818	68,983	7,136	22,191	38,319	32,429
14	58,989	27,370	565	78,302	3,267	5,355
15	2,739	67,453	3,211	3,112	34,758	35,533
16	22,987	48,997	1,143	27,751	20,892	24,484
17	29,881	36,349	6,600	33,531	14,671	24,628
18	5,128	65,284	2,352	6,656	40,947	25,160
19	28,481	26,568	369	32,228	7,188	16,003
20	54,217	6,302	070	50,334	1,065	2,896
21	33,824	0	0	33,824	0	0
22	33,786	0	0	33,786	0	0
X	77,630	45,100	4,941	83,796	14,056	29,820
Y	18,169	3,221	363	20,222	333	1,198
NA	2,434	1,858	844	2,446	122	2,568
UL	2,056	6,182	1,020	2,395	1,969	4,894

The table presents summary statistics for the draft genome sequence over the entire genome and by individual chromosome. NA, clones that could not be placed into the sequenced clone layout. UL, clones that could be placed in the layout, but that could not reliably be placed on a chromosome. First three columns, data from finished clones, draft clones and predraft clones. The last three columns break the data down according to the type of sequence contig. Contigs containing finished clones represent sequence contigs that consist of finished sequence plus any (small) extensions from merged sequence contigs that arise from overlap with flanking draft clones. Deep coverage sequence contigs include sequence from two or more overlapping unfinished clones; they consist of roughly full shotgun coverage and thus are longer than the average unfinished sequence contig. Draft/predraft sequence contigs are all of the other sequence contigs in unfinished clones. Thus, the draft genome sequence consists of approximately one-third finished sequence, one-third deep coverage sequence and one-third draft/pre-draft coverage sequence. In all of the statistics, we count only nonoverlapping bases in the draft genome sequence.

contigs in 942 fingerprint clone contigs.

The hierarchy of contigs is summarized in Fig. 7. Initial sequence contigs are integrated to create merged sequence contigs, which are then linked to form sequence-contig scaffolds. These scaffolds reside within sequenced-clone contigs, which in turn reside within fingerprint clone contigs.

The draft genome sequence

The result of the assembly process is an integrated draft sequence of the human genome. Several features of the draft genome sequence are reported in Tables 5–7, including the proportion represented by finished, draft and predraft categories. The Tables also show the numbers and lengths of different types of contig, for each chromosome and for the genome as a whole.

The contiguity of the draft genome sequence at each level is an important feature. Two commonly used statistics have significant drawbacks for describing contiguity. The 'average length' of a contig is deflated by the presence of many small contigs comprising only a small proportion of the genome, whereas the 'length-weighted average length' is inflated by the presence of large segments of finished sequence. Instead, we chose to describe the contiguity as a property of the 'typical' nucleotide. We used a statistic called the 'N50 length', defined as the largest length L such that 50% of all nucleotides are contained in contigs of size at least L.

The continuity of the draft genome sequence reported here and the effectiveness of assembly can be readily seen from the following: half of all nucleotides reside within an initial sequence contig of at least 21.7 kb, a sequence contig of at least 82 kb, a sequence-contig scaffold of at least 274 kb, a sequenced-clone contig of at least 826 kb and a fingerprint clone contig of at least 8.4 Mb (Tables 6, 7). The cumulative distributions for each of these measures of contiguity are shown in Fig. 8, in which the N50 values for each measure can be seen as the value at which the cumulative distributions cross 50%. We have also estimated the size of each chromosome, by estimating the gap sizes (see below) and the extent of missing heterochromatic sequence[93,94,105–108] (Table 8). This is undoubtedly an oversimplification and does not adequately take into account the sequence status of each chromosome. Nonetheless, it provides a useful way to relate the draft sequence to the chromosomes.

Quality assessment

The draft genome sequence already covers the vast majority of the genome, but it remains an incomplete, intermediate product that is regularly updated as we work towards a complete finished sequence. The current version contains many gaps and errors. We therefore sought to evaluate the quality of various aspects of the current draft genome sequence, including the sequenced clones themselves, their assignment to a position in the fingerprint clone contigs, and the assembly of initial sequence contigs from the individual clones into sequence-contig scaffolds.

Nucleotide accuracy is reflected in a PHRAP score assigned to each base in the draft genome sequence and available to users through the Genome Browsers (see below) and public database entries. A summary of these scores for the unfinished portion of the genome is shown in Table 9. About 91% of the unfinished draft genome sequence has an error rate of less than 1 per 10,000 bases (PHRAP score > 40), and about 96% has an error rate of less than 1 in 1,000 bases (PHRAP > 30). These values are based only on the quality scores for the bases in the sequenced clones; they do not reflect additional confidence in the sequences that are represented in overlapping clones. The finished portion of the draft genome sequence has an error rate of less than 1 per 10,000 bases.

Individual sequenced clones. We assessed the frequency of misassemblies, which can occur when the assembly program PHRAP joins two nonadjacent regions in the clone into a single initial sequence contig. The frequency of misassemblies depends heavily on the depth and quality of coverage of each clone and the nature of the underlying sequence; thus it may vary among genomic regions and among individual centres. Most clone misassemblies are readily corrected as coverage is added during finishing, but they may have been propagated into the current version of the draft genome sequence and they justify caution for certain applications.

We estimated the frequency of misassembly by examining instances in which there was substantial overlap between a draft clone and a finished clone. We studied 83 Mb of such overlaps, involving about 9,000 initial sequence contigs. We found 5.3 instances per Mb in which the alignment of an initial sequence contig to the finished sequence failed to extend to within 200 bases

Table 6 Clone level contiguity of the draft genome sequence

Chromosome	Sequenced-clone contigs		Sequenced-clone-contig scaffolds		Fingerprint clone contigs with sequence	
	Number	N50 length (kb)	Number	N50 length (kb)	Number	N50 length (kb)
All	4,884	826	2,191	2,279	942	8,398
1	453	650	197	1,915	106	3,537
2	348	1,028	127	3,140	52	10,628
3	409	672	201	1,550	73	5,077
4	384	606	163	1,659	41	6,918
5	385	623	164	1,642	48	5,747
6	292	814	98	3,292	17	24,680
7	224	1.074	86	3,527	29	20,401
8	292	542	115	1,742	43	6,236
9	143	1,242	78	2,411	21	29,108
10	179	1,097	105	1,952	16	30,284
11	224	887	89	3,024	31	9,414
12	196	1,138	76	2,717	28	9,546
13	128	1,151	56	3,257	13	25,256
14	54	3,079	27	8,489	14	22,128
15	123	797	56	2,095	19	8,274
16	159	620	92	1,317	57	2,716
17	138	831	58	2,138	43	2,816
18	137	709	47	2,572	24	4,887
19	159	569	79	1,200	51	1,534
20	42	2,318	20	6,862	9	23,489
21	5	28,515	5	28,515	5	28,515
22	11	23,048	11	23,048	11	23,048
X	325	572	181	1,082	143	1,436
Y	27	1,539	20	3,290	8	5,135
UL	47	227	40	281	40	281

Number and size of sequenced-clone contigs, sequenced-clone-contig scaffolds and those fingerprint clone contigs (see Box 1) that contain sequenced clones; some small fingerprint clone contigs do not as yet have associated sequence. UL, fingerprint clone contigs that could not reliably be placed on a chromosome. These length estimates are from the draft genome sequence, in which gaps between sequence contigs are arbitrarily represented with 100 Ns and gaps between sequence clone contigs with 50,000 Ns for 'bridged gaps' and 100,000 Ns for 'unbridged gaps'. These arbitrary values differ minimally from empirical estimates of gap size (see text), and using the empirically derived estimates would change the N50 lengths presented here only slightly. For unfinished chromosomes, the N50 length ranges from 1.5 to 3 times the arithmetic mean for sequenced-clone contigs, 1.5 to 3 times for sequenced-clone-contig scaffolds, and 1.5 to 6 times for fingerprint clone contigs with sequence.

of the end of the contig, suggesting a possible false join in the assembly of the initial sequence contig. In about half of these cases, the potential misassembly involved fewer than 400 bases, suggesting that a single raw sequence read may have been incorrectly joined. We found 1.9 instances per Mb in which the alignment showed an internal gap, again suggesting a possible misassembly; and 0.5 instances per Mb in which the alignment indicated that two initial sequence contigs that overlapped by at least 150 bp had not been merged by PHRAP. Finally, there were another 0.9 instances per Mb with various other problems. This gives a total of 8.6 instances per Mb of possible misassembly, with about half being relatively small issues involving a few hundred bases.

Some of the potential problems might not result from misassembly, but might reflect sequence polymorphism in the population,

small rearrangements during growth of the large-insert clones, regions of low-quality sequence or matches between segmental duplications. Thus, the frequency of misassemblies may be overstated. On the other hand, the criteria for recognizing overlap between draft and finished clones may have eliminated some misassemblies.

Layout of the sequenced clones. We assessed the accuracy of the layout of sequenced clones onto the fingerprinted clone contigs by calculating the concordance between the positions assigned to a sequenced clone on the basis of *in silico* digestion and the position assigned on the basis of BAC end sequence data. The positions agreed in 98% of cases in which independent assignments could be made by both methods. The results were also compared with well studied regions containing both finished and draft genome sequence. These results indicated that sequenced clone order in the fingerprint map was reliable to within about half of one clone length ($\sim$100 kb).

A direct test of the layout is also provided by the draft genome sequence assembly itself. With extensive coverage of the genome, a correctly placed clone should usually (although not always) show sequence overlap with its neighbours in the map. We found only 421 instances of 'singleton' clones that failed to overlap a neighbouring clone. Close examination of the data suggests that most of these are correctly placed, but simply do not yet overlap an adjacent sequenced clone. About 150 clones appeared to be candidates for being incorrectly placed.

Alignment of the fingerprint clone contigs. The alignment of the fingerprint clone contigs with the chromosomes was based on the radiation hybrid, YAC and genetic maps of STSs. The positions of most of the STSs in the draft genome sequence were consistent with these previous maps, but the positions of about 1.7% differed from one or more of them. Some of these disagreements may be due to errors in the layout of the sequenced clones or in the underlying

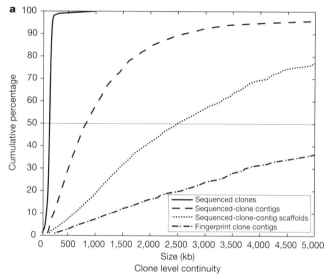

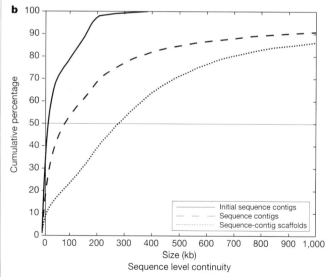

Figure 8 Cumulative distributions of several measures of clone level contiguity and sequence contiguity. The figures represent the proportion of the draft genome sequence contained in contigs of at most the indicated size. **a**, Clone level contiguity. The clones have a tight size distribution with an N50 of $\sim$ 160 kb (corresponding to 50% on the cumulative distribution). Sequenced-clone contigs represent the next level of continuity, and are linked by mRNA sequences or pairs of BAC end sequences to yield the sequenced-clone-contig scaffolds. The underlying contiguity of the layout of sequenced clones against the fingerprinted clone contigs is only partially shown at this scale.
b, Sequence contiguity. The input fragments have low continuity (N50 = 21.7 kb). After merging, the sequence contigs grow to an N50 length of about 82 kb. After linking, sequence-contig scaffolds with an N50 length of about 274 kb are created.

Figure 9 Overview of features of draft human genome. The Figure shows the occurrences of twelve important types of feature across the human genome. Large grey blocks represent centromeres and centromeric heterochromatin (size not precisely to scale). Each of the feature types is depicted in a track, from top to bottom as follows. (1) Chromosome position in Mb. (2) The approximate positions of Giemsa-stained chromosome bands at the 800 band resolution. (3) Level of coverage in the draft genome sequence. Red, areas covered by finished clones; yellow, areas covered by predraft sequence. Regions covered by draft sequenced clones are in orange, with darker shades reflecting increasing shotgun sequence coverage. (4) GC content. Percentage of bases in a 20,000 base window that are C or G. (5) Repeat density. Red line, density of SINE class repeats in a 100,000-base window; blue line, density of LINE class repeats in a 100,000-base window. (6) Density of SNPs in a 50,000-base window. The SNPs were detected by sequencing and alignments of random genomic reads. Some of the heterogeneity in SNP density reflects the methods used for SNP discovery. Rigorous analysis of SNP density requires comparing the number of SNPs identified to the precise number of bases surveyed. (7) Non-coding RNA genes. Brown, functional RNA genes such as tRNAs, snoRNAs and rRNAs; light orange, RNA pseudogenes. (8) CpG islands. Green ticks represent regions of $\sim$ 200 bases with CpG levels significantly higher than in the genome as a whole, and GC ratios of at least 50%. (9) Exofish ecores. Regions of homology with the pufferfish *T. nigroviridis*[292] are blue. (10) ESTs with at least one intron when aligned against genomic DNA are shown as black tick marks. (11) The starts of genes predicted by Genie or Ensembl are shown as red ticks. The starts of known genes from the RefSeq database[110] are shown in blue. (12) The names of genes that have been uniquely located in the draft genome sequence, characterized and named by the HGM Nomenclature Committee. Known disease genes from the OMIM database are red, other genes blue. This Figure is based on an earlier version of the draft genome sequence than analysed in the text, owing to production constraints. We are aware of various errors in the Figure, including omissions of some known genes and misplacements of others. Some genes are mapped to more than one location, owing to errors in assembly, close paralogues or pseudogenes. Manual review was performed to select the most likely location in these cases and to correct other regions. For updated information, see http://genome.ucsc.edu/ and http://www.ensembl.org/.

Table 7 Sequence level contiguity of the draft genome sequence

Chromosome	Initial sequence contigs		Sequence contigs		Sequence-contig scaffolds	
	Number	N50 length (kb)	Number	N50 length (kb)	Number	N50 length (kb)
All	396,913	21.7	149,821	81.9	87,757	274.3
1	37,656	16.5	12,256	59.1	5,457	278.4
2	32,280	19.9	13,228	57.3	6,959	248.5
3	38,848	15.6	15,098	37.7	8,964	167.4
4	28,600	16.0	13,152	33.0	7,402	158.9
5	30,096	20.4	10,689	72.9	6,378	241.2
6	17,472	43.6	5,547	180.3	2,554	485.0
7	12,733	86.4	4,562	335.7	2,726	591.3
8	19,042	18.1	8,984	38.2	4,631	198.9
9	15,955	20.1	6,226	55.6	3,766	216.2
10	21,762	18.7	9,126	47.9	6,886	133.0
11	29,723	14.3	8,503	40.0	4,684	193.2
12	22,050	19.1	8,422	63.4	5,526	217.0
13	13,737	21.7	5,193	70.5	2,659	300.1
14	4,470	161.4	829	1,371.0	541	2,009.5
15	13,134	15.3	5,840	30.3	3,229	149.7
16	10,297	34.4	4,916	119.5	3,337	356.3
17	10,369	22.9	4,339	90.6	2,616	248.9
18	16,266	15.3	4,461	51.4	2,540	216.1
19	6,009	38.4	2,503	134.4	1,551	375.5
20	2,884	108.6	511	1,346.7	312	813.8
21	103	340.0	5	28,515.3	5	28,515.3
22	526	113.9	11	23,048.1	11	23,048.1
X	11,062	58.8	4,607	218.6	2,610	450.7
Y	557	154.3	140	1,388.6	106	1,439.7
UL	1,282	21.4	613	46.0	297	166.4

This Table is similar to Table 6 but shows the number and N50 length for various types of sequence contig (see Box 1). See legend to Table 6 concerning treatment of gaps. For sequence contigs in the draft genome sequence, the N50 length ranges from 1.7 to 5.5 times the arithmetic mean for initial sequence contigs, 2.5 to 8.2 times for merged sequence contigs, and 6.1 to 10 times for sequence-contig scaffolds.

Table 8 Chromosome size estimates

Chromosome*	Sequenced bases† (Mb)	FCC gaps‡		SCC gaps‖		Sequence gaps#		Heterochromatin and short arm adjustments**(Mb)	Total estimated chromosome size (including artefactual duplication in draft genome sequence)†† (Mb)	Previously estimated chromosome size‡‡ (Mb)
		Number	Total bases in gaps§ (Mb)	Number	Total bases in gaps¶ (Mb)	Number	Total bases in gaps☆ (Mb)			
All	2,692.9	897	152.0	4,076	142.7	145,514	80.6	212	3,289	3,286
1	212.2	104	17.7	347	12.1	11,803	6.5	30	279	263
2	221.6	50	8.5	296	10.4	12,880	7.1	3	251	255
3	186.2	71	12.1	336	11.8	14,689	8.1	3	221	214
4	168.1	39	6.6	343	12.0	12,768	7.1	3	197	203
5	169.7	46	7.8	337	11.8	10,304	5.7	3	198	194
6	158.1	15	2.6	275	9.6	5,225	2.9	3	176	183
7	146.2	27	4.6	195	6.8	4,338	2.4	3	163	171
8	124.3	41	7.0	249	8.7	8,692	4.8	3	148	155
9	106.9	19	3.2	122	4.3	6,083	3.4	22	140	145
10	127.1	14	2.4	163	5.7	8,947	5.0	3	143	144
11	128.6	29	4.9	193	6.8	8,279	4.6	3	148	144
12	124.5	26	4.4	168	5.9	8,226	4.6	3	142	143
13	92.9	12	2.0	115	4.0	5,065	2.8	16	118	114
14	86.9	13	2.2	40	1.4	775	0.4	16	107	109
15	73.4	18	3.1	104	3.6	5,717	3.2	17	100	106
16	73.1	55	9.4	102	3.6	4,757	2.6	15	104	98
17	72.8	41	7.0	95	3.3	4,261	2.4	3	88	92
18	72.9	22	3.7	113	4.0	4,324	2.4	3	86	85
19	55.4	49	8.3	108	3.8	2,344	1.3	3	72	67
20	60.5	7	1.2	33	1.2	469	0.3	3	66	72
21	33.8	4	0.1	0	0.0	0	0.0	11	45	50
22	33.8	10	1.0	0	0.0	0	0.0	13	48	56
X	127.7	141	24.0	182	6.4	4,282	2.4	3	163	164
Y	21.8	6	1.0	19	0.7	113	0.1	27	51	59
NA	5.1	0	0	134	0.0	577	0.3	0	0	0
UL	9.3	38	0	7	0.0	566	0.3	0	0	0

* NA, sequenced clones that could not be associated with fingerprint clone contigs. UL, clone contigs that could not be reliably placed on a chromosome.
† Total number of bases in the draft genome sequence, excluding gaps. Total length of scaffold (including gaps contained within clones) is 2.916 Gb.
‡ Gaps between those fingerprint clone contigs that contain sequenced clones excluding gaps for centromeres.
§ For unfinished chromosomes, we estimate an average size of 0.17 Mb per FCC gap, based on retrospective estimates of the clone coverage of chromosomes 21 and 22. Gap estimates for chromosomes 21 and 22 are taken from refs 93, 94.
‖ Gaps between sequenced-clone contigs within a fingerprint clone contig.
¶ For unfinished chromosomes, we estimate sequenced clone gaps at 0.035 Mb each, based on evaluation of a sample of these gaps.
# Gaps between two sequence contigs within a sequenced-clone contig.
☆ We estimate the average number of bases in sequence gaps from alignments of the initial sequence contigs of unfinished clones (see text) and extrapolation to the whole chromosome.
** Including adjustments for estimates of the sizes of the short arms of the acrocentric chromosomes 13, 14, 15, 21 and 22 (ref. 105), estimates for the centromere and heterochromatic regions of chromosomes 1, 9 and 16 (refs 106, 107) and estimates of 3 Mb for the centromere and 24 Mb for telomeric heterochromatin for the Y chromosome[108].
†† The sum of the five lengths in the preceding columns. This is an overestimate, because the draft genome sequence contains some artefactual sequence owing to inability to correctly to merge all underlying sequence contigs. The total amount of artefactual duplication varies among chromosomes; the overall amount is estimated by computational analysis to be about 100 Mb, or about 3% of the total length given, yielding a total estimated size of about 3,200 Mb for the human genome.
‡‡ Including heterochromatic regions and acrocentric short arm(s)[105].

Figure 9

Note: The diagrams of the chromosomes span more than one page. For ease of reference, the end of a chromosome segment on one page is repeated on the following page, so that the segments overlap by 2–4 megabases at each end.

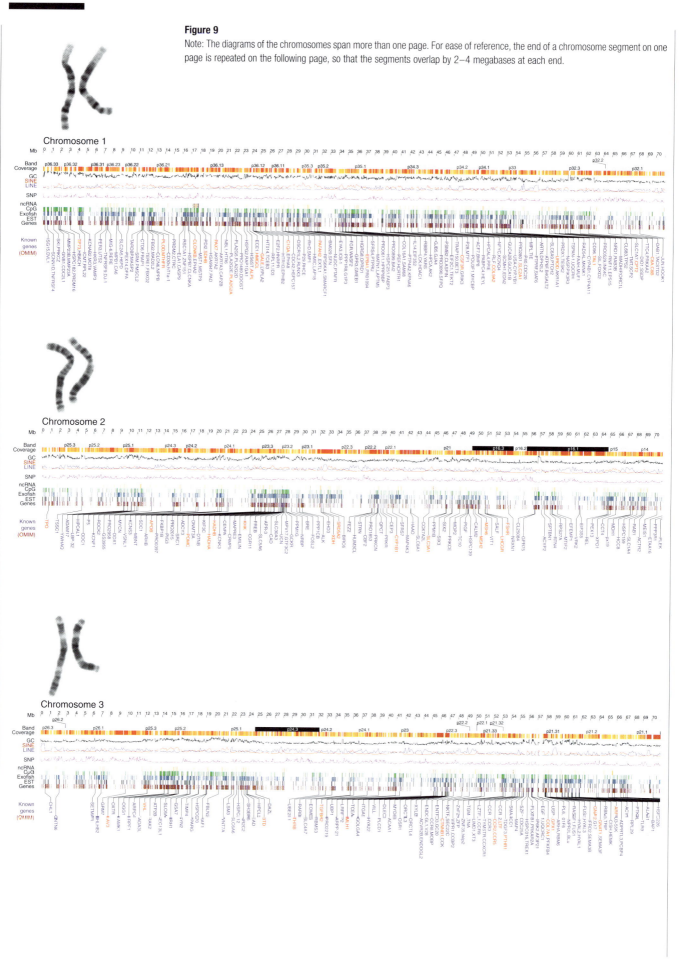

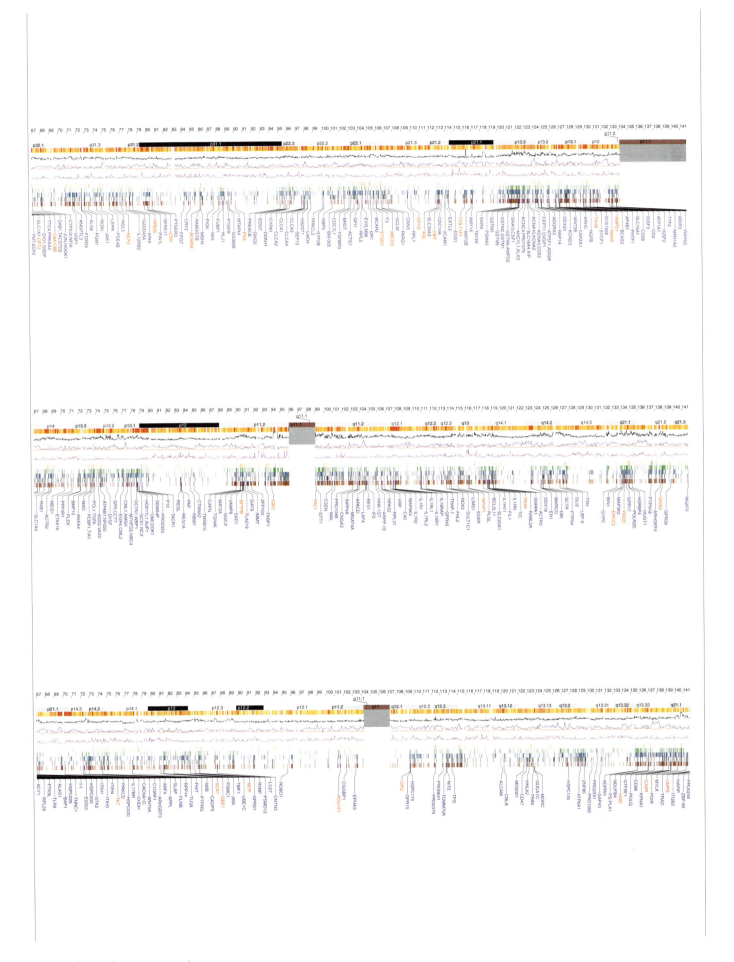

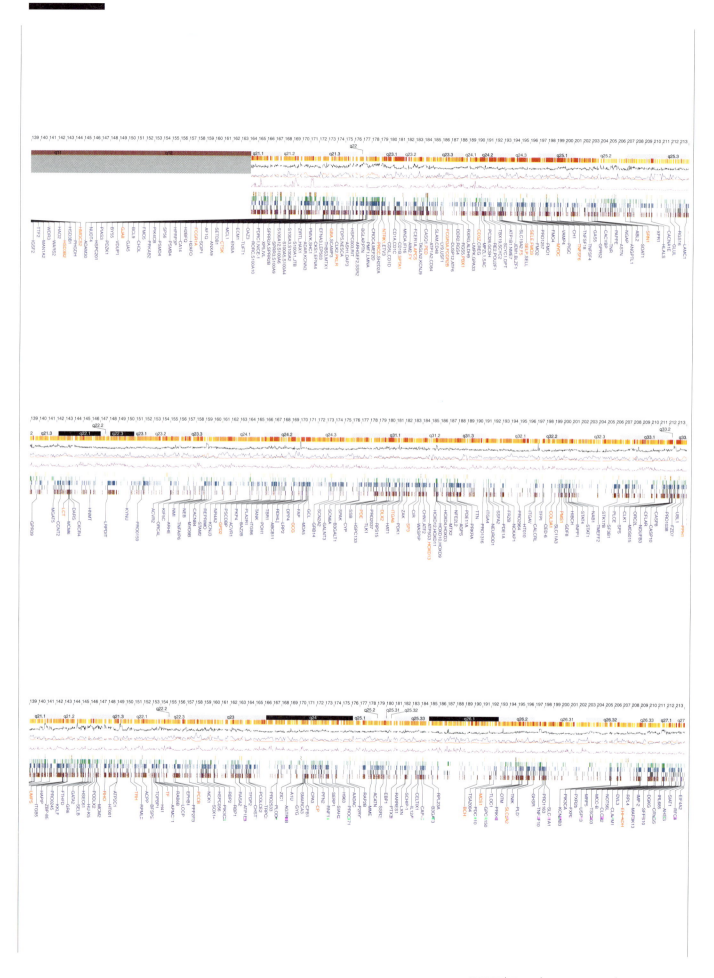

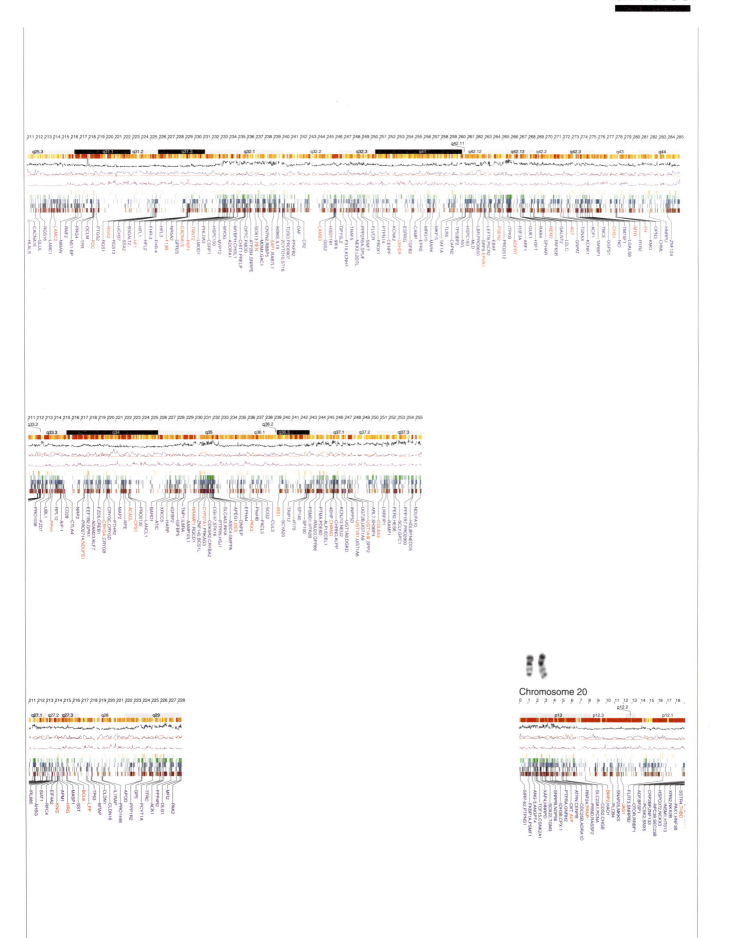

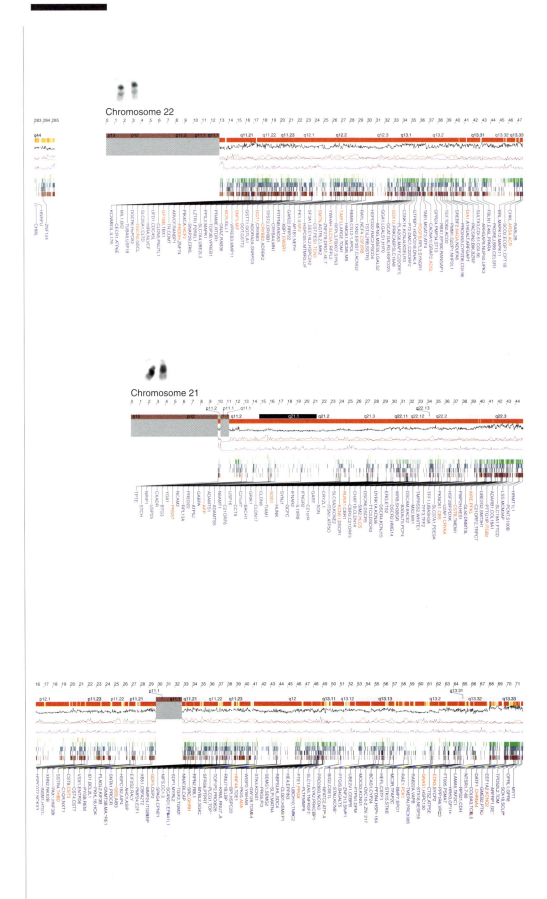

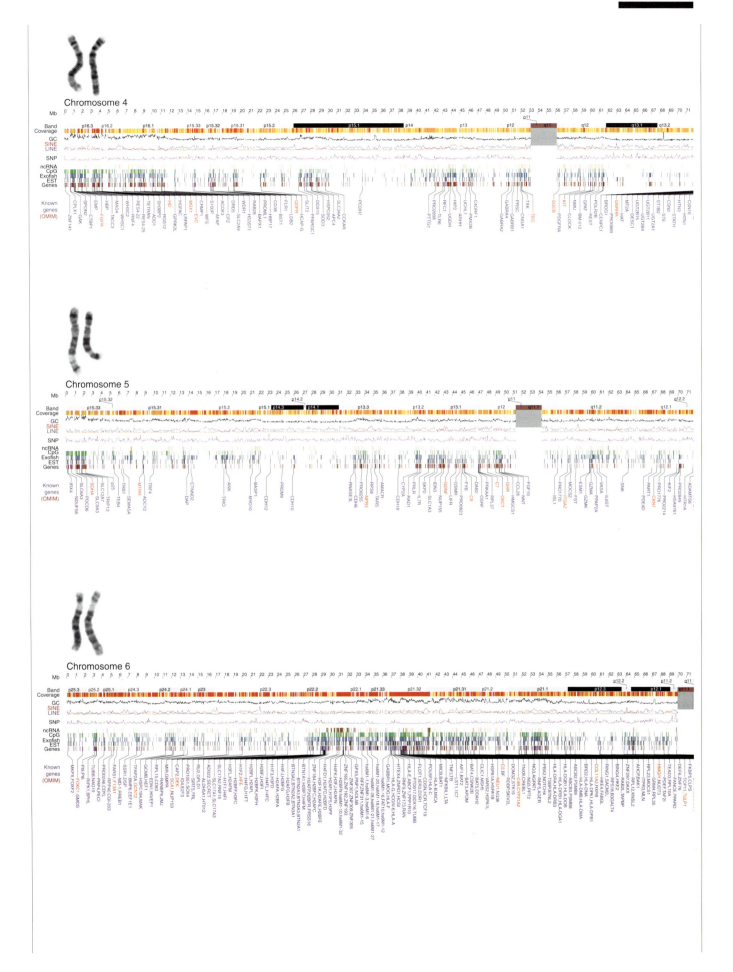

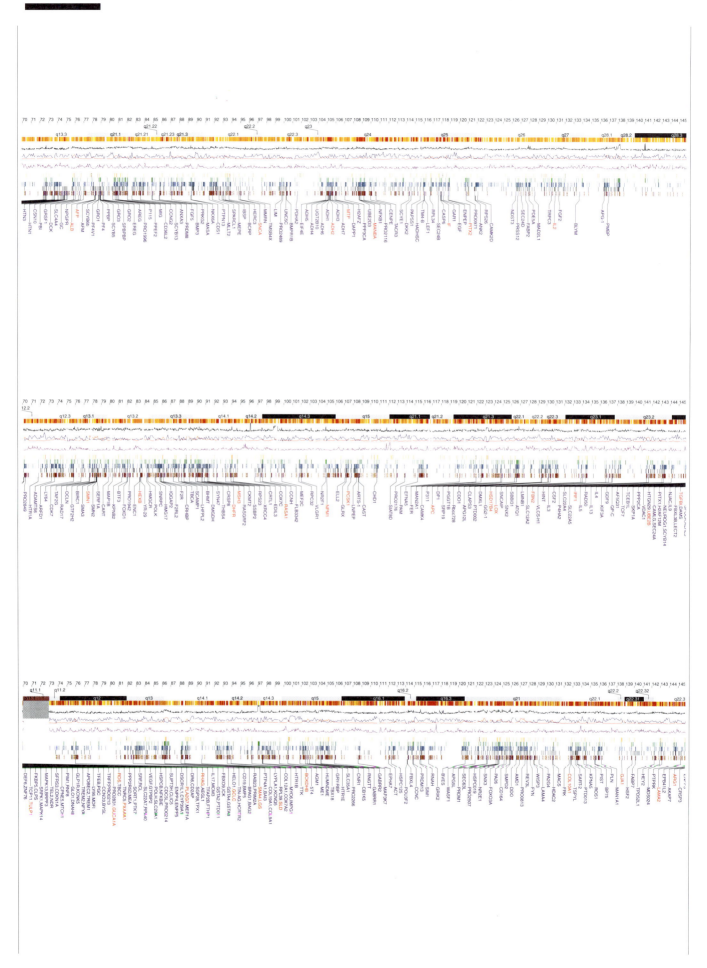

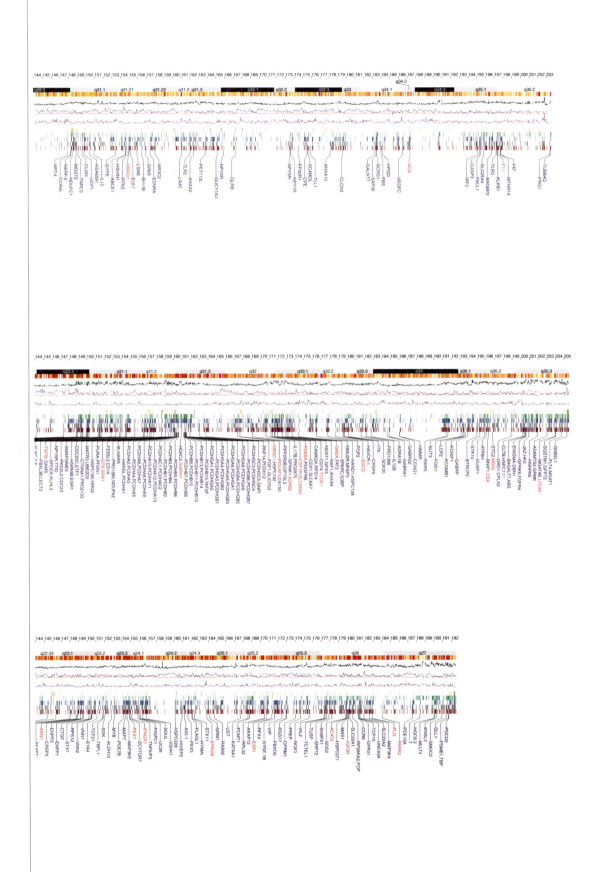

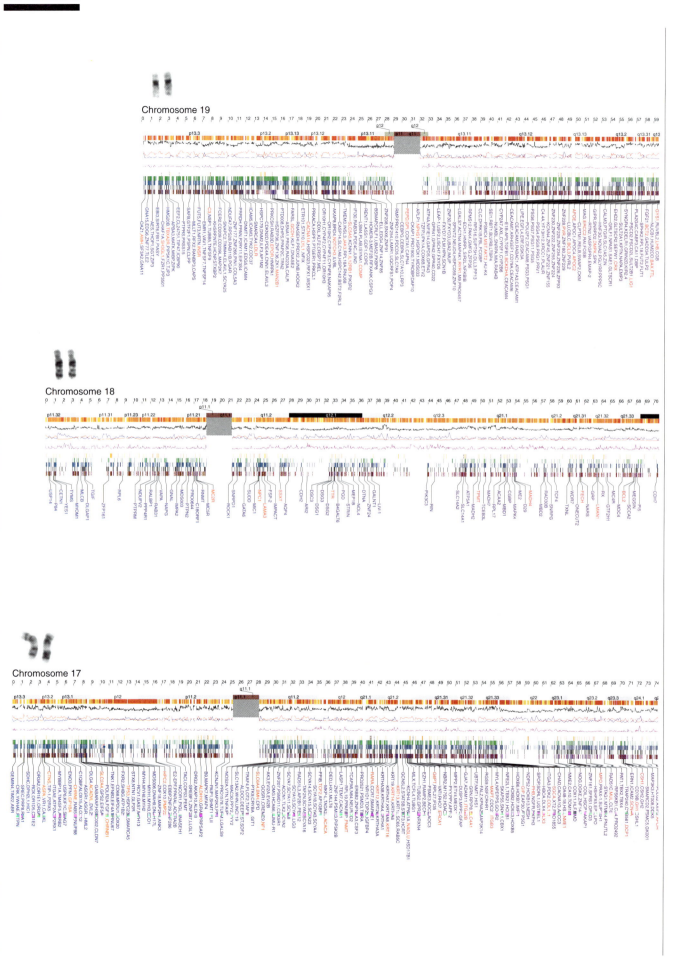

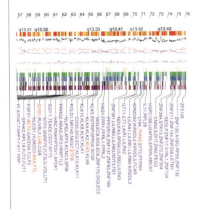

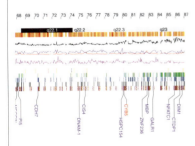

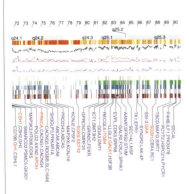

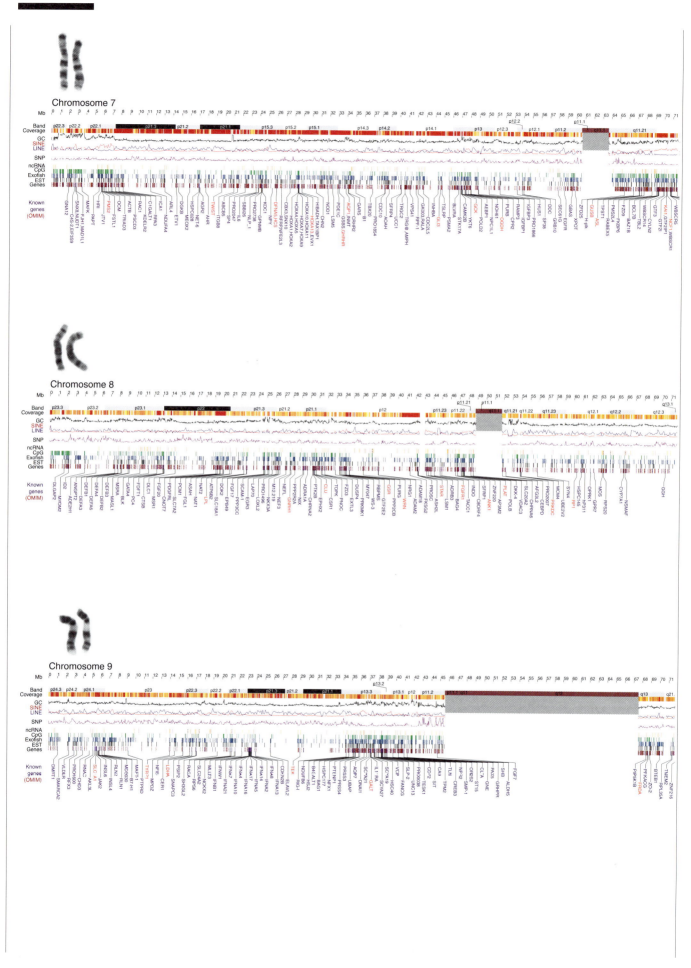

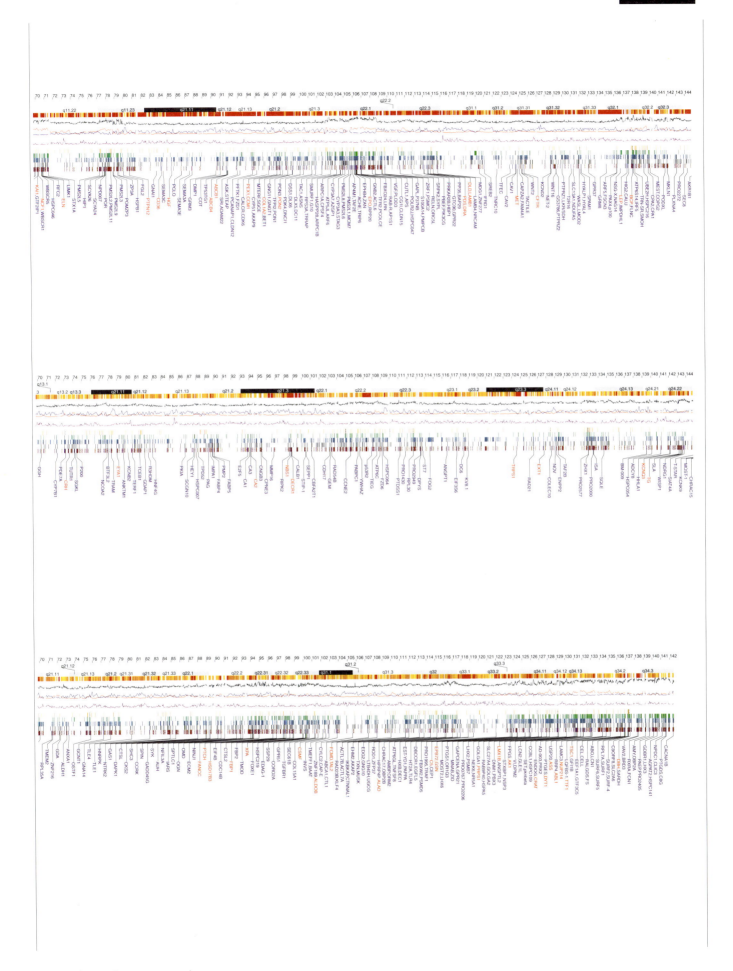

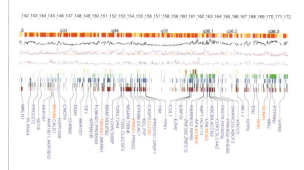

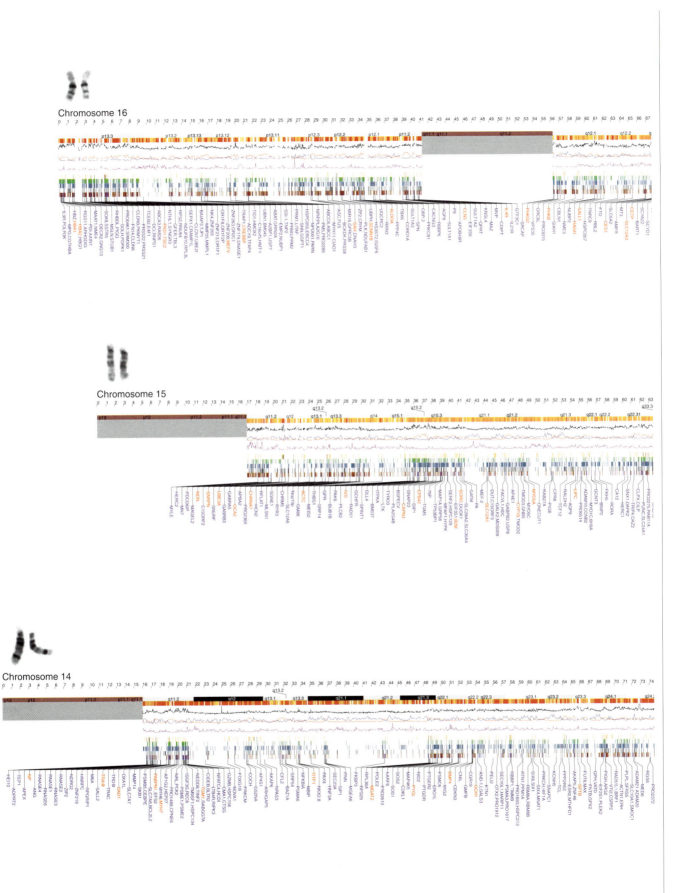

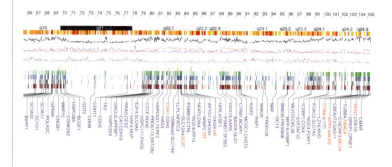

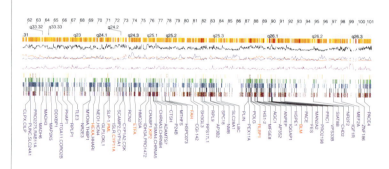

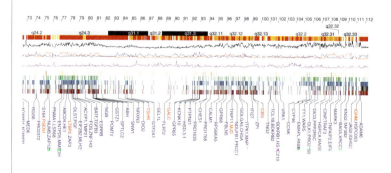

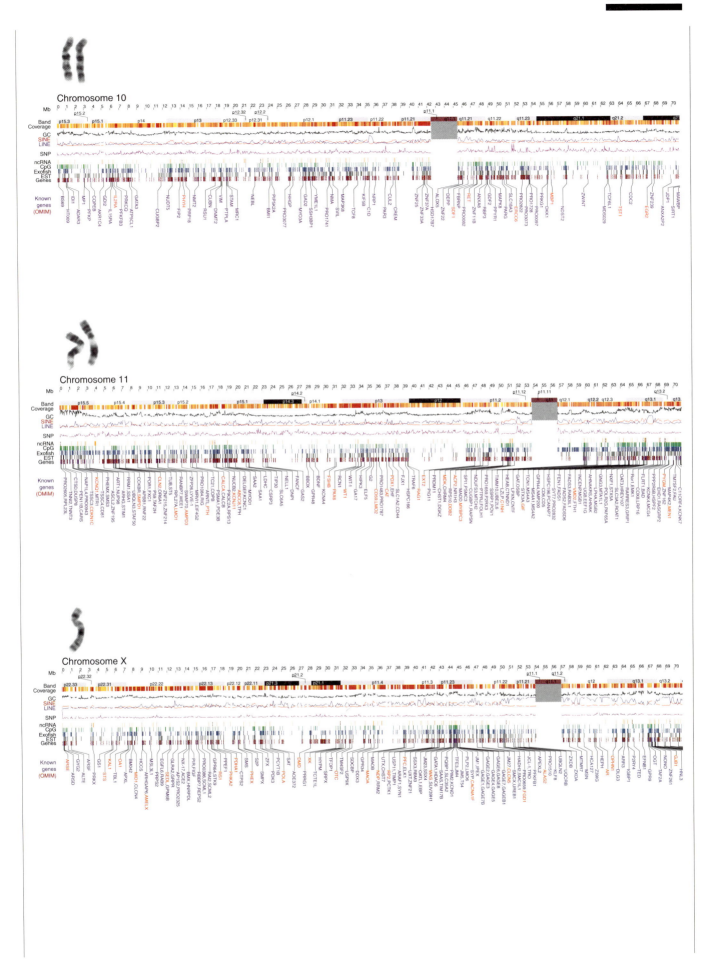

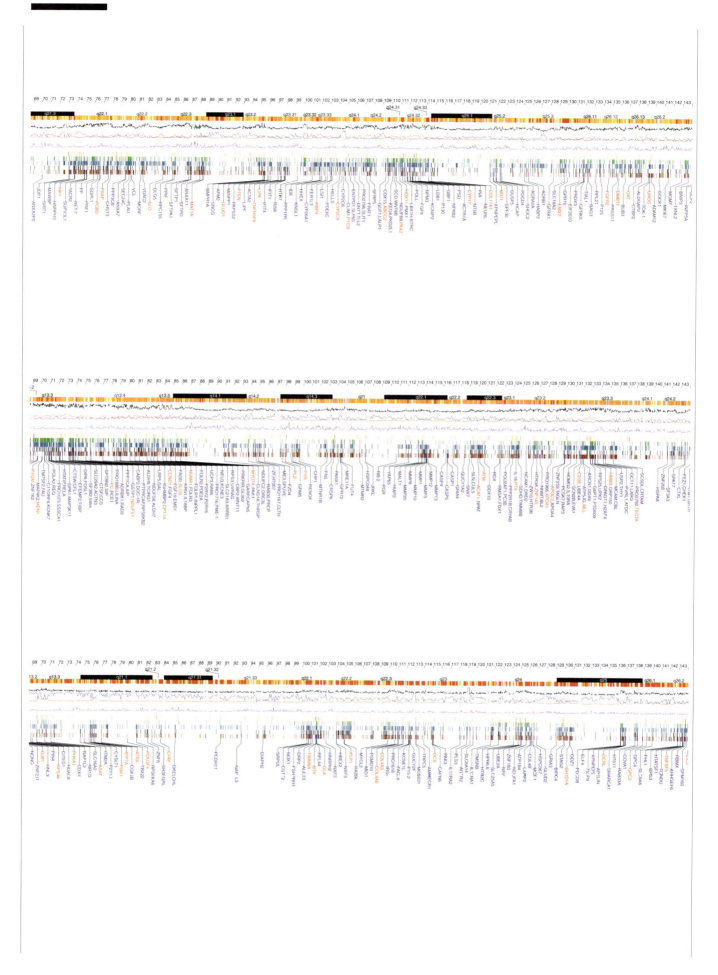

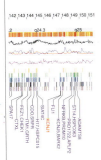

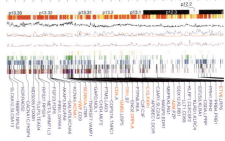

Chromosome 12

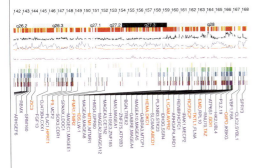

Chromosome 13

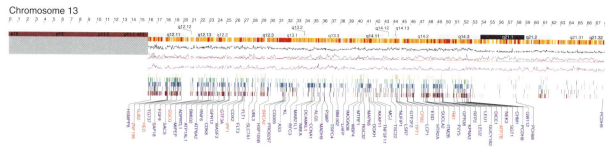

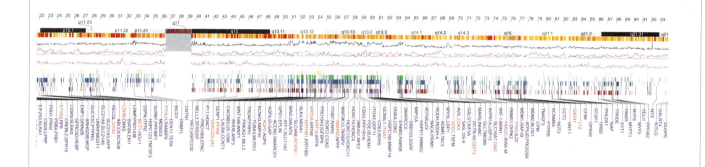

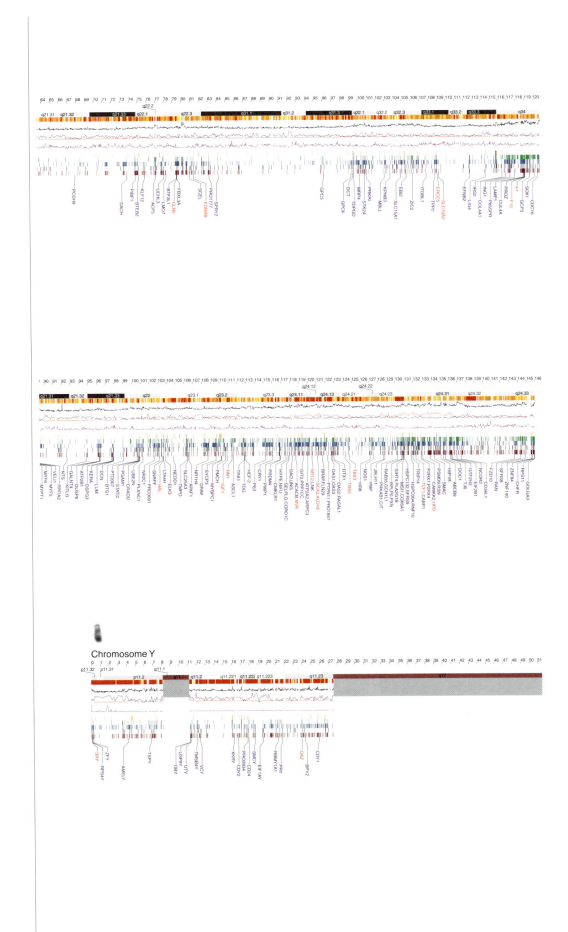

Chromosome Y

fingerprint map. However, many involve STSs that have been localized on only one or two of the previous maps or that occur as isolated discrepancies in conflict with several flanking STSs. Many of these cases are probably due to errors in the previous maps (with error rates for individual maps estimated at 1–2%[100]). Others may be due to incorrect assignment of the STSs to the draft genome sequence (by the electronic polymerase chain reaction (e-PCR) computer program) or to database entries that contain sequence data from more than one clone (owing to cross-contamination).

Graphical views of the independent data sets were particularly useful in detecting problems with order or orientation (Fig. 5). Areas of conflict were reviewed and corrected if supported by the underlying data. In the version discussed here, there were 41 sequenced clones falling in 14 sequenced-clone contigs with STS content information from multiple maps that disagreed with the flanking clones or sequenced-clone contigs; the placement of these clones thus remains suspect. Four of these instances suggest errors in the fingerprint map, whereas the others suggest errors in the layout of sequenced clones. These cases are being investigated and will be corrected in future versions.

Assembly of the sequenced clones. We assessed the accuracy of the assembly by using a set of 148 draft clones comprising 22.4 Mb for which finished sequence subsequently became available[104]. The initial sequence contigs lack information about order and orientation, and GigAssembler attempts to use linking data to infer such information as far as possible[104]. Starting with initial sequence contigs that were unordered and unoriented, the program placed 90% of the initial sequence contigs in the correct orientation and 85% in the correct order with respect to one another. In a separate test, GigAssembler was tested on simulated draft data produced from finished sequence on chromosome 22 and similar results were obtained.

Some problems remain at all levels. First, errors in the initial sequence contigs persist in the merged sequence contigs built from them and can cause difficulties in the assembly of the draft genome sequence. Second, GigAssembler may fail to merge some overlapping sequences because of poor data quality, allelic differences or misassemblies of the initial sequence contigs; this may result in apparent local duplication of a sequence. We have estimated by various methods the amount of such artefactual duplication in the assembly from these and other sources to be about 100 Mb. On the other hand, nearby duplicated sequences may occasionally be incorrectly merged. Some sequenced clones remain incorrectly placed on the layout, as discussed above, and others (< 0.5%) remain unplaced. The fingerprint map has undoubtedly failed to resolve some closely related duplicated regions, such as the Williams region and several highly repetitive subtelomeric and pericentric regions (see below). Detailed examination and sequence finishing may be required to sort out these regions precisely, as has been done with chromosome Y[89]. Finally, small sequenced-clone contigs with limited or no STS

landmark content remain difficult to place. Full utilization of the higher resolution radiation hybrid map (the TNG map) may help in this[95]. Future targeted FISH experiments and increased map continuity will also facilitate positioning of these sequences.

Genome coverage

We next assessed the nature of the gaps within the draft genome sequence, and attempted to estimate the fraction of the human genome not represented within the current version.

Gaps in draft genome sequence coverage. There are three types of gap in the draft genome sequence: gaps within unfinished sequenced clones; gaps between sequenced-clone contigs, but within fingerprint clone contigs; and gaps between fingerprint clone contigs. The first two types are relatively straightforward to close simply by performing additional sequencing and finishing on already identified clones. Closing the third type may require screening of additional large-insert clone libraries and possibly new technologies for the most recalcitrant regions. We consider these three cases in turn.

We estimated the size of gaps within draft clones by studying instances in which there was substantial overlap between a draft clone and a finished clone, as described above. The average gap size in these draft sequenced clones was 554 bp, although the precise estimate was sensitive to certain assumptions in the analysis. Assuming that the sequence gaps in the draft genome sequence are fairly represented by this sample, about 80 Mb or about 3% (likely range 2–4%) of sequence may lie in the 145,514 gaps within draft sequenced clones.

The gaps between sequenced-clone contigs but within fingerprint clone contigs are more difficult to evaluate directly, because the draft genome sequence flanking many of the gaps is often not precisely aligned with the fingerprinted clones. However, most are much smaller than a single BAC. In fact, nearly three-quarters of these gaps are bridged by one or more individual BACs, as indicated by linking information from BAC end sequences. We measured the sizes of a subset of gaps directly by examining restriction fragment fingerprints of overlapping clones. A study of 157 'bridged' gaps and 55 'unbridged' gaps gave an average gap size of 25 kb. Allowing for the possibility that these gaps may not be fully representative and that some restriction fragments are not included in the calculation, a more conservative estimate of gap size would be 35 kb. This would indicate that about 150 Mb or 5% of the human genome may reside in the 4,076 gaps between sequenced-clone contigs. This sequence should be readily obtained as the clones spanning them are sequenced.

The size of the gaps between fingerprint clone contigs was estimated by comparing the fingerprint maps to the essentially completed chromosomes 21 and 22. The analysis shows that the fingerprinted BAC clones in the global database cover 97–98% of the sequenced portions of those chromosomes[86]. The published sequences of these chromosomes also contain a few small gaps (5 and 11, respectively) amounting to some 1.6% of the euchromatic sequence, and do not include the heterochromatic portion. This suggests that the gaps between contigs in the fingerprint map contain about 4% of the euchromatic genome. Experience with closure of such gaps on chromosomes 20 and 7 suggests that many of these gaps are less than one clone in length and will be closed by clones from other libraries. However, recovery of sequence from these gaps represents the most challenging aspect of producing a complete finished sequence of the human genome.

As another measure of the representation of the BAC libraries, Riethman[109] has found BAC or cosmid clones that link to telomeric half-YACs or to the telomeric sequence itself for 40 of the 41 non-satellite telomeres. Thus, the fingerprint map appears to have no substantial gaps in these regions. Many of the pericentric regions are also represented, but analysis is less complete here (see below).

Representation of random raw sequences. In another approach to measuring coverage, we compared a collection of random raw sequence reads to the existing draft genome sequence. In principle,

Table 9 Distribution of PHRAP scores in the draft genome sequence

PHRAP score	Percentage of bases in the draft genome sequence
0–9	0.6
10–19	1.3
20–29	2.2
30–39	4.8
40–49	8.1
50–59	8.7
60–69	9.0
70–79	12.1
80–89	17.3
>90	35.9

PHRAP scores are a logarithmically based representation of the error probability. A PHRAP score of X corresponds to an error probability of $10^{-X/10}$. Thus, PHRAP scores of 20, 30 and 40 correspond to accuracy of 99%, 99.9% and 99.99%, respectively. PHRAP scores are derived from quality scores of the underlying sequence reads used in sequence assembly. See http://www.genome.washington.edu/UWGC/analysistools/phrap.htm.

the fraction of reads matching the draft genome sequence should provide an estimate of genome coverage. In practice, the comparison is complicated by the need to allow for repeat sequences, the imperfect sequence quality of both the raw sequence and the draft genome sequence, and the possibility of polymorphism. Nonetheless, the analysis provides a reasonable view of the extent to which the genome is represented in the draft genome sequence and the public databases.

We compared the raw sequence reads against both the sequences used in the construction of the draft genome sequence and all of GenBank using the BLAST computer program. Of the 5,615 raw sequence reads analysed (each containing at least 100 bp of contiguous non-repetitive sequence), 4,924 had a match of ≥97% identity with a sequenced clone, indicating that 88 ± 1.5% of the genome was represented in sequenced clones. The estimate is subject to various uncertainties. Most serious is the proportion of repeat sequence in the remainder of the genome. If the unsequenced portion of the genome is unusually rich in repeated sequence, we would underestimate its size (although the excess would be comprised of repeated sequence).

We examined those raw sequences that failed to match by comparing them to the other publicly available sequence resources. Fifty (0.9%) had matches in public databases containing cDNA sequences, STSs and similar data. An additional 276 (or 43% of the remaining raw sequence) had matches to the whole-genome shotgun reads discussed above (consistent with the idea that these reads cover about half of the genome).

We also examined the extent of genome coverage by aligning the cDNA sequences for genes in the RefSeq dataset[110] to the draft genome sequence. We found that 88% of the bases of these cDNAs could be aligned to the draft genome sequence at high stringency (at least 98% identity). (A few of the alignments with either the random raw sequence reads or the cDNAs may be to a highly similar region in the genome, but such matches should affect the estimate of genome coverage by considerably less than 1%, based on the estimated extent of duplication within the genome (see below).)

These results indicate that about 88% of the human genome is represented in the draft genome sequence and about 94% in the combined publicly available sequence databases. The figure of 88% agrees well with our independent estimates above that about 3%, 5% and 4% of the genome reside in the three types of gap in the draft genome sequence.

Finally, a small experimental check was performed by screening a large-insert clone library with probes corresponding to 16 of the whole genome shotgun reads that failed to match the draft genome sequence. Five hybridized to many clones from different fingerprint clone contigs and were discarded as being repetitive. Of the remaining eleven, two fell within sequenced clones (presumably within sequence gaps of the first type), eight fell in fingerprint clone

contigs but between sequenced clones (gaps of the second type) and one failed to identify clones in the fingerprint map (gaps of the third type) but did identify clones in another large-insert library. Although these numbers are small, they are consistent with the view that the much of the remaining genome sequence lies within already identified clones in the current map.

Estimates of genome and chromosome sizes. Informed by this analysis of genome coverage, we proceeded to estimate the sizes of the genome and each of the chromosomes (Table 8). Beginning with the current assigned sequence for each chromosome, we corrected for the known gaps on the basis of their estimated sizes (see above). We attempted to account for the sizes of centromeres and heterochromatin, neither of which are well represented in the draft sequence. Finally, we corrected for around 100 Mb of artefactual duplication in the assembly. We arrived at a total human genome size estimate of around 3,200 Mb, which compares favourably with previous estimates based on DNA content.

We also independently estimated the size of the euchromatic portion of the genome by determining the fraction of the 5,615 random raw sequences that matched the finished portion of the human genome (whose total length is known with greater precision). Twenty-nine per cent of these raw sequences found a match among 835 Mb of nonredundant finished sequence. This leads to an estimate of the euchromatic genome size of 2.9 Gb. This agrees reasonably with the prediction above based on the length of the draft genome sequence (Table 8).

Update. The results above reflect the data on 7 October 2000. New data are continually being added, with improvements being made to the physical map, new clones being sequenced to close gaps and draft clones progressing to full shotgun coverage and finishing. The draft genome sequence will be regularly reassembled and publicly released.

Currently, the physical map has been refined such that the number of fingerprint clone contigs has fallen from 1,246 to 965; this reflects the elimination of some artefactual contigs and the closure of some gaps. The sequence coverage has risen such that 90% of the human genome is now represented in the sequenced clones and more than 94% is represented in the combined publicly available sequence databases. The total amount of finished sequence is now around 1 Gb.

Broad genomic landscape

What biological insights can be gleaned from the draft sequence? In this section, we consider very large-scale features of the draft genome sequence: the distribution of GC content, CpG islands and recombination rates, and the repeat content and gene content of the human genome. The draft genome sequence makes it possible to integrate these features and others at scales ranging from individual

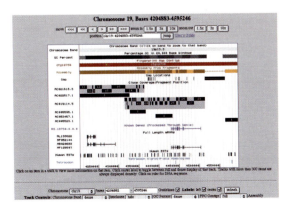

Figure 10 Screen shot from UCSC Draft Human Genome Browser. See http://genome.ucsc.edu/.

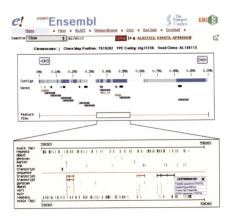

Figure 11 Screen shot from the Genome Browser of Project Ensembl. See http://www.ensembl.org.

87

nucleotides to collections of chromosomes. Unless noted, all analyses were conducted on the assembled draft genome sequence described above.

Figure 9 provides a high-level view of the contents of the draft genome sequence, at a scale of about 3.8 Mb per centimetre. Of course, navigating information spanning nearly ten orders of magnitude requires computational tools to extract the full value. We have created and made freely available various 'Genome Browsers'. Browsers were developed and are maintained by the University of California at Santa Cruz (Fig. 10) and the EnsEMBL project of the European Bioinformatics Institute and the Sanger Centre (Fig. 11). Additional browsers have been created; URLs are listed at www.nhgri.nih.gov/genome_hub. These web-based computer tools allow users to view an annotated display of the draft genome sequence, with the ability to scroll along the chromosomes and zoom in or out to different scales. They include: the nucleotide sequence, sequence contigs, clone contigs, sequence coverage and finishing status, local GC content, CpG islands, known STS markers from previous genetic and physical maps, families of repeat sequences, known genes, ESTs and mRNAs, predicted genes, SNPs and sequence similarities with other organisms (currently the pufferfish *Tetraodon nigroviridis*). These browsers will be updated as the draft genome sequence is refined and corrected as additional annotations are developed.

In addition to using the Genome Browsers, one can download

from these sites the entire draft genome sequence together with the annotations in a computer-readable format. The sequences of the underlying sequenced clones are all available through the public sequence databases. URLs for these and other genome websites are listed in Box 2. A larger list of useful URLs can be found at www.nhgri.nih.gov/genome_hub. An introduction to using the draft genome sequence, as well as associated databases and analytical tools, is provided in an accompanying paper[111].

In addition, the human cytogenetic map has been integrated with the draft genome sequence as part of a related project. The BAC Resource Consortium[103] established dense connections between the maps using more than 7,500 sequenced large-insert clones that had been cytogenetically mapped by FISH; the average density of the map is 2.3 clones per Mb. Although the precision of the integration is limited by the resolution of FISH, the links provide a powerful tool for the analysis of cytogenetic aberrations in inherited diseases and cancer. These cytogenetic links can also be accessed through the Genome Browsers.

Long-range variation in GC content

The existence of GC-rich and GC-poor regions in the human genome was first revealed by experimental studies involving density gradient separation, which indicated substantial variation in average GC content among large fragments. Subsequent studies have indicated that these GC-rich and GC-poor regions may have different biological properties, such as gene density, composition of repeat sequences, correspondence with cytogenetic bands and recombination rate[112–117]. Many of these studies were indirect, owing to the lack of sufficient sequence data.

The draft genome sequence makes it possible to explore the variation in GC content in a direct and global manner. Visual inspection (Fig. 9) confirms that local GC content undergoes substantial long-range excursions from its genome-wide average of 41%. If the genome were drawn from a uniform distribution of GC content, the local GC content in a window of size n bp should be $41 \pm \sqrt{(41)(59)/n}$%. Fluctuations would be modest, with the standard deviation being halved as the window size is quadrupled— for example, 0.70%, 0.35%, 0.17% and 0.09% for windows of size 5, 20, 80 and 320 kb.

The draft genome sequence, however, contains many regions with much more extreme variation. There are huge regions (> 10 Mb) with GC content far from the average. For example, the most distal 48 Mb of chromosome 1p (from the telomere to about STS marker D1S3279) has an average GC content of 47.1%, and chromosome 13 has a 40-Mb region (roughly between STS marker A005X38 and

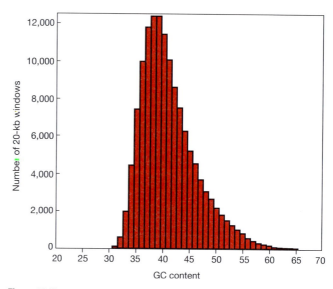

Figure 12 Histogram of GC content of 20-kb windows in the draft genome sequence.

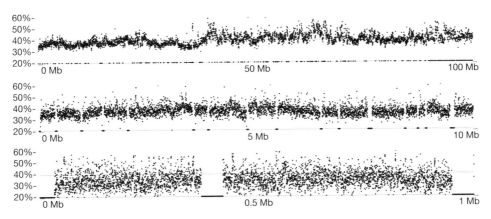

Figure 13 Variation in GC content at various scales. The GC content in subregions of a 100-Mb region of chromosome 1 is plotted, starting at about 83 Mb from the beginning of the draft genome sequence. This region is AT-rich overall. Top, the GC content of the entire 100-Mb region analysed in non-overlapping 20-kb windows. Middle, GC content of the first 10 Mb, analysed in 2-kb windows. Bottom, the GC content of the first 1 Mb, analysed in 200-bp windows. At this scale, gaps in the sequence can be seen.

stsG30423) with only 36% GC content. There are also examples of large shifts in GC content between adjacent multimegabase regions. For example, the average GC content on chromosome 17q is 50% for the distal 10.3 Mb but drops to 38% for the adjacent 3.9 Mb. There are regions of less than 300 kb with even wider swings in GC content, for example, from 33.1% to 59.3%.

Long-range variation in GC content is evident not just from extreme outliers, but throughout the genome. The distribution of average GC content in 20-kb windows across the draft genome sequence is shown in Fig. 12. The spread is 15-fold larger than predicted by a uniform process. Moreover, the standard deviation barely decreases as window size increases by successive factors of four—5.9%, 5.2%, 4.9% and 4.6% for windows of size 5, 20, 80 and 320 kb. The distribution is also notably skewed, with 58% below the average and 42% above the average of 41%, with a long tail of GC-rich regions.

Bernardi and colleagues[118,119] proposed that the long-range variation in GC content may reflect that the genome is composed of a mosaic of compositionally homogeneous regions that they dubbed 'isochores'. They suggested that the skewed distribution is composed of five normal distributions, corresponding to five distinct types of isochore (L1, L2, H1, H2 and H3, with GC contents of < 38%, 38–42%, 42–47%, 47–52% and > 52%, respectively).

We studied the draft genome sequence to see whether strict isochores could be identified. For example, the sequence was divided into 300-kb windows, and each window was subdivided into 20-kb subwindows. We calculated the average GC content for each window and subwindow, and investigated how much of the variance in the GC content of subwindows across the genome can be statistically 'explained' by the average GC content in each window. About three-quarters of the genome-wide variance among 20-kb windows can be statistically explained by the average GC content of 300-kb windows that contain them, but the residual variance among subwindows (standard deviation, 2.4%) is still far too large to be consistent with a homogeneous distribution. In fact, the hypothesis of homogeneity could be rejected for each 300-kb window in the draft genome sequence.

Similar results were obtained with other window and subwindow sizes. Some of the local heterogeneity in GC content is attributable to transposable element insertions (see below). Such repeat elements typically have a higher GC content than the surrounding sequence, with the effect being strongest for the most recent insertions.

These results rule out a strict notion of isochores as compositionally homogeneous. Instead, there is substantial variation at many different scales, as illustrated in Fig. 13. Although isochores do not appear to merit the prefix 'iso', the genome clearly does contain large regions of distinctive GC content and it is likely to be worth redefining the concept so that it becomes possible rigorously to partition the genome into regions. In the absence of a precise definition, we will loosely refer to such regions as 'GC content domains' in the context of the discussion below.

Fickett *et al.*[120] have explored a model in which the underlying preference for a particular GC content drifts continuously throughout the genome, an approach that bears further examination. Churchill[121] has proposed that the boundaries between GC content domains can in some cases be predicted by a hidden Markov model, with one state representing a GC-rich region and one representing an AT-rich region. We found that this approach tended to identify only very short domains of less than a kilobase (data not shown), but variants of this approach deserve further attention.

The correlation between GC content domains and various biological properties is of great interest, and this is likely to be the most fruitful route to understanding the basis of variation in GC content. As described below, we confirm the existence of strong correlations with both repeat content and gene density. Using the integration between the draft genome sequence and the cytogenetic map described above, it is possible to confirm a statistically significant correlation between GC content and Giemsa bands (G-bands). For example, 98% of large-insert clones mapping to the darkest G-bands are in 200-kb regions of low GC content (average 37%), whereas more than 80% of clones mapping to the lightest G-bands are in regions of high GC content (average 45%)[103]. Estimated band locations can be seen in Fig. 9 and viewed in the context of other genome annotation at http://genome.ucsc.edu/goldenPath/mapPlots/ and http://genome.ucsc.edu/goldenPath/hgTracks.html.

CpG islands

A related topic is the distribution of so-called CpG islands across the genome. The dinucleotide CpG is notable because it is greatly under-represented in human DNA, occurring at only about one-fifth of the roughly 4% frequency that would be expected by simply multiplying the typical fraction of Cs and Gs (0.21×0.21). The deficit occurs because most CpG dinucleotides are methylated on the cytosine base, and spontaneous deamination of methyl-C residues gives rise to T residues. (Spontaneous deamination of ordinary cytosine residues gives rise to uracil residues that are readily recognized and repaired by the cell.) As a result, methyl-CpG dinucleotides steadily mutate to TpG dinucleotides. However, the genome contains many 'CpG islands' in which CpG dinucleotides are not methylated and occur at a frequency closer to that predicted by the local GC content. CpG islands are of particular interest because many are associated with the 5′ ends of genes[122–127].

We searched the draft genome sequence for CpG islands. Ideally, they should be defined by directly testing for the absence of cytosine methylation, but that was not practical for this report. There are

Table 10 Number of CpG islands by GC content

GC content of island	Number of islands	Percentage of islands	Nucleotides in islands	Percentage of nucleotides in islands
Total	28,890	100	19,818,547	100
>80%	22	0.08	5,916	0.03
70–80%	5,884	20	3,111,965	16
60–70%	18,779	65	13,110,924	66
50–60%	4,205	15	3,589,742	18

Potential CpG islands were identified by searching the draft genome sequence one base at a time, scoring each dinucleotide (+17 for GC, −1 for others) and identifying maximally scoring segments. Each segment was then evaluated to determine GC content (≥50%), length (>200) and ratio of observed proportion of GC dinucleotides to the expected proportion on the basis of the GC content of the segment (>0.60), using a modification of a program developed by G. Micklem (personal communication).

various computer programs that attempt to identify CpG islands on the basis of primary sequence alone. These programs differ in some important respects (such as how aggressively they subdivide long CpG-containing regions), and the precise correspondence with experimentally undermethylated islands has not been validated. Nevertheless, there is a good correlation, and computational analysis thus provides a reasonable picture of the distribution of CpG islands in the genome.

To identify CpG islands, we used the definition proposed by Gardiner-Garden and Frommer[128] and embodied in a computer program. We searched the draft genome sequence for CpG islands, using both the full sequence and the sequence masked to eliminate repeat sequences. The number of regions satisfying the definition of a CpG island was 50,267 in the full sequence and 28,890 in the repeat-masked sequence. The difference reflects the fact that some repeat elements (notably Alu) are GC-rich. Although some of these repeat elements may function as control regions, it seems unlikely that most of the apparent CpG islands in repeat sequences are functional. Accordingly, we focused on those in the non-repeated sequence. The count of 28,890 CpG islands is reasonably close to the previous estimate of about 35,000 (ref. 129, as modified by ref. 130). Most of the islands are short, with 60–70% GC content (Table 10). More than 95% of the islands are less than 1,800 bp long, and more than 75% are less than 850 bp. The longest CpG island (on chromosome 10) is 36,619 bp long, and 322 are longer than 3,000 bp. Some of the larger islands contain ribosomal pseudogenes, although RNA genes and pseudogenes account for only a small proportion of all islands (< 0.5%). The smaller islands are consistent with their previously hypothesized function, but the role of these larger islands is uncertain.

The density of CpG islands varies substantially among some of the chromosomes. Most chromosomes have 5–15 islands per Mb, with a mean of 10.5 islands per Mb. However, chromosome Y has an unusually low 2.9 islands per Mb, and chromosomes 16, 17 and 22 have 19–22 islands per Mb. The extreme outlier is chromosome 19, with 43 islands per Mb. Similar trends are seen when considering the percentage of bases contained in CpG islands. The relative density of CpG islands correlates reasonably well with estimates of relative gene density on these chromosomes, based both on previous mapping studies involving ESTs (Fig. 14) and on the distribution of gene predictions discussed below.

Comparison of genetic and physical distance

The draft genome sequence makes it possible to compare genetic and physical distances and thereby to explore variation in the rate of recombination across the human chromosomes. We focus here on large-scale variation. Finer variation is examined in an accompanying paper[131].

The genetic and physical maps are integrated by 5,282 polymorphic loci from the Marshfield genetic map[102], whose positions are known in terms of centimorgans (cM) and Mb along the chromosomes. Figure 15 shows the comparison of the draft genome sequence for chromosome 12 with the male, female and sex-averaged maps. One can calculate the approximate ratio of cM per Mb across a chromosome (reflected in the slopes in Fig. 15) and the average recombination rate for each chromosome arm.

Two striking features emerge from analysis of these data. First, the average recombination rate increases as the length of the chromosome arm decreases (Fig. 16). Long chromosome arms have an average recombination rate of about 1 cM per Mb, whereas the shortest arms are in the range of 2 cM per Mb. A similar trend has been seen in the yeast genome[132,133], despite the fact that the physical scale is nearly 200 times as small. Moreover, experimental studies have shown that lengthening or shortening yeast chromosomes results in a compensatory change in recombination rate[132].

The second observation is that the recombination rate tends to be suppressed near the centromeres and higher in the distal portions of most chromosomes, with the increase largely in the terminal

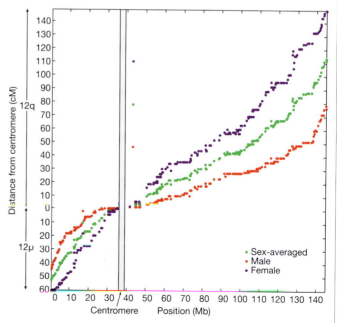

Figure 15 Distance in cM along the genetic map of chromosome 12 plotted against position in Mb in the draft genome sequence. Female, male and sex-averaged maps are shown. Female recombination rates are much higher than male recombination rates. The increased slopes at either end of the chromosome reflect the increased rates of recombination per Mb near the telomeres. Conversely, the flatter slope near the centromere shows decreased recombination there, especially in male meiosis. This is typical of the other chromosomes as well (see http://genome.ucsc.edu/goldenPath/mapPlots). Discordant markers may be map, marker placement or assembly errors.

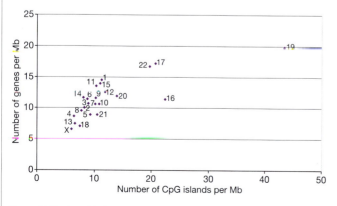

Figure 14 Number of CpG islands per Mb for each chromosome, plotted against the number of genes per Mb (the number of genes was taken from GeneMap98 (ref. 100)). Chromosomes 16, 17, 22 and particularly 19 are clear outliers, with a density of CpG islands that is even greater than would be expected from the high gene counts for these four chromosomes.

20–35 Mb. The increase is most pronounced in the male meiotic map. The effect can be seen, for example, from the higher slope at both ends of chromosome 12 (Fig. 15). Regional and sex-specific effects have been observed for chromosome 21 (refs 110, 134).

Why is recombination higher on smaller chromosome arms? A higher rate would increase the likelihood of at least one crossover during meiosis on each chromosome arm, as is generally observed in human chiasmata counts[135]. Crossovers are believed to be necessary for normal meiotic disjunction of homologous chromosome pairs in eukaryotes. An extreme example is the pseudoautosomal regions on chromosomes Xp and Yp, which pair during male meiosis; this physical region of only 2.6 Mb has a genetic length of 50 cM (corresponding to 20 cM per Mb), with the result that a crossover is virtually assured.

Mechanistically, the increased rate of recombination on shorter chromosome arms could be explained if, once an initial recombination event occurs, additional nearby events are blocked by positive crossover interference on each arm. Evidence from yeast mutants in which interference is abolished shows that interference plays a key role in distributing a limited number of crossovers among the various chromosome arms in yeast[136]. An alternative possibility is that a checkpoint mechanism scans for and enforces the presence of at least one crossover on each chromosome arm.

Variation in recombination rates along chromosomes and between the sexes is likely to reflect variation in the initiation of meiosis-induced double-strand breaks (DSBs) that initiate recombination. DSBs in yeast have been associated with open chromatin[137,138], rather than with specific DNA sequence motifs. With the availability of the draft genome sequence, it should be possible to explore in an analogous manner whether variation in human recombination rates reflects systematic differences in chromosome accessibility during meiosis.

Repeat content of the human genome

A puzzling observation in the early days of molecular biology was that genome size does not correlate well with organismal complexity. For example, *Homo sapiens* has a genome that is 200 times as large as that of the yeast *S. cerevisiae*, but 200 times as small as that of *Amoeba dubia*[139,140]. This mystery (the C-value paradox) was largely resolved with the recognition that genomes can contain a large quantity of repetitive sequence, far in excess of that devoted to protein-coding genes (reviewed in refs 140, 141).

In the human, coding sequences comprise less than 5% of the genome (see below), whereas repeat sequences account for at least 50% and probably much more. Broadly, the repeats fall into five classes: (1) transposon-derived repeats, often referred to as interspersed repeats; (2) inactive (partially) retroposed copies of cellular genes (including protein-coding genes and small structural RNAs), usually referred to as processed pseudogenes; (3) simple sequence repeats, consisting of direct repetitions of relatively short k-mers such as $(A)_n$, $(CA)_n$ or $(CGG)_n$; (4) segmental duplications, consisting of blocks of around 10–300 kb that have been copied from one region of the genome into another region; and (5) blocks of tandemly repeated sequences, such as at centromeres, telomeres, the short arms of acrocentric chromosomes and ribosomal gene clusters. (These regions are intentionally under-represented in the draft genome sequence and are not discussed here.)

Repeats are often described as 'junk' and dismissed as uninteresting. However, they actually represent an extraordinary trove of information about biological processes. The repeats constitute a rich palaeontological record, holding crucial clues about evolutionary events and forces. As passive markers, they provide assays for studying processes of mutation and selection. It is possible to recognize cohorts of repeats 'born' at the same time and to follow their fates in different regions of the genome or in different species. As active agents, repeats have reshaped the genome by causing ectopic rearrangements, creating entirely new genes, modifying and reshuffling existing genes, and modulating overall GC content. They also shed light on chromosome structure and dynamics, and provide tools for medical genetic and population genetic studies.

The human is the first repeat-rich genome to be sequenced, and so we investigated what information could be gleaned from this majority component of the human genome. Although some of the general observations about repeats were suggested by previous studies, the draft genome sequence provides the first comprehensive view, allowing some questions to be resolved and new mysteries to emerge.

Transposon-derived repeats

Most human repeat sequence is derived from transposable elements[142,143]. We can currently recognize about 45% of the genome as belonging to this class. Much of the remaining 'unique' DNA must also be derived from ancient transposable element copies that have diverged too far to be recognized as such. To describe our analyses of interspersed repeats, it is necessary briefly to review the relevant features of human transposable elements.

Classes of transposable elements. In mammals, almost all transposable elements fall into one of four types (Fig. 17), of which three transpose through RNA intermediates and one transposes directly as DNA. These are long interspersed elements (LINEs), short interspersed elements (SINEs), LTR retrotransposons and DNA transposons.

LINEs are one of the most ancient and successful inventions in eukaryotic genomes. In humans, these transposons are about 6 kb long, harbour an internal polymerase II promoter and encode two open reading frames (ORFs). Upon translation, a LINE RNA assembles with its own encoded proteins and moves to the nucleus, where an endonuclease activity makes a single-stranded nick and the reverse transcriptase uses the nicked DNA to prime reverse transcription from the 3′ end of the LINE RNA. Reverse transcription frequently fails to proceed to the 5′ end, resulting in many truncated, nonfunctional insertions. Indeed, most LINE-derived repeats are short, with an average size of 900 bp for all LINE1 copies, and a median size of 1,070 bp for copies of the currently active LINE1 element (L1Hs). New insertion sites are flanked by a small

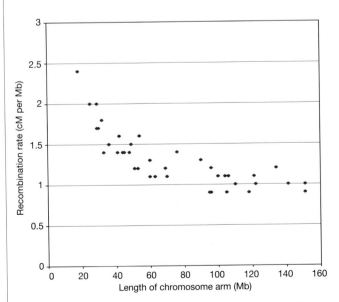

Figure 16 Rate of recombination averaged across the euchromatic portion of each chromosome arm plotted against the length of the chromosome arm in Mb. For large chromosomes, the average recombination rates are very similar, but as chromosome arm length decreases, average recombination rates rise markedly.

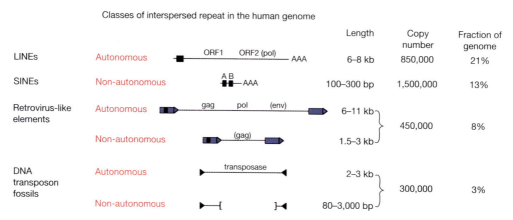

Figure 17 Almost all transposable elements in mammals fall into one of four classes. See text for details.

target site duplication of 7–20 bp. The LINE machinery is believed to be responsible for most reverse transcription in the genome, including the retrotransposition of the non-autonomous SINEs[144] and the creation of processed pseudogenes[145,146]. Three distantly related LINE families are found in the human genome: LINE1, LINE2 and LINE3. Only LINE1 is still active.

SINEs are wildly successful freeloaders on the backs of LINE elements. They are short (about 100–400 bp), harbour an internal polymerase III promoter and encode no proteins. These non-autonomous transposons are thought to use the LINE machinery for transposition. Indeed, most SINEs 'live' by sharing the 3′ end with a resident LINE element[144]. The promoter regions of all known SINEs are derived from tRNA sequences, with the exception of a single monophyletic family of SINEs derived from the signal recognition particle component 7SL. This family, which also does not share its 3′ end with a LINE, includes the only active SINE in the human genome: the Alu element. By contrast, the mouse has both tRNA-derived and 7SL-derived SINEs. The human genome contains three distinct monophyletic families of SINEs: the active Alu, and the inactive MIR and Ther2/MIR3.

LTR retroposons are flanked by long terminal direct repeats that contain all of the necessary transcriptional regulatory elements. The autonomous elements (retrotransposons) contain *gag* and *pol* genes, which encode a protease, reverse transcriptase, RNAse H and integrase. Exogenous retroviruses seem to have arisen from endogenous retrotransposons by acquisition of a cellular *envelope* gene (*env*)[147]. Transposition occurs through the retroviral mechanism with reverse transcription occurring in a cytoplasmic virus-like particle, primed by a tRNA (in contrast to the nuclear location and chromosomal priming of LINEs). Although a variety of LTR retrotransposons exist, only the vertebrate-specific endogenous retroviruses (ERVs) appear to have been active in the mammalian genome. Mammalian retroviruses fall into three classes (I–III), each comprising many families with independent origins. Most (85%) of the LTR retroposon-derived 'fossils' consist only of an isolated LTR, with the internal sequence having been lost by homologous recombination between the flanking LTRs.

DNA transposons resemble bacterial transposons, having terminal inverted repeats and encoding a transposase that binds near the inverted repeats and mediates mobility through a 'cut-and-paste' mechanism. The human genome contains at least seven major classes of DNA transposon, which can be subdivided into many families with independent origins[148] (see RepBase, http://www.girinst.org/~server/repbase.html). DNA transposons tend to have short life spans within a species. This can be explained by contrasting the modes of transposition of DNA transposons and LINE elements. LINE transposition tends to involve only functional elements, owing to the *cis*-preference by which LINE proteins assemble with the RNA from which they were translated. By

contrast, DNA transposons cannot exercise a *cis*-preference: the encoded transposase is produced in the cytoplasm and, when it returns to the nucleus, it cannot distinguish active from inactive elements. As inactive copies accumulate in the genome, transposition becomes less efficient. This checks the expansion of any DNA transposon family and in due course causes it to die out. To survive, DNA transposons must eventually move by horizontal transfer to virgin genomes, and there is considerable evidence for such transfer[149–153].

Transposable elements employ different strategies to ensure their evolutionary survival. LINEs and SINEs rely almost exclusively on vertical transmission within the host genome[154] (but see refs 148, 155). DNA transposons are more promiscuous, requiring relatively frequent horizontal transfer. LTR retroposons use both strategies, with some being long-term active residents of the human genome (such as members of the ERVL family) and others having only short residence times.

Table 11 Number of copies and fraction of genome for classes of interspersed repeat

	Number of copies (× 1,000)	Total number of bases in the draft genome sequence (Mb)	Fraction of the draft genome sequence (%)	Number of families (subfamilies)
SINEs	1,558	359.6	13.14	
Alu	1,090	290.1	10.60	1 (~20)
MIR	393	60.1	2.20	1 (1)
MIR3	75	9.3	0.34	1 (1)
LINEs	868	558.8	20.42	
LINE1	516	462.1	16.89	1 (~55)
LINE2	315	88.2	3.22	1 (2)
LINE3	37	8.4	0.31	1 (2)
LTR elements	443	227.0	8.29	
ERV-class I	112	79.2	2.89	72 (132)
ERV(K)-class II	8	8.5	0.31	10 (20)
ERV (L)-class III	83	39.5	1.44	21 (42)
MaLR	240	99.8	3.65	1 (31)
DNA elements	294	77.6	2.84	
hAT group				
MER1-Charlie	182	30.1	1.39	25 (50)
Zaphod	13	4.3	0.16	4 (10)
Tc-1 group				
MER2-Tigger	57	28.0	1.02	12 (28)
Tc2	4	0.9	0.03	1 (5)
Mariner	14	2.6	0.10	4 (5)
PiggyBac-like	2	0.5	0.02	10 (20)
Unclassified	22	3.2	0.12	7 (7)
Unclassified	3	3.8	0.14	3 (4)
Total interspersed repeats		1,226.8	44.83	

The number of copies and base pair contributions of the major classes and subclasses of transposable elements in the human genome. Data extracted from a RepeatMasker analysis of the draft genome sequence (RepeatMasker version 09092000, sensitive settings, using RepBase Update 5.08). In calculating percentages, RepeatMasker excluded the runs of Ns linking the contigs in the draft genome sequence. In the last column, separate consensus sequences in the repeat databases are considered subfamilies, rather than families, when the sequences are closely related or related through intermediate subfamilies.

Census of human repeats. We began by taking a census of the transposable elements in the draft genome sequence, using a recently updated version of the RepeatMasker program (version 09092000) run under sensitive settings (see http://repeatmasker. genome.washington.edu). This program scans sequences to identify full-length and partial members of all known repeat families represented in RepBase Update (version 5.08; see http://www. girinst.org/~server/repbase.html and ref. 156). Table 11 shows the number of copies and fraction of the draft genome sequence occupied by each of the four major classes and the main subclasses.

The precise count of repeats is obviously underestimated because the genome sequence is not finished, but their density and other properties can be stated with reasonable confidence. Currently recognized SINEs, LINEs, LTR retroposons and DNA transposon copies comprise 13%, 20%, 8% and 3% of the sequence, respectively. We expect these densities to grow as more repeat families are recognized, among which will be lower copy number LTR elements and DNA transposons, and possibly high copy number ancient (highly diverged) repeats.

Age distribution. The age distribution of the repeats in the human genome provides a rich 'fossil record' stretching over several hundred million years. The ancestry and approximate age of each fossil can be inferred by exploiting the fact that each copy is derived from, and therefore initially carried the sequence of, a then-active transposon and, being generally under no functional constraint, has accumulated mutations randomly and independently of other copies. We can infer the sequence of the ancestral active elements by clustering the modern derivatives into phylogenetic trees and building a consensus based on the multiple sequence alignment of a cluster of copies. Using available consensus sequences for known repeat subfamilies, we calculated the per cent divergence from the inferred ancestral active transposon for each of three million interspersed repeats in the draft genome sequence.

The percentage of sequence divergence can be converted into an approximate age in millions of years (Myr) on the basis of evolutionary information. Care is required in calibrating the clock, because the rate of sequence divergence may not be constant over time or between lineages[139]. The relative-rate test[157] can be used to calculate the sequence divergence that accumulated in a lineage after a given timepoint, on the basis of comparison with a sibling species that diverged at that time and an outgroup species. For example, the substitution rate over roughly the last 25 Myr in the human lineage can be calculated by using old world monkeys (which diverged about 25 Myr ago) as a sibling species and new world monkeys as an outgroup. We have used currently available calibrations for the human lineage, but the issue should be revisited as sequence information becomes available from different mammals.

Figure 18a shows the representation of various classes of trans-

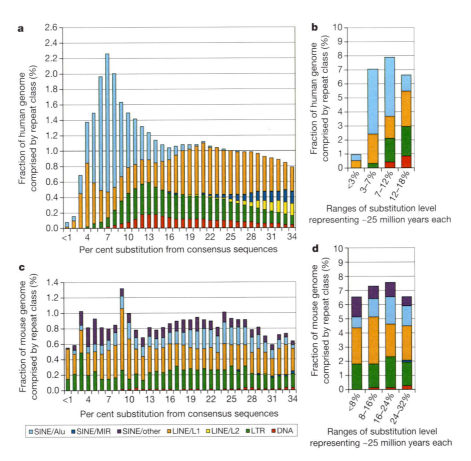

Figure 18 Age distribution of interspersed repeats in the human and mouse genomes. Bases covered by interspersed repeats were sorted by their divergence from their consensus sequence (which approximates the repeat's original sequence at the time of insertion). The average number of substitutions per 100 bp (substitution level, K) was calculated from the mismatch level p assuming equal frequency of all substitutions (the one-parameter Jukes–Cantor model, $K = -3/4\ln(1 - 4/3p)$). This model tends to underestimate higher substitution levels. CpG dinucleotides in the consensus were excluded from the substitution level calculations because the C→T transition rate in CpG pairs is about tenfold higher than other transitions and causes distortions in comparing

transposable elements with high and low CpG content. **a**, The distribution, for the human genome, in bins corresponding to 1% increments in substitution levels. **b**, The data grouped into bins representing roughly equal time periods of 25 Myr. **c,d**, Equivalent data for available mouse genomic sequence. There is a different correspondence between substitution levels and time periods owing to different rates of nucleotide substitution in the two species. The correspondence between substitution levels and time periods was largely derived from three-way species comparisons (relative rate test[139,157]) with the age estimates based on fossil data. Human divergence from gibbon 20–30 Myr; old world monkey 25–35 Myr; prosimians 55–80 Myr; eutherian mammalian radiation ~100 Myr.

posable elements in categories reflecting equal amounts of sequence divergence. In Fig. 18b the data are grouped into four bins corresponding to successive 25-Myr periods, on the basis of an approximate clock. Figure 19 shows the mean ages of various subfamilies of DNA transposons. Several facts are apparent from these graphs. First, most interspersed repeats in the human genome predate the eutherian radiation. This is a testament to the extremely slow rate with which nonfunctional sequences are cleared from vertebrate genomes (see below concerning comparison with the fly).

Second, LINE and SINE elements have extremely long lives. The monophyletic LINE1 and Alu lineages are at least 150 and 80 Myr old, respectively. In earlier times, the reigning transposons were LINE2 and MIR[148,158]. The SINE MIR was perfectly adapted for reverse transcription by LINE2, as it carried the same 50-base sequence at its 3′ end. When LINE2 became extinct 80–100 Myr ago, it spelled the doom of MIR.

Third, there were two major peaks of DNA transposon activity (Fig. 19). The first involved Charlie elements and occurred long before the eutherian radiation; the second involved Tigger elements and occurred after this radiation. Because DNA transposons can produce large-scale chromosome rearrangements[159–162], it is possible that widespread activity could be involved in speciation events.

Fourth, there is no evidence for DNA transposon activity in the past 50 Myr in the human genome. The youngest two DNA transposon families that we can identify in the draft genome sequence (MER75 and MER85) show 6–7% divergence from their respective consensus sequences representing the ancestral element (Fig. 19), indicating that they were active before the divergence of humans and new world monkeys. Moreover, these elements were relatively unsuccessful, together contributing just 125 kb to the draft genome sequence.

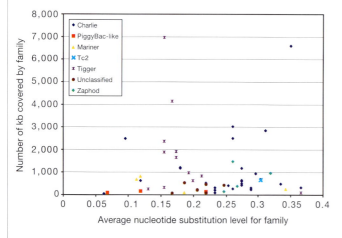

Figure 19 Median ages and per cent of the genome covered by subfamilies of DNA transposons. The Charlie and Zaphod elements were hobo-Activator-Tam3 (hAT) DNA transposons; Mariner, Tc2 and Tigger were Tc1-like elements. Unlike retroposons, DNA transposons are thought to have a short life span in a genome. Thus, the average or median divergence of copies from the consensus is a particularly accurate measure of the age of the DNA transposon copies.

Finally, LTR retroposons appear to be teetering on the brink of extinction, if they have not already succumbed. For example, the most prolific elements (ERVL and MaLRs) flourished for more than 100 Myr but appear to have died out about 40 Myr ago[163,164]. Only a single LTR retroposon family (HERVK10) is known to have transposed since our divergence from the chimpanzee 7 Myr ago, with only one known copy (in the HLA region) that is not shared between all humans[165]. In the draft genome sequence, we can identify only three full-length copies with all ORFs intact (the final total may be slightly higher owing to the imperfect state of the draft genome sequence).

More generally, the overall activity of all transposons has declined markedly over the past 35–50 Myr, with the possible exception of LINE1 (Fig. 18). Indeed, apart from an exceptional burst of activity of Alus peaking around 40 Myr ago, there would appear to have been a fairly steady decline in activity in the hominid lineage since the mammalian radiation. The extent of the decline must be even greater than it appears because old repeats are gradually removed by random deletion and because old repeat families are harder to recognize and likely to be under-represented in the repeat databases. (We confirmed that the decline in transposition is not an artefact arising from errors in the draft genome sequence, which, in principle, could increase the divergence level in recent elements. First, the sequence error rate (Table 9) is far too low to have a significant effect on the apparent age of recent transposons; and second, the same result is seen if one considers only finished sequence.)

What explains the decline in transposon activity in the lineage leading to humans? We return to this question below, in the context of the observation that there is no similar decline in the mouse genome.

Comparison with other organisms. We compared the complement of transposable elements in the human genome with those of the other sequenced eukaryotic genomes. We analysed the fly, worm and mustard weed genomes for the number and nature of repeats (Table 12) and the age distribution (Fig. 20). (For the fly, we analysed the 114 Mb of unfinished 'large' contigs produced by the whole-genome shotgun assembly[166], which are reported to represent euchromatic sequence. Similar results were obtained by analysing 30 Mb of finished euchromatic sequence.) The human genome stands in stark contrast to the genomes of the other organisms.

(1) The euchromatic portion of the human genome has a much higher density of transposable element copies than the euchromatic DNA of the other three organisms. The repeats in the other organisms may have been slightly underestimated because the repeat databases for the other organisms are less complete than for the human, especially with regard to older elements; on the other hand, recent additions to these databases appear to increase the repeat content only marginally.

(2) The human genome is filled with copies of ancient transposons, whereas the transposons in the other genomes tend to be of more recent origin. The difference is most marked with the fly, but is clear for the other genomes as well. The accumulation of old repeats is likely to be determined by the rate at which organisms engage in 'housecleaning' through genomic deletion. Studies of pseudogenes

Table 12 Number and nature of interspersed repeats in eukaryotic genomes

	Human		Fly		Worm		Mustard weed	
	Percentage of bases	Approximate number of families	Percentage of bases	Approximate number of families	Percentage of bases	Approximate number of families	Percentage of bases	Approximate number of families
LINE/SINE	33.40%	6	0.70%	20	0.40%	10	0.50%	10
LTR	8.10%	100	1.50%	50	0.00%	4	4.80%	70
DNA	2.80%	60	0.70%	20	5.30%	80	5.10%	80
Total	44.40%	170	3.10%	90	6.50%	90	10.50%	160

The complete genomes of fly, worm, and chromosomes 2 and 4 of mustard weed (as deposited at ncbi.nlm.nih.gov/genbank/genomes) were screened against the repeats in RepBase Update 5.02 (September 2000) with RepeatMasker at sensitive settings.

have suggested that small deletions occur at a rate that is 75-fold higher in flies than in mammals; the half-life of such nonfunctional DNA is estimated at 12 Myr for flies and 800 Myr for mammals[167]. The rate of large deletions has not been systematically compared, but seems likely also to differ markedly.

(3) Whereas in the human two repeat families (LINE1 and Alu) account for 60% of all interspersed repeat sequence, the other organisms have no dominant families. Instead, the worm, fly and mustard weed genomes all contain many transposon families, each consisting of typically hundreds to thousands of elements. This difference may be explained by the observation that the vertically transmitted, long-term residential LINE and SINE elements represent 75% of interspersed repeats in the human genome, but only 5–25% in the other genomes. In contrast, the horizontally transmitted and shorter-lived DNA transposons represent only a small portion of all interspersed repeats in humans (6%) but a much larger fraction in fly, mustard weed and worm (25%, 49% and 87%, respectively). These features of the human genome are probably general to all mammals. The relative lack of horizontally transmitted elements may have its origin in the well developed immune system of mammals, as horizontal transfer requires infectious vectors, such as viruses, against which the immune system guards.

We also looked for differences among mammals, by comparing the transposons in the human and mouse genomes. As with the human genome, care is required in calibrating the substitution clock for the mouse genome. There is considerable evidence that the rate of substitution per Myr is higher in rodent lineages than in the hominid lineages[139,168,169]. In fact, we found clear evidence for different rates of substitution by examining families of transposable elements whose insertions predate the divergence of the human and mouse lineages. In an analysis of 22 such families, we found that the substitution level was an average of 1.7-fold higher in mouse than human (not shown). (This is likely to be an underestimate because of an ascertainment bias against the most diverged copies.) The faster clock in mouse is also evident from the fact that the ancient LINE2 and MIR elements, which transposed before the mammalian radiation and are readily detectable in the human genome, cannot be readily identified in available mouse genomic sequence (Fig. 18).

We used the best available estimates to calibrate substitution levels and time[169]. The ratio of substitution rates varied from about 1.7-fold higher over the past 100 Myr to about 2.6-fold higher over the past 25 Myr.

The analysis shows that, although the overall density of the four transposon types in human and mouse is similar, the age distribution is strikingly different (Fig. 18). Transposon activity in the mouse genome has not undergone the decline seen in humans and proceeds at a much higher rate. In contrast to their possible extinction in humans, LTR retroposons are alive and well in the mouse with such representatives as the active IAP family and putatively active members of the long-lived ERVL and MaLR families. LINE1 and a variety of SINEs are quite active. These evolutionary findings are consistent with the empirical observations that new spontaneous mutations are 30 times more likely to be caused by LINE insertions in mouse than in human (~3% versus 0.1%)[170] and 60 times more likely to be caused by transposable elements in general. It is estimated that around 1 in 600 mutations in human are due to transpositions, whereas 10% of mutations in mouse are due to transpositions (mostly IAP insertions).

The contrast between human and mouse suggests that the explanation for the decline of transposon activity in humans may lie in some fundamental difference between hominids and rodents. Population structure and dynamics would seem to be likely suspects. Rodents tend to have large populations, whereas hominid populations tend to be small and may undergo frequent bottlenecks. Evolutionary forces affected by such factors include inbreeding and genetic drift, which might affect the persistence of active transposable elements[171]. Studies in additional mammalian lineages may shed light on the forces responsible for the differences in the activity of transposable elements[172].

Variation in the distribution of repeats. We next explored variation in the distribution of repeats across the draft genome sequence, by calculating the repeat density in windows of various sizes across the genome. There is striking variation at smaller scales.

Some regions of the genome are extraordinarily dense in repeats. The prizewinner appears to be a 525-kb region on chromosome Xp11, with an overall transposable element density of 89%. This region contains a 200-kb segment with 98% density, as well as a segment of 100 kb in which LINE1 sequences alone comprise 89% of the sequence. In addition, there are regions of more than 100 kb with extremely high densities of Alu (> 56% at three loci, including one on 7q11 with a 50-kb stretch of > 61% Alu) and the ancient transposons MIR (> 15% on chromosome 1p36) and LINE2 (> 18% on chromosome 22q12).

In contrast, some genomic regions are nearly devoid of repeats. The absence of repeats may be a sign of large-scale *cis*-regulatory elements that cannot tolerate being interrupted by insertions. The four regions with the lowest density of interspersed repeats in the human genome are the four homeobox gene clusters, HOXA, HOXB, HOXC and HOXD (Fig. 21). Each locus contains regions of around 100 kb containing less than 2% interspersed repeats. Ongoing sequence analysis of the four HOX clusters in mouse, rat and baboon shows a similar absence of transposable elements, and reveals a high density of conserved noncoding elements (K. Dewar and B. Birren, manuscript in preparation). The presence of a complex collection of regulatory regions may explain why individual HOX genes carried in transgenic mice fail to show proper regulation.

It may be worth investigating other repeat-poor regions, such as a region on chromosome 8q21 (1.5% repeat over 63 kb) containing a gene encoding a homeodomain zinc-finger protein (homologous to mouse pID 9663936), a region on chromosome 1p36 (5% repeat over 100 kb) with no obvious genes and a region on chromosome 18q22 (4% over 100 kb) containing three genes of unknown function (among which is KIAA0450). It will be interesting to see whether the homologous regions in the mouse genome have

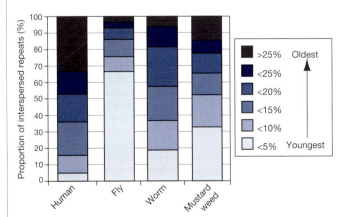

Figure 20 Comparison of the age of interspersed repeats in eukaryotic genomes. The copies of repeats were pooled by their nucleotide substitution level from the consensus.

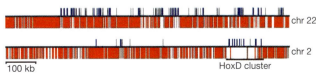

Figure 21 Two regions of about 1 Mb on chromosomes 2 and 22. Red bars, interspersed repeats; blue bars, exons of known genes. Note the deficit of repeats in the HoxD cluster, which contains a collection of genes with complex, interrelated regulation.

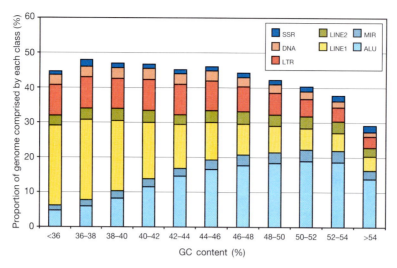

Figure 22 Density of the major repeat classes as a function of local GC content, in windows of 50 kb.

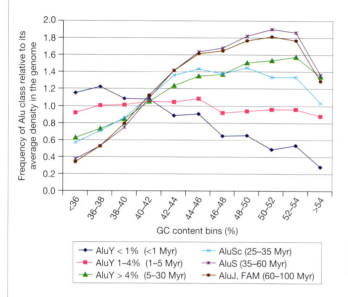

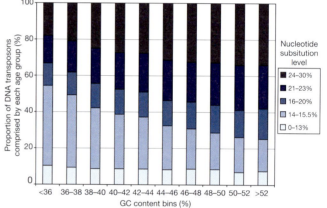

Figure 24 DNA transposon copies in AT-rich DNA tend to be younger than those in more GC-rich DNA. DNA transposon families were grouped into five age categories by their median substitution level (see Fig. 19). The proportion attributed to each age class is shown as a function of GC content. Similar patterns are seen for LINE1 and LTR elements.

Figure 23 Alu elements target AT-rich DNA, but accumulate in GC-rich DNA. This graph shows the relative distribution of various Alu cohorts as a function of local GC content. The divergence levels (including CpG sites) and ages of the cohorts are shown in the key.

similarly resisted the insertion of transposable elements during rodent evolution.

Distribution by GC content. We next focused on the correlation between the nature of the transposons in a region and its GC content. We calculated the density of each repeat type as a function of the GC content in 50-kb windows (Fig. 22). As has been reported[142,173–176], LINE sequences occur at much higher density in AT-rich regions (roughly fourfold enriched), whereas SINEs (MIR, Alu) show the opposite trend (for Alu, up to fivefold lower in AT-rich DNA). LTR retroposons and DNA transposons show a more uniform distribution, dipping only in the most GC-rich regions.

The preference of LINEs for AT-rich DNA seems like a reasonable way for a genomic parasite to accommodate its host, by targeting gene-poor AT-rich DNA and thereby imposing a lower mutational burden. Mechanistically, selective targeting is nicely explained by the fact that the preferred cleavage site of the LINE endonuclease is TTTT/A (where the slash indicates the point of cleavage), which is used to prime reverse transcription from the poly(A) tail of LINE RNA[177].

The contrary behaviour of SINEs, however, is baffling. How do

SINEs accumulate in GC-rich DNA, particularly if they depend on the LINE transposition machinery[178]? Notably, the same pattern is seen for the Alu-like B1 and the tRNA-derived SINEs in mouse and for MIR in human[142]. One possibility is that SINEs somehow target GC-rich DNA for insertion. The alternative is that SINEs initially insert with the same proclivity for AT-rich DNA as LINEs, but that the distribution is subsequently reshaped by evolutionary forces[142,179].

We used the draft genome sequence to investigate this mystery by comparing the proclivities of young, adolescent, middle-aged and old Alus (Fig. 23). Strikingly, recent Alus show a preference for AT-rich DNA resembling that of LINEs, whereas progressively older Alus show a progressively stronger bias towards GC-rich DNA. These results indicate that the GC bias must result from strong pressure: Fig. 23 shows that a 13-fold enrichment of Alus in GC-rich DNA has occurred within the last 30 Myr, and possibly more recently.

These results raise a new mystery. What is the force that produces the great and rapid enrichment of Alus in GC-rich DNA? One explanation may be that deletions are more readily tolerated in gene-poor AT-rich regions than in gene-rich GC-rich regions, resulting in older elements being enriched in GC-rich regions. Such an enrichment is seen for transposable elements such as

DNA transposons (Fig. 24). However, this effect seems too slow and too small to account for the observed remodelling of the Alu distribution. This can be seen by performing a similar analysis for LINE elements (Fig. 25). There is no significant change in the LINE distribution over the past 100 Myr, in contrast to the rapid change seen for Alu. There is an eventual shift after more than 100 Myr, although its magnitude is still smaller than seen for Alus.

These observations indicate that there may be some force acting particularly on Alus. This could be a higher rate of random loss of Alus in AT-rich DNA, negative selection against Alus in AT-rich DNA or positive selection in favour of Alus in GC-rich DNA. The first two possibilities seem unlikely because AT-rich DNA is

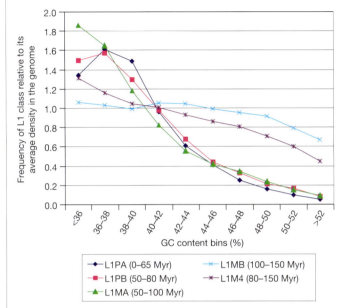

Figure 25 Distribution of various LINE cohorts as a function of local GC content. The divergence levels and ages of the cohorts are shown in the key. (The divergence levels were measured for the 3′ UTR of the LINE1 element only, which is best characterized evolutionarily. This region contains almost no CpG sites, and thus 1% divergence level corresponds to a much longer time than for CpG-rich Alu copies).

gene-poor and tolerates the accumulation of other transposable elements. The third seems more feasible, in that it involves selecting in favour of the minority of Alus in GC-rich regions rather than against the majority that lie in AT-rich regions. But positive selection for Alus in GC-rich regions would imply that they benefit the organism.

Schmid[180] has proposed such a function for SINEs. This hypothesis is based on the observation that in many species SINEs are transcribed under conditions of stress, and the resulting RNAs specifically bind a particular protein kinase (PKR) and block its ability to inhibit protein translation[181–183]. SINE RNAs would thus promote protein translation under stress. SINE RNA may be well suited to such a role in regulating protein translation, because it can be quickly transcribed in large quantities from thousands of elements and it can function without protein translation. Under this theory, there could be positive selection for SINEs in readily transcribed open chromatin such as is found near genes. This could explain the retention of Alus in gene-rich GC-rich regions. It is also consistent with the observation that SINE density in AT-rich DNA is higher near genes[142].

Further insight about Alus comes from the relationship between Alu density and GC content on individual chromosomes (Fig. 26). There are two outliers. Chromosome 19 is even richer in Alus than predicted by its (high) GC content; the chromosome comprises 2% of the genome, but contains 5% of Alus. On the other hand, chromosome Y shows the lowest density of Alus relative to its GC content, being higher than average for GC content less than 40% and lower than average for GC content over 40%. Even in AT-rich DNA, Alus are under-represented on chromosome Y compared with other young interspersed repeats (see below). These phenomena may be related to an unusually high gene density on chromosome 19 and an unusually low density of somatically active genes on chromosome Y (both relative to GC content). This would be consistent with the idea that Alu correlates not with GC content but with actively transcribed genes.

Our results may support the controversial idea that SINEs actually earn their keep in the genome. Clearly, much additional work will be needed to prove or disprove the hypothesis that SINEs are genomic symbionts.

Biases in human mutation. Indirect studies have suggested that nucleotide substitution is not uniform across mammalian

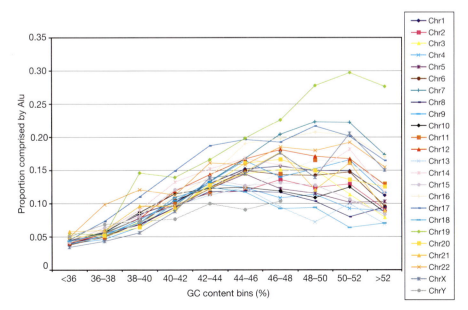

Figure 26 Comparison of the Alu density of each chromosome as a function of local GC content. At higher GC levels, the Alu density varies widely between chromosomes, with chromosome 19 being a particular outlier. In contrast, the LINE1 density pattern is quite uniform for most chromosomes, with the exception of a 1.5 to 2-fold over-representation in AT-rich regions of the X and Y chromosomes (not shown).

genomes[184–187]. By studying sets of repeat elements belonging to a common cohort, one can directly measure nucleotide substitution rates in different regions of the genome. We find strong evidence that the pattern of neutral substitution differs as a function of local GC content (Fig. 27). Because the results are observed in repetitive elements throughout the genome, the variation in the pattern of nucleotide substitution seems likely to be due to differences in the underlying mutational process rather than to selection.

The effect can be seen most clearly by focusing on the substitution process $\gamma \leftrightarrow \alpha$, where γ denotes GC or CG base pairs and α denotes AT or TA base pairs. If K is the equilibrium constant in the direction of α base pairs (defined by the ratio of the forward and reverse rates), then the equilibrium GC content should be $1/(1 + K)$. Two observations emerge.

First, there is a regional bias in substitution patterns. The equilibrium constant varies as a function of local GC content: γ base pairs are more likely to mutate towards α base pairs in AT-rich regions than in GC-rich regions. For the analysis in Fig. 27, the equilibrium constant K is 2.5, 1.9 and 1.2 when the draft genome sequence is partitioned into three bins with average GC content of 37, 43 and 50%, respectively. This bias could be due to a reported tendency for GC-rich regions to replicate earlier in the cell cycle than AT-rich regions and for guanine pools, which are limiting for

DNA replication, to become depleted late in the cell cycle, thereby resulting in a small but significant shift in substitution towards α base pairs[186,188]. Another theory proposes that many substitutions are due to differences in DNA repair mechanisms, possibly related to transcriptional activity and thereby to gene density and GC content[185,189,190].

There is also an absolute bias in substitution patterns resulting in directional pressure towards lower GC content throughout the human genome. The genome is not at equilibrium with respect to the pattern of nucleotide substitution: the expected equilibrium GC content corresponding to the values of K above is 29, 35 and 44% for regions with average GC contents of 37, 43 and 50%, respectively. Recent observations on SNPs[190] confirm that the mutation pattern in GC-rich DNA is biased towards α base pairs; it should be possible to perform similar analyses throughout the genome with the availability of 1.4 million SNPs[97,191]. On the basis solely of nucleotide substitution patterns, the GC content would be expected to be about 7% lower throughout the genome.

What accounts for the higher GC content? One possible explanation is that in GC-rich regions, a considerable fraction of the nucleotides is likely to be under functional constraint owing to the high gene density. Selection on coding regions and regulatory CpG islands may maintain the higher-than-predicted GC content. Another is that throughout the rest of the genome, a constant influx of transposable elements tends to increase GC content (Fig. 28). Young repeat elements clearly have a higher GC content than their surrounding regions, except in extremely GC-rich regions. Moreover, repeat elements clearly shift with age towards a lower GC content, closer to that of the neighbourhood in which they reside. Much of the 'non-repeat' DNA in AT-rich regions probably consists of ancient repeats that are not detectable by current methods and that have had more time to approach the local equilibrium value.

The repeats can also be used to study how the mutation process is affected by the immediately adjacent nucleotide. Such 'context effects' will be discussed elsewhere (A. Kas and A. F. A. Smit, unpublished results).

Fast living on chromosome Y. The pattern of interspersed repeats can be used to shed light on the unusual evolutionary history of chromosome Y. Our analysis shows that the genetic material on

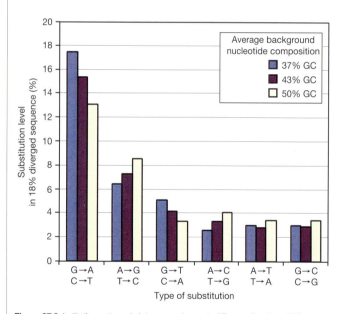

Figure 27 Substitution patterns in interspersed repeats differ as a function of GC content. We collected all copies of five DNA transposons (Tigger1, Tigger2, Charlie3, MER1 and HSMAR2), chosen for their high copy number and well defined consensus sequences. DNA transposons are optimal for the study of neutral substitutions: they do not segregate into subfamilies with diagnostic differences, presumably because they are short-lived and new active families do not evolve in a genome (see text). Duplicates and close paralogues resulting from duplication after transposition were eliminated. The copies were grouped on the basis of GC content of the flanking 1,000 bp on both sides and aligned to the consensus sequence (representing the state of the copy at integration). Recursive efforts using parameters arising from this study did not change the alignments significantly. Alignments were inspected by hand, and obvious misalignments caused by insertions and duplications were eliminated. Substitutions ($n = 80,000$) were counted for each position in the consensus, excluding those in CpG dinucleotides, and a substitution frequency matrix was defined. From the matrices for each repeat (which corresponded to different ages), a single rate matrix was calculated for these bins of GC content (< 40% GC, 40–47% GC and > 47% GC). Data are shown for a repeat with an average divergence (in non-CpG sites) of 18% in 43% GC content (the repeat has slightly higher divergence in AT-rich DNA and lower in GC-rich DNA). From the rate matrix, we calculated log-likelihood matrices with different entropies (divergence levels), which are theoretically optimal for alignments of neutrally diverged copies to their common ancestral state (A. Kas and A. F. A. Smit, unpublished). These matrices are in use by the RepeatMasker program.

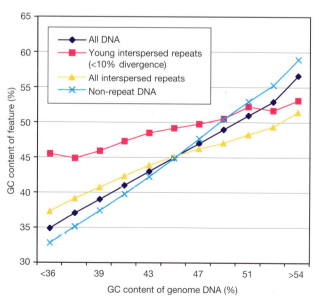

Figure 28 Interspersed repeats tend to diminish the differences between GC bins, despite the fact that GC-rich transposable elements (specifically Alu) accumulate in GC-rich DNA, and AT-rich elements (LINE1) in AT-rich DNA. The GC content of particular components of the sequence (repeats, young repeats and non-repeat sequence) was calculated as a function of overall GC content.

chromosome Y is unusually young, probably owing to a high tolerance for gain of new material by insertion and loss of old material by deletion. Several lines of evidence support this picture. For example, LINE elements on chromosome Y are on average much younger than those on autosomes (not shown). Similarly, MaLR-family retroposons on chromosome Y are younger than those on autosomes, with the representation of subfamilies showing a strong inverse correlation with the age of the subfamily. Moreover, chromosome Y has a relative over-representation of the younger retroviral class II (ERVK) and a relative under-representation of the primarily older class III (ERVL) compared with other chromosomes. Overall, chromosome Y seems to maintain a youthful appearance by rapid turnover.

Interspersed repeats on chromosome Y can also be used to estimate the relative mutation rates, α_m and α_f, in the male and female germlines. Chromosome Y always resides in males, whereas chromosome X resides in females twice as often as in males. The substitution rates, μ_Y and μ_X, on these two chromosomes should thus be in the ratio $\mu_Y:\mu_X = (\alpha_m):(\alpha_m + 2\alpha_f)/3$, provided that one considers equivalent neutral sequences. Several authors have estimated the mutation rate in the male germline to be fivefold higher than in the female germline, by comparing the rates of evolution of X- and Y-linked genes in humans and primates. However, Page and colleagues[192] have challenged these estimates as too high. They studied a 39-kb region that is apparently devoid of genes and resides within a large segmental duplication from X to Y that occurred 3–4 Myr ago in the human lineage. On the basis of phylogenetic analysis of the sequence on human Y and human, chimp and gorilla X, they obtained a much lower estimate of $\mu_Y:\mu_X = 1.36$, corresponding to $\alpha_m:\alpha_f = 1.7$. They suggested that the other estimates may have been higher because they were based on much longer evolutionary periods or because the genes studied may have been under selection.

Our database of human repeats provides a powerful resource for addressing this question. We identified the repeat elements from recent subfamilies (effectively, birth cohorts dating from the past 50 Myr) and measured the substitution rates for subfamily members on chromosomes X and Y (Fig. 29). There is a clear linear relationship with a slope of $\mu_Y:\mu_X = 1.57$ corresponding to $\alpha_m:\alpha_f = 2.1$. The estimate is in reasonable agreement with that of Page et al., although it is based on much more total sequence (360 kb on Y, 1.6 Mb on X) and a much longer time period. In particular, the discrepancy with earlier reports is not explained by recent changes in the human lineage. Various theories have been proposed for the higher mutation rate in the male germline, including the greater number of cell divisions in the formation of sperm than eggs and different repair mechanisms in sperm and eggs.

Active transposons. We were interested in identifying the youngest retrotransposons in the draft genome sequence. This set should contain the currently active retrotransposons, as well as the insertion sites that are still polymorphic in the human population.

The youngest branch in the phylogenetic tree of human LINE1 elements is called L1Hs (ref. 158); it differs in its 3′ untranslated region (UTR) by 12 diagnostic substitutions from the next oldest subfamily (L1PA2). Within the L1Hs family, there are two subsets referred to as Ta and pre-Ta, defined by a diagnostic trinucleotide[193,194]. All active L1 elements are thought to belong to these two subsets, because they account for all 14 known cases of human disease arising from new L1 transposition (with 13 belonging to the Ta subset and one to the pre-Ta subset)[195,196]. These subsets are also of great interest for population genetics because at least 50% are still segregating as polymorphisms in the human population[194,197]; they provide powerful markers for tracing population history because they represent unique (non-recurrent and non-revertible) genetic events that can be used (along with similarly polymorphic Alus) for reconstructing human migrations.

LINE1 elements that are retrotransposition-competent should consist of a full-length sequence and should have both ORFs intact. Eleven such elements from the Ta subset have been identified, including the likely progenitors of mutagenic insertions into the factor VIII and dystrophin genes[198–202]. A cultured cell retrotransposition assay has revealed that eight of these elements remain retrotransposition-competent[200,202,203].

We searched the draft genome sequence and identified 535 LINEs belonging to the Ta subset and 415 belonging to the pre-Ta subset. These elements provide a large collection of tools for probing human population history. We also identified those consisting of full-length elements with intact ORFs, which are candidate active LINEs. We found 39 such elements belonging to the Ta subset and 22 belonging to the pre-Ta subset; this substantially increases the number in the first category and provides the first known examples in the second category. These elements can now be tested for retrotransposition competence in the cell culture assay. Preliminary analysis resulted in the identification of two of these elements as the likely progenitors of mutagenic insertions into the β-globin and RP2 genes (R. Badge and J. V. Moran, unpublished data). Similar analyses should allow the identification of the progenitors of most, if not all, other known mutagenic L1 insertions.

L1 elements can carry extra DNA if transcription extends through the native transcriptional termination site into flanking genomic DNA. This process, termed L1-mediated transduction, provides a means for the mobilization of DNA sequences around the genome and may be a mechanism for 'exon shuffling'[204]. Twenty-one per cent of the 71 full-length L1s analysed contained non-L1-derived sequences before the 3′ target-site duplication site, in cases in which the site was unambiguously recognizable. The length of the transduced sequence was 30–970 bp, supporting the suggestion that 0.5–1.0% of the human genome may have arisen by LINE-based transduction of 3′ flanking sequences[205,206].

Our analysis also turned up two instances of 5′ transduction (145 bp and 215 bp). Although this possibility had been suggested on the basis of cell culture models[195,203], these are the first documented examples. Such events may arise from transcription initiating in a cellular promoter upstream of the L1 elements. L1 transcription is generally confined to the germline[207,208], but transcription from other promoters could explain a somatic L1 retrotransposition event that resulted in colon cancer[206].

Transposons as a creative force. The primary force for the origin and expansion of most transposons has been selection for their ability to create progeny, and not a selective advantage for the host. However, these selfish pieces of DNA have been responsible for important innovations in many genomes, for example by contributing regulatory elements and even new genes.

Twenty human genes have been recognized as probably derived

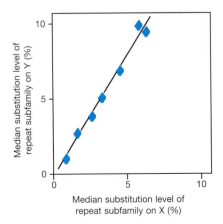

Figure 29 Higher substitution rate on chromosome Y than on chromosome X. We calculated the median substitution level (excluding CpG sites) for copies of the most recent L1 subfamilies (L1Hs–L1PA8) on the X and Y chromosomes. Only the 3′ UTR of the L1 element was considered because its consensus sequence is best established.

from transposons[142,209]. These include the RAG1 and RAG2 recombinases and the major centromere-binding protein CENPB. We scanned the draft genome sequence and identified another 27 cases, bringing the total to 47 (Table 13; refs 142, 209). All but four are derived from DNA transposons, which give rise to only a small proportion of the interspersed repeats in the genome. Why there are so many DNA transposase-like genes, many of which still contain the critical residues for transposase activity, is a mystery.

To illustrate this concept, we describe the discovery of one of the new examples. We searched the draft genome sequence to identify the autonomous DNA transposon responsible for the distribution of the non-autonomous MER85 element, one of the most recently (40–50 Myr ago) active DNA transposons. Most non-autonomous elements are internal deletion products of a DNA transposon. We identified one instance of a large (1,782 bp) ORF flanked by the 5′ and 3′ halves of a MER85 element. The ORF encodes a novel protein (partially published as pID 6453533) whose closest homologue is the transposase of the piggyBac DNA transposon, which is found in insects and has the same characteristic TTAA target-site duplications[210] as MER85. The ORF is actively transcribed in fetal brain and in cancer cells. That it has not been lost to mutation in 40–50 Myr of evolution (whereas the flanking, noncoding, MER85-

like termini show the typical divergence level of such elements) and is actively transcribed provides strong evidence that it has been adopted by the human genome as a gene. Its function is unknown.

LINE1 activity clearly has also had fringe benefits. We mentioned above the possibility of exon reshuffling by cotranscription of neighbouring DNA. The LINE1 machinery can also cause reverse transcription of genic mRNAs, which typically results in nonfunctional processed pseudogenes but can, occasionally, give rise to functional processed genes. There are at least eight human and eight mouse genes for which evidence strongly supports such an origin[211] (see http://www-ifi.uni-muenster.de/exapted-retrogenes/tables.html). Many other intronless genes may have been created in the same way.

Transposons have made other creative contributions to the genome. A few hundred genes, for example, use transcriptional terminators donated by LTR retroposons (data not shown). Other genes employ regulatory elements derived from repeat elements[211].

Simple sequence repeats

Simple sequence repeats (SSRs) are a rather different type of repetitive structure that is common in the human genome—perfect or slightly imperfect tandem repeats of a particular k-mer. SSRs with a short repeat unit ($n = 1$–13 bases) are often termed microsa-

Table 13 Human genes derived from transposable elements

GenBank ID*	Gene name	Related transposon family†	Possible fusion gene§	Newly recognized derivation‖
nID 3150436	BC200	FLAM Alu‡		
pID 2330017	Telomerase	non-LTR retrotransposon		
pID 1196425	HERV-3 env	Retroviridae/HERV-R‡		
pID 4773880	Syncytin	Retroviridae/HERV-W‡		
pID 131827	RAG1 and 2	Tc1-like		
pID 29863	CENP-B	Tc1/Pogo		
EST 2529718		Tc1/Pogo		+
PID 10047247		Tc1/Pogo/Pogo		+
EST 4524463		Tc1/Pogo/Pogo		+
pID 4504807	Jerky	Tc1/Pogo/Tigger		
pID 7513096	JRKL	Tc1/Pogo/Tigger		
EST 5112721		Tc1/Pogo/Tigger		
EST 11097233		Tc1/Pogo/Tigger		+
EST 6986275	Sancho	Tc1/Pogo/Tigger		+
EST 8616450		Tc1/Pogo/Tigger		
EST 8750408		Tc1/Pogo/Tigger		+
EST 5177004		Tc1/Pogo/Tigger		+
PID 3413884	KIAA0461	Tc1/Pogo/Tc2	+	
PID 7959287	KIAA1513	Tc1/Pogo/Tc2		
PID 2231380		Tc1/Mariner/Hsmar1‡	+	+
EST 10219887		hAT/Hobo	+	+
PID 6581095	Buster1	hAT/Charlie	+	
PID 7243087	Buster2	hAT/Charlie	+	
PID 6581097	Buster3	hAT/Charlie		
PID 7662294	KIAA0766	hAT/Charlie	+	
PID 10439678		hAT/Charlie		
PID 7243087	KIAA1353	hAT/Charlie		+
PID 7021900		hAT/Charlie/Charlie3‡		+
PID 4263748		hAT/Charlie/Charlie8‡	+	
EST 8161741		hAT/Charlie/Charlie9‡		+
pID 4758872	DAP4,pP52[rIPK]	hAT/Tip100/Zaphod		
EST 10990063		hAT/Tip100/Zaphod		+
EST 10101591		hAT/Tip100/Zaphod		+
pID 7513011	KIAA0543	hAT/Tip100/Tip100	+	
pID 10439744		hAT/Tip100/Tip100		+
pID 10047247	KIAA1586	hAT/Tip100/Tip100		+
pID 10439762		hAT/Tip100	+	‖
EST 10459804		hAT/Tip100		+
pID 4160548	Tramp	hAT/Tam3		+
BAC 3522927		hAT/Tam3		+
pID 3327088	KIAA0637	hAT/Tam3	+	
EST 1928552		hAT/Tam3		+
pID 6453533		piggyBac/MER85‡	+	
EST 3594004		piggyBac/MER85‡		
BAC 4309921		piggyBac/MER85‡		+
EST 4073914		piggyBac/MER75‡		+
EST 1963278		piggyBac		+

The Table lists 47 human genes, with a likely origin in up to 38 different transposon copies.
* Where available, the GenBank ID numbers are given for proteins, otherwise a representative EST or a clone name is shown. Six groups (two or three genes each) have similarity at the DNA level well beyond that observed between different DNA transposon families in the genome; they are indicated in italics, with all but the initial member of each group indented. This could be explained if the genes were paralogous (derived from a single inserted transposon and subsequently duplicated).
† Classification of the transposon.
‡ Indicates that the transposon from which the gene is derived is precisely known.
§ Proteins probably formed by fusion of a cellular and transposon gene; many have acquired zinc-finger domains.
‖ Not previously reported as being derived from transposable element genes. The remaining genes can be found in refs 142, 209.

Table 14 SSR content of the human genome

Length of repeat unit	Average bases per Mb	Average number of SSR elements per Mb
1	1,660	36.7
2	5,046	43.1
3	1,013	11.8
4	3,383	32.5
5	2,686	17.6
6	1,376	15.2
7	906	8.4
8	1,139	11.1
9	900	8.6
10	1,576	8.6
11	770	8.7

SSRs were identified by using the computer program Tandem Repeat Finder with the following parameters: match score 2, mismatch score 3, indel 5, minimum alignment 50, maximum repeat length 500, minimum repeat length 1.

tellites, whereas those with longer repeat units ($n = 14–500$ bases) are often termed minisatellites. With the exception of poly(A) tails from reverse transcribed messages, SSRs are thought to arise by slippage during DNA replication[212,213].

We compiled a catalogue of all SSRs over a given length in the human draft genome sequence, and studied their properties (Table 14). SSRs comprise about 3% of the human genome, with the greatest single contribution coming from dinucleotide repeats (0.5%). (The precise criteria for the number of repeat units and the extent of divergence allowed in an SSR affect the exact census, but not the qualitative conclusions.)

There is approximately one SSR per 2 kb (the number of non-overlapping tandem repeats is 437 per Mb). The catalogue confirms various properties of SSRs that have been inferred from sampling approaches (Table 15). The most frequent dinucleotide repeats are AC and AT (50 and 35% of dinucleotide repeats, respectively), whereas AG repeats (15%) are less frequent and GC repeats (0.1%) are greatly under-represented. The most frequent trinucleotides are AAT and AAC (33% and 21%, respectively), whereas ACC (4.0%), AGC (2.2%), ACT (1.4%) and ACG (0.1%) are relatively rare. Overall, trinucleotide SSRs are much less frequent than dinucleotide SSRs[214].

SSRs have been extremely important in human genetic studies, because they show a high degree of length polymorphism in the human population owing to frequent slippage by DNA polymerase during replication. Genetic markers based on SSRs—particularly $(CA)_n$ repeats—have been the workhorse of most human disease-mapping studies[101,102]. The availability of a comprehensive catalogue of SSRs is thus a boon for human genetic studies.

The SSR catalogue also allowed us to resolve a mystery regarding mammalian genetic maps. Such genetic maps in rat, mouse and human have a deficit of polymorphic $(CA)_n$ repeats on chromosome X[30,101]. There are two possible explanations for this deficit. There may simply be fewer $(CA)_n$ repeats on chromosome X; or $(CA)_n$ repeats may be as dense on chromosome X but less polymorphic in

Table 15 SSRs by repeat unit

Repeat unit	Number of SSRs per Mb
AC	27.7
AT	19.4
AG	8.2
GC	0.1
AAT	4.1
AAC	2.6
AGG	1.5
AAG	1.4
ATG	0.7
CGG	0.6
ACC	0.4
AGC	0.3
ACT	0.2
ACG	0.0

SSRs were identified as in Table 14.

the population. In fact, analysis of the draft genome sequence shows that chromosome X has the same density of $(CA)_n$ repeats per Mb as the autosomes (data not shown). Thus, the deficit of polymorphic markers relative to autosomes results from population genetic forces. Possible explanations include that chromosome X has a smaller effective population size, experiences more frequent selective sweeps reducing diversity (owing to its hemizygosity in males), or has a lower mutation rate (owing to its more frequent passage through the less mutagenic female germline). The availability of the draft genome sequence should provide ways to test these alternative explanations.

Segmental duplications

A remarkable feature of the human genome is the segmental duplication of portions of genomic sequence[215–217]. Such duplications involve the transfer of 1–200-kb blocks of genomic sequence to one or more locations in the genome. The locations of both donor and recipient regions of the genome are often not tandemly arranged, suggesting mechanisms other than unequal crossing-over for their origin. They are relatively recent, inasmuch as strong sequence identity is seen in both exons and introns (in contrast to regions that are considered to show evidence of ancient duplications, characterized by similarities only in coding regions). Indeed, many such duplications appear to have arisen in very recent evolutionary time, as judged by high sequence identity and by their absence in closely related species.

Segmental duplications can be divided into two categories. First, interchromosomal duplications are defined as segments that are duplicated among nonhomologous chromosomes. For example, a 9.5-kb genomic segment of the adrenoleukodystrophy locus from Xq28 has been duplicated to regions near the centromeres of chromosomes 2, 10, 16 and 22 (refs 218, 219). Anecdotal observations suggest that many interchromosomal duplications map near the centromeric and telomeric regions of human chromosomes[218–233].

The second category is intrachromosomal duplications, which occur within a particular chromosome or chromosomal arm. This category includes several duplicated segments, also known as low copy repeat sequences, that mediate recurrent chromosomal structural rearrangements associated with genetic disease[215,217]. Examples on chromosome 17 include three copies of a roughly 200-kb repeat separated by around 5 Mb and two copies of a roughly 24-kb repeat separated by 1.5 Mb. The copies are so similar (99% identity) that paralogous recombination events can occur, giving rise to contiguous gene syndromes: Smith–Magenis syndrome and Charcot–Marie–Tooth syndrome 1A, respectively[34,234]. Several other examples are known and are also suspected to be responsible for recurrent microdeletion syndromes (for example, Prader–Willi/Angelman,

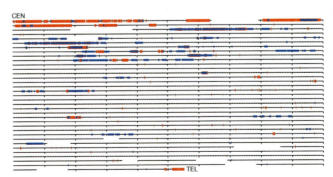

Figure 30 Duplication landscape of chromosome 22. The size and location of intrachromosomal (blue) and interchromosomal (red) duplications are depicted for chromosome 22q, using the PARASIGHT computer program (Bailey and Eichler, unpublished). Each horizontal line represents 1 Mb (ticks, 100-kb intervals). The chromosome sequence is oriented from centromere (top left) to telomere (bottom right). Pairwise alignments with > 90% nucleotide identity and > 1 kb long are shown. Gaps within the chromosomal sequence are of known size and shown as empty space.

velocardiofacial/DiGeorge and Williams' syndromes[215,235–240]).

Until now, the identification and characterization of segmental duplications have been based on anecdotal reports—for example, finding that certain probes hybridize to multiple chromosomal sites or noticing duplicated sequence at certain recurrent chromosomal breakpoints. The availability of the entire genomic sequence will make it possible to explore the nature of segmental duplications more systematically. This analysis can begin with the current state of the draft genome sequence, although caution is required because some apparent duplications may arise from a failure to merge sequence contigs from overlapping clones. Alternatively, erroneous

assembly of closely related sequences from nonoverlapping clones may underestimate the true frequency of such features, particularly among those segments with the highest sequence similarity. Accordingly, we adopted a conservative approach for estimating such duplication from the available draft genome sequence.

Pericentromeres and subtelomeres. We began by re-evaluating the finished sequences of chromosomes 21 and 22. The initial papers on these chromosomes[93,94] noted some instances of interchromosomal duplication near each centromere. With the ability now to compare these chromosomes to the vast majority of the genome, it is apparent that the regions near the centromeres consist almost entirely of interchromosomal duplicated segments, with little or no unique sequence. Smaller regions of interchromosomal duplication are also observed near the telomeres.

Chromosome 22 contains a region of 1.5 Mb adjacent to the centromere in which 90% of sequence can now be recognized to consist of interchromosomal duplication (Fig. 30). Conversely, 52% of the interchromosomal duplications on chromosome 22 were located in this region, which comprises only 5% of the chromosome. Also, the subtelomeric end consists of a 50-kb region consisting almost entirely of interchromosomal duplications.

Chromosome 21 presents a similar landscape (Fig. 31). The first 1 Mb after the centromere is composed of interchromosomal repeats, as well as the largest (> 200 kb) block of intrachromosomally duplicated material. Again, most interchromosomal duplications on the chromosome map to this region and the most subtelomeric region (30 kb) shows extensive duplication among nonhomologous chromosomes.

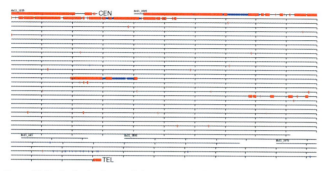

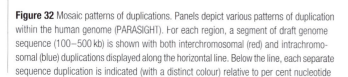

Figure 31 Duplication landscape of chromosome 21. The size and location of intrachromosomal (blue) and interchromosomal (red) duplications are depicted along the sequence of the long arm of chromosome 21. Gaps between finished sequence are denoted by empty space but do not represent actual gap size.

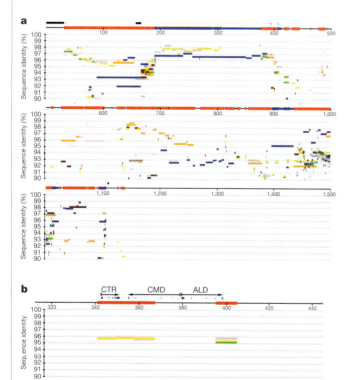

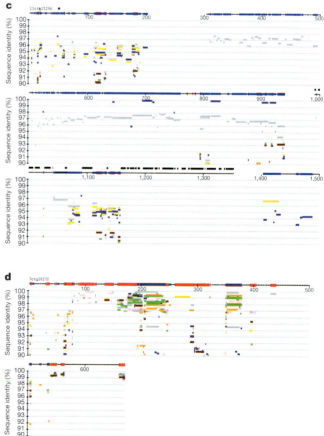

Figure 32 Mosaic patterns of duplications. Panels depict various patterns of duplication within the human genome (PARASIGHT). For each region, a segment of draft genome sequence (100–500 kb) is shown with both interchromosomal (red) and intrachromosomal (blue) duplications displayed along the horizontal line. Below the line, each separate sequence duplication is indicated (with a distinct colour) relative to per cent nucleotide

identity for the duplicated segment (y axis). Black bars show the relative locations of large blocks of heterochromatic sequences (alpha, gamma and HSAT sequence). **a**, An active pericentromeric region on chromosome 21. **b**, An ancestral region from Xq28 that has contributed various 'genic' segments to pericentromeric regions. **c**, A pericentromeric region from chromosome 11. **d**, A subtelomeric region from chromosome 7p.

The pericentromeric regions are structurally very complex, as illustrated for chromosome 21 in Fig. 32a. The pericentromeric regions appear to have been bombarded by successive insertions of duplications; the insertion events must be fairly recent because the degree of sequence conservation with the genomic source loci is fairly high (90–100%, with an apparent peak around 96%). Distinct insertions are typically separated by AT-rich or GC-rich minisatellite-like repeats that have been hypothesized to have a functional role in targeting duplications to these regions[233,241].

A single genomic source locus often gives rise to pericentromeric copies on multiple chromosomes, with each having essentially the same breakpoints and the same degree of divergence. An example of such a source locus on Xq28 is shown in Fig. 32b. Phylogenetic analysis has suggested a two-step mechanism for the origin and dispersal of these segments, whereby an initial segmental duplication in the pericentromeric region of one chromosome occurs and is then redistributed as part of a larger cassette to other such regions[242].

A comprehensive analysis for all chromosomes will have to await complete sequencing of the genome, but the evidence from the draft genome sequence indicates that the same picture is likely to be seen throughout the genome. Several papers have analysed finished segments within pericentromeric regions of chromosomes 2 (160 kb), 10 (400 kb) and 16 (300 kb), all of which show extensive interchromosomal segmental duplication[215,219,232,233]. An example from another pericentromeric region on chromosome 11 is shown in Fig. 32c. Interchromosomal duplications in subtelomeric regions also appear to be a fairly general phenomenon, as illustrated by a large tract (~500 kb) of complex duplication on chromosome 7 (Fig. 32d).

The explanation for the clustering of segmental duplications may be that the genome has a damage-control mechanism whereby chromosomal breakage products are preferentially inserted into pericentromeric and, to a lesser extent, subtelomeric regions. The possibility of a specific mechanism for the insertion of these sequences has been suggested on the basis of the unusual sequences found flanking the insertions. Although it is also possible that these regions simply have greater tolerance for large insertions, many large gene-poor 'deserts' have been identified[93] and there is no accumulation of duplicated segments within these regions. Along with the fact that transitions between duplicons (from different regions of the genome) occur at specific sequences, this suggests that active recruitment of duplications to such regions may occur. In any case, the duplicated regions are in general young (with many duplications showing <6% nucleotide divergence from their source loci) and in constant flux, both through additional duplications and by large-scale exchange among similar chromosomal environments. There is evidence of structural polymorphism in the human population, such as the presence or absence of olfactory receptor segments located within the telomeric regions of several human chromosomes[226,227].

Genome-wide analysis of segmental duplications. We also performed a global genome-wide analysis to characterize the amount of segmental duplication in the genome. We 'repeat-masked' the known interspersed repeats in the draft genome sequence and compared the remaining draft genomic sequence with itself in a massive all-by-all BLASTN similarity search. We excluded matches in which the sequence identity was so high that it might reflect artefactual duplications resulting from a failure to overlap sequence contigs correctly in assembling the draft genome sequence. Specifically, we considered only matches with less than 99.5% identity for finished sequence and less than 98% identity for unfinished sequence.

We took several approaches to avoid counting artefactual duplications in the sequence. In the first approach, we studied only finished sequence. We compared the finished sequence with itself, to identify segments of at least 1 kb and 90–99.5% sequence identity. This analysis will underestimate the extent of segmental

duplication, because it requires that at least two copies of the segment are present in the finished sequence and because some true duplications have over 99.5% identity.

The finished sequence consists of at least 3.3% segmental duplication (Table 16). Interchromosomal duplication accounts for about 1.5% and intrachromosomal duplication for about 2%, with some overlap (0.2%) between these categories. We analysed the lengths and divergence of the segmental duplications (Fig. 33). The duplications tend to be large (10–50 kb) and highly homologous, especially for the interchromosomal segments. The sequence divergence for the interchromosomal duplications appears to peak between 96.5% and 97.5%. This may indicate that interchromosomal duplications occurred in a punctuated manner. It will be intriguing to investigate whether such genomic upheaval has a role in speciation events.

In a second approach, we compared the entire human draft genome sequence (finished and unfinished) with itself to identify duplications with 90–98% sequence identity (Table 17). The draft genome sequence contains at least 3.6% segmental duplication. The actual proportion will be significantly higher, because we excluded many true matches with more than 98% sequence identity (at least 1.1% of the finished sequence). Although exact measurement must await a finished sequence, the human genome seems likely to contain about 5% segmental duplication, with most of this sequence in large blocks (>10 kb). Such a high proportion of large duplications clearly distinguishes the human genome from other sequenced genomes, such as the fly and worm (Table 18).

The structure of large highly paralogous regions presents one of the 'serious and unanticipated challenges' to producing a finished sequence of the genome[46]. The absence of unique STS or fingerprint signatures over large genomic distances (~1 Mb) and the high degree of sequence similarity makes the distinction between paralogous sequence variation and allelic polymorphism problematic. Furthermore, the fact that such regions frequently harbour intron–exon structures of genuine unique sequence will complicate efforts to generate a genome-wide SNP map. The data indicate that a modest portion of the human genome may be relatively recalcitrant to genomic-based methods for SNP detection. Owing to their repetitive nature and their location in the genome, segmental

Table 16 Fraction of finished sequence in inter- and intrachromosomal duplications

Chromosome	Intrachromosomal (%)	Interchromosomal (%)	All (%)
1	1.4	0.5	1.9
2	0.1	0.6	0.7
3	0.3	1.1	1.1
4	0.0	1.0	1.0
5	0.6	0.3	0.9
6	0.8	0.4	1.1
7	3.4	1.3	4.1
8	0.3	0.1	0.3
9	0.8	2.9	3.7
10	2.1	0.8	2.9
11	1.2	2.1	2.3
12	1.5	0.3	1.8
13	0.0	0.5	0.5
14	0.6	0.4	1.0
15	3.0	6.9	6.9
16	4.5	2.0	5.8
17	1.6	0.3	1.8
18	0.0	0.7	0.7
19	3.6	0.3	3.8
20	0.2	0.3	0.5
21	1.4	1.6	3.0
22	6.1	2.6	7.5
X	1.8	3.2	5.0
Y	12.1	16.0	27.4
Un	0.0	0.5	0.5
Total	2.0	1.5	3.3

Excludes duplications with identities >99.5% to avoid artefactual duplication due to incomplete merger in the assembly process. Calculation was performed on the finished sequence available in September 2000 and reflects the duplications found within the total amount of finished sequence then. Note that there is some overlap between the interchromosomal and intrachromosomal sets.

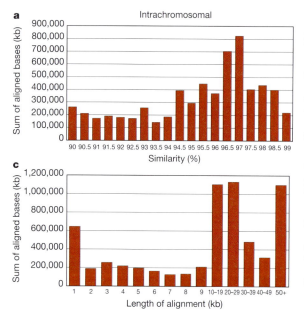

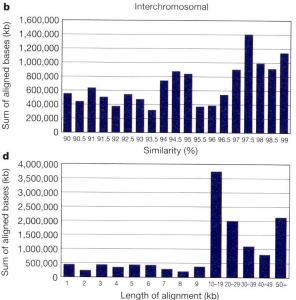

Figure 33 a–d, Sequence properties of segmental duplications. Distributions of length and per cent nucleotide identity for segmental duplications are shown as a function of the number of aligned bp, for the subset of finished genome sequence. Intrachromosomal, red; interchromosomal, blue.

duplications may well be underestimated by the current analysis. An understanding of the biology, pathology and evolution of these duplications will require specialized efforts within these exceptional regions of the human genome. The presence and distribution of such segments may provide evolutionary fodder for processes of exon shuffling and a general increase in protein diversity associated with domain accretion. It will be important to consider both genome-wide duplication events and more restricted punctuated events of genome duplication as forces in the evolution of vertebrate genomes.

Table 17 Fraction of the draft genome sequence in inter- and intrachromosomal duplications

Chromosome	Intrachromosomal (%)	Interchromosomal (%)	All (%)
1	2.1	1.7	3.4
2	1.6	1.6	2.6
3	1.8	1.4	2.7
4	1.5	2.2	3.0
5	1.0	0.9	1.8
6	1.5	1.4	2.7
7	3.6	1.8	4.5
8	1.2	1.5	2.1
9	2.1	2.3	3.8
10	3.3	2.0	4.7
11	2.7	1.4	3.7
12	2.1	1.2	2.8
13	1.7	1.6	3.0
14	0.6	0.6	1.2
15	4.1	4.4	6.7
16	3.4	3.4	5.5
17	4.4	1.7	5.7
18	0.9	1.0	1.9
19	5.4	1.6	6.3
20	0.8	1.4	2.0
21	1.9	4.0	4.8
22	6.8	7.7	11.9
X	1.2	1.1	2.2
Y	10.9	13.1	20.8
NA	2.3	7.8	8.3
UL	11.6	20.8	22.2
Total	2.3	2.0	3.6

Excludes duplications with identities >98% to avoid artefactual duplication due to incomplete merger in the assembly process. Calculation was performed on an earlier version of the draft genome sequence based on data available in July 2000 and reflects the duplications found within the total amount of finished sequence then. Note that there is some overlap between the interchromosomal and intrachromosomal sets.

Gene content of the human genome

Genes (or at least their coding regions) comprise only a tiny fraction of human DNA, but they represent the major biological function of the genome and the main focus of interest by biologists. They are also the most challenging feature to identify in the human genome sequence.

The ultimate goal is to compile a complete list of all human genes and their encoded proteins, to serve as a 'periodic table' for biomedical research[243]. But this is a difficult task. In organisms with small genomes, it is straightforward to identify most genes by the presence of long ORFs. In contrast, human genes tend to have small exons (encoding an average of only 50 codons) separated by long introns (some exceeding 10 kb). This creates a signal-to-noise problem, with the result that computer programs for direct gene prediction have only limited accuracy. Instead, computational prediction of human genes must rely largely on the availability of cDNA sequences or on sequence conservation with genes and proteins from other organisms. This approach is adequate for strongly conserved genes (such as histones or ubiquitin), but may be less sensitive to rapidly evolving genes (including many crucial to speciation, sex determination and fertilization).

Here we describe our efforts to recognize both the RNA genes and protein-coding genes in the human genome. We also study the properties of the predicted human protein set, attempting to discern how the human proteome differs from those of invertebrates such as worm and fly.

Noncoding RNAs

Although biologists often speak of a tight coupling between 'genes

Table 18 Cross-species comparison for large, highly homologous segmental duplications

	Percentage of genome (%)		
	Fly	Worm	Human (finished)*
> 1 kb	1.2	4.25	3.25
> 5 kb	0.37	1.50	2.86
> 10 kb	0.08	0.66	2.52

* This is an underestimate of the total amount of segmental duplication in the human genome because it only reflects duplication detectable with available finished sequence. The proportion of segmental duplications of > 1 kb is probably about 5% (see text).

and their encoded protein products', it is important to remember that thousands of human genes produce noncoding RNAs (ncRNAs) as their ultimate product[244]. There are several major classes of ncRNA. (1) Transfer RNAs (tRNAs) are the adapters that translate the triplet nucleic acid code of RNA into the amino-acid sequence of proteins; (2) ribosomal RNAs (rRNAs) are also central to the translational machinery, and recent X-ray crystallography results strongly indicate that peptide bond formation is catalysed by rRNA, not protein[245,246]; (3) small nucleolar RNAs (snoRNAs) are required for rRNA processing and base modification in the nucleolus[247,248]; and (4) small nuclear RNAs (snRNAs) are critical components of spliceosomes, the large ribonucleoprotein (RNP) complexes that splice introns out of pre-mRNAs in the nucleus. Humans have both a major, U2 snRNA-dependent spliceosome that splices most introns, and a minor, U12 snRNA-dependent spliceosome that splices a rare class of introns that often have AT/AC dinucleotides at the splice sites instead of the canonical GT/AG splice site consensus[249].

Other ncRNAs include both RNAs of known biochemical function (such as telomerase RNA and the 7SL signal recognition particle RNA) and ncRNAs of enigmatic function (such as the large Xist transcript implicated in X dosage compensation[250], or the small vault RNAs found in the bizarre vault ribonucleoprotein complex[251], which is three times the mass of the ribosome but has unknown function).

ncRNAs do not have translated ORFs, are often small and are not polyadenylated. Accordingly, novel ncRNAs cannot readily be found by computational gene-finding techniques (which search for features such as ORFs) or experimental sequencing of cDNA or EST libraries (most of which are prepared by reverse transcription using a primer complementary to a poly(A) tail). Even if the complete finished sequence of the human genome were available, discovering novel ncRNAs would still be challenging. We can, however, identify genomic sequences that are homologous to known ncRNA genes, using BLASTN or, in some cases, more specialized methods.

It is sometimes difficult to tell whether such homologous genes are orthologues, paralogues or closely related pseudogenes (because inactivating mutations are much less obvious than for protein-coding genes). For tRNA, there is sufficiently detailed information about the cloverleaf secondary structure to allow true genes and pseudogenes to be distinguished with high sensitivity. For many other ncRNAs, there is much less structural information and so we employ an operational criterion of high sequence similarity (> 95% sequence identity and > 95% full length) to distinguish true genes from pseudogenes. These assignments will eventually need to be reconciled with experimental data.

Transfer RNA genes. The classical experimental estimate of the number of human tRNA genes is 1,310 (ref. 252). In the draft genome sequence, we find only 497 human tRNA genes (Tables 19, 20). How do we account for this discrepancy? We believe that the original estimate is likely to have been inflated in two respects. First, it came from a hybridization experiment that probably counted closely related pseudogenes; by analysis of the draft genome sequence, there are in fact 324 tRNA-derived putative pseudogenes

(Table 20). Second, the earlier estimate assumed too high a value for the size of the human genome; repeating the calculation using the correct value yields an estimate of about 890 tRNA-related loci, which is in reasonable accord with our count of 821 tRNA genes and pseudogenes in the draft genome sequence.

The human tRNA gene set predicted from the draft genome sequence appears to include most of the known human tRNA species. The draft genome sequence contains 37 of 38 human tRNA species listed in a tRNA database[253], allowing for up to one mismatch. This includes one copy of the known gene for a specialized selenocysteine tRNA, one of several components of a baroque translational mechanism that reads UGA as a selenocysteine codon in certain rare mRNAs that carry a specific *cis*-acting RNA regulatory site (a so-called SECIS element) in their 3' UTRs. The one tRNA gene in the database not found in the draft genome sequence is DE9990, a tRNAGlu species, which differs in two positions from the most related tRNA gene in the human genome. Possible explanations are that the database version of this tRNA contains two errors, the gene is polymorphic or this is a genuine functional tRNA that is missing from the draft genome sequence. (The database also lists one additional tRNA gene (DS9994), but this is apparently a contaminant, most similar to bacterial tRNAs; the parent entry (Z13399) was withdrawn from the DNA database, but the tRNA entry has not yet been removed from the tRNA database.) Although the human set appears substantially complete by this test, the tRNA gene numbers in Table 19 should be considered tentative and used with caution. The human and fly (but not the worm) are known to be missing significant amounts of heterochromatic DNA, and additional tRNA genes could be located there.

With this caveat, the results indicate that the human has fewer tRNA genes than the worm, but more than the fly. This may seem surprising, but tRNA gene number in metazoans is thought to be related not to organismal complexity, but more to idiosyncrasies of the demand for tRNA abundance in certain tissues or stages of embryonic development. For example, the frog *Xenopus laevis*, which must load each oocyte with a remarkable 40 ng of tRNA, has thousands of tRNA genes[254].

The degeneracy of the genetic code has allowed an inspired economy of tRNA anticodon usage. Although 61 sense codons need to be decoded, not all 61 different anticodons are present in tRNAs. Rather, tRNAs generally follow stereotyped and conserved wobble rules[255–257]. Wobble reduces the number of required anticodons substantially, and provides a connection between the genetic code and the hybridization stability of modified and unmodified RNA bases. In eukaryotes, the rules proposed by Guthrie and Abelson[256] predict that about 46 tRNA species will be sufficient to read the 61 sense codons (counting the initiator and elongator methionine tRNAs as two species). According to these rules, in the codon's third (wobble) position, U and C are generally decoded by a single tRNA species, whereas A and G are decoded by two separate tRNA species.

In 'two-codon boxes' of the genetic code (where codons ending with U/C encode a different amino acid from those ending with A/G), the U/C wobble position should be decoded by a G at position 34 in the tRNA anticodon. Thus, in the top left of Fig. 34, there is no tRNA with an AAA anticodon for Phe, but the GAA anticodon can recognize both UUU and UUC codons in the mRNA. In 'four-codon boxes' of the genetic code (where U, C, A and G in the wobble position all encode the same amino acid), the U/C wobble position is almost always decoded by I34 (inosine) in the tRNA, where the inosine is produced by post-transcriptional modification of an adenine (A). In the bottom left of Fig. 34, for example, the GUU and GUC codons of the four-codon Val box are decoded by a tRNA with an anticodon of AAC, which is no doubt modified to IAC. Presumably this pattern, which is strikingly conserved in eukaryotes, has to do with the fact that IA base pairs are also possible; thus

Table 19 Number of tRNA genes in various organisms

Organism	Number of canonical tRNAs	SeCys tRNA
Human	497	1
Worm	584	1
Fly	284	1
Yeast	273	0
Methanococcus jannaschii	36	1
Escherichia coli	86	1

Number of tRNA genes in each of six genome sequences, according to analysis by the computer program tRNAscan-SE . Canonical tRNAs read one of the standard 61 sense codons; this category excludes pseudogenes, undetermined anticodons, putative supressors and selenocysteine tRNAs. Most organisms have a selenocysteine (SeCys) tRNA species, but some unicellular eukaryotes do not (such as the yeast S. cerevisiae).

the IAC anticodon for a Val tRNA could recognize GUU, GUC and even GUA codons. Were this same I34 to be utilized in two-codon boxes, however, misreading of the NNA codon would occur, resulting in translational havoc. Eukaryotic glycine tRNAs represent a conserved exception to this last rule; they use a GCC anticodon to decode GGU and GGC, rather than the expected ICC anticodon.

Satisfyingly, the human tRNA set follows these wobble rules almost perfectly (Fig. 34). Only three unexpected tRNA species are found: single genes for a tRNATyr-AUA, tRNAIle-GAU, and tRNAAsn-AUU. Perhaps these are pseudogenes, but they appear to be plausible tRNAs. We also checked the possibility of sequencing errors in their anticodons, but each of these three genes is in a region of high sequence accuracy, with PHRAP quality scores higher than 70 for every base in their anticodons.

As in all other organisms, human protein-coding genes show codon bias—preferential use of one synonymous codon over another[258] (Fig. 34). In less complex organisms, such as yeast or bacteria, highly expressed genes show the strongest codon bias. Cytoplasmic abundance of tRNA species is correlated with both codon bias and overall amino-acid frequency (for example, tRNAs for preferred codons and for more common amino acids are more abundant). This is presumably driven by selective pressure for efficient or accurate translation[259]. In many organisms, tRNA abundance in turn appears to be roughly correlated with tRNA gene copy number, so tRNA gene copy number has been used as a proxy for tRNA abundance[260]. In vertebrates, however, codon bias is not so obviously correlated with gene expression level. Differing codon biases between human genes is more a function of their location in regions of different GC composition[261]. In agreement with the literature, we see only a very rough correlation of human tRNA gene number with either amino-acid frequency or codon bias (Fig. 34). The most obvious outliers in these weak correlations are

the strongly preferred CUG leucine codon, with a mere six tRNA-Leu-CAG genes producing a tRNA to decode it, and the relatively rare cysteine UGU and UGC codons, with 30 tRNA genes to decode them.

The tRNA genes are dispersed throughout the human genome. However, this dispersal is nonrandom. tRNA genes have sometimes been seen in clusters at small scales[262,263] but we can now see striking clustering on a genome-wide scale. More than 25% of the tRNA genes (140) are found in a region of only about 4 Mb on chromosome 6. This small region, only about 0.1% of the genome, contains an almost sufficient set of tRNA genes all by itself. The 140 tRNA genes contain a representative for 36 of the 49 anticodons found in the complete set; and of the 21 isoacceptor types, only tRNAs to decode Asn, Cys, Glu and selenocysteine are missing. Many of these tRNA genes, meanwhile, are clustered elsewhere; 18 of the 30 Cys tRNAs are found in a 0.5-Mb stretch of chromosome 7 and many of the Asn and Glu tRNA genes are loosely clustered on chromosome 1. More than half of the tRNA genes (280 out of 497) reside on either chromosome 1 or chromosome 6. Chromosomes 3, 4, 8, 9, 10, 12, 18, 20, 21 and X appear to have fewer than 10 tRNA genes each; and chromosomes 22 and Y have none at all (each has a single pseudogene).

Ribosomal RNA genes. The ribosome, the protein synthetic machine of the cell, is made up of two subunits and contains four rRNA species and many proteins. The large ribosomal subunit contains 28S and 5.8S rRNAs (collectively called 'large subunit' (LSU) rRNA) and also a 5S rRNA. The small ribosomal subunit contains 18S rRNA ('small subunit' (SSU) rRNA). The genes for LSU and SSU rRNA occur in the human genome as a 44-kb tandem repeat unit[264]. There are thought to be about 150–200 copies of this repeat unit arrayed on the short arms of acrocentric chromosomes 13, 14, 15, 21 and 22 (refs 254, 264). There are no true complete

	Codon	Freq	Anticodon	tRNA		Codon	Freq	Anticodon	tRNA		Codon	Freq	Anticodon	tRNA		Codon	Freq	Anticodon	tRNA
Phe	UUU	171	AAA	0	Ser	UCU	147	AGA	10	Tyr	UAU	124	AUA	1	Cys	UGU	99	ACA	0
	UUC	203	GAA	14		UCC	172	GGA	0		UAC	158	GUA	11		UGC	119	GCA	30
Leu	UUA	73	UAA	8		UCA	118	UGA	5	stop	UAA	0	UUA	0	stop	UGA	0	UCA	0
	UUG	125	CAA	6		UCG	45	CGA	4	stop	UAG	0	CUA	0	Trp	UGG	122	CCA	7
Leu	CUU	127	AAG	13	Pro	CCU	175	AGG	11	His	CAU	104	AUG	0	Arg	CGU	47	ACG	9
	CUC	187	GAG	0		CCC	197	GGG	0		CAC	147	GUG	12		CGC	107	GCG	0
	CUA	69	UAG	2		CCA	170	UGG	10	Gln	CAA	121	UUG	11		CGA	63	UCG	7
	CUG	392	CAG	6		CCG	69	CGG	4		CAG	343	CUG	21		CGG	115	CCG	5
Ile	AUU	165	AAU	13	Thr	ACU	131	AGU	8	Asn	AAU	174	AUU	1	Ser	AGU	121	ACU	0
	AUC	218	GAU	1		ACC	192	GGU	0		AAC	199	GUU	33		AGC	191	GCU	7
	AUA	71	UAU	5		ACA	150	UGU	10	Lys	AAA	248	UUU	16	Arg	AGA	113	UCU	5
Met	AUG	221	CAU	17		ACG	63	CGU	7		AAG	331	CUU	22		AGG	110	CCU	4
Val	GUU	111	AAC	20	Ala	GCU	185	AGC	25	Asp	GAU	230	AUC	0	Gly	GGU	112	ACC	0
	GUC	146	GAC	0		GCC	282	GGC	0		GAC	262	GUC	10		GGC	230	GCC	11
	GUA	72	UAC	5		GCA	160	UGC	10	Glu	GAA	301	UUC	14		GGA	168	UCC	5
	GUG	288	CAC	19		GCG	74	CGC	5		GAG	404	CUC	8		GGG	160	CCC	8

Figure 34 The human genetic code and associated tRNA genes. For each of the 64 codons, we show: the corresponding amino acid; the observed frequency of the codon per 10,000 codons; the codon; predicted wobble pairing to a tRNA anticodon (black lines); an unmodified tRNA anticodon sequence; and the number of tRNA genes found with this anticodon. For example, phenylalanine is encoded by UUU or UUC; UUC is seen more frequently, 203 to 171 occurrences per 10,000 total codons; both codons are expected to be decoded by a single tRNA anticodon type, GAA, using a G/U wobble; and there are 14 tRNA genes found with this anticodon. The modified anticodon sequence in the mature tRNA is not shown, even where post-transcriptional modifications can be confidently predicted (for example, when an A is used to decode a U/C third position, the A is almost certainly an inosine in the mature tRNA). The Figure also does not show the number of distinct tRNA species (such as distinct sequence families) for each anticodon; often there is more than one species for each anticodon.

copies of the rDNA tandem repeats in the draft genome sequence, owing to the deliberate bias in the initial phase of the sequencing effort against sequencing BAC clones whose restriction fragment fingerprints showed them to contain primarily tandemly repeated sequence. Sequence similarity analysis with the BLASTN computer program does, however, detect hundreds of rDNA-derived sequence fragments dispersed throughout the complete genome, including one 'full-length' copy of an individual 5.8S rRNA gene not associated with a true tandem repeat unit (Table 20).

The 5S rDNA genes also occur in tandem arrays, the largest of which is on chromosome 1 between 1q41.11 and 1q42.13, close to the telomere[265,266]. There are 200–300 true 5S genes in these arrays[265,267]. The number of 5S-related sequences in the genome, including numerous dispersed pseudogenes, is classically cited as 2,000 (refs 252, 254). The long tandem array on chromosome 1 is not yet present in the draft genome sequence because there are no *Eco*RI or *Hin*dIII sites present, and thus it was not cloned in the most heavily utilized BAC libraries (Table 1). We expect to recover it during the finishing stage. We do detect four individual copies of 5S rDNA by our search criteria ($\geq 95\%$ identity and $\geq 95\%$ full length). We also find many more distantly related dispersed sequences (520 at $P \leq 0.001$), which we interpret as probable pseudogenes (Table 20).

Small nucleolar RNA genes. Eukaryotic rRNA is extensively processed and modified in the nucleolus. Much of this activity is directed by numerous snoRNAs. These come in two families: C/D box snoRNAs (mostly involved in guiding site-specific 2'-O-ribose methylations of other RNAs) and H/ACA snoRNAs (mostly involved in guiding site-specific pseudouridylations)[247,248]. We compiled a set of 97 known human snoRNA gene sequences; 84 of these (87%) have at least one copy in the draft genome sequence (Table 20), almost all as single-copy genes.

It is thought that all 2'-O-ribose methylations and pseudouridylations in eukaryotic rRNA are guided by snoRNAs. There are 105–107 methylations and around 95 pseudouridylations in human rRNA[268]. Only about half of these have been tentatively assigned to known guide snoRNAs. There are also snoRNA-directed modifications on other stable RNAs, such as U6 (ref. 269), and the extent of this is just beginning to be explored. Sequence similarity has so far proven insufficient to recognize all snoRNA genes. We therefore expect that there are many unrecognized snoRNA genes that are not detected by BLAST queries.

Spliceosomal RNAs and other ncRNA genes. We also looked for copies of other known ncRNA genes. We found at least one copy of 21 (95%) of 22 known ncRNAs, including the spliceosomal snRNAs. There were multiple copies for several ncRNAs, as expected; for example, we find 44 dispersed genes for U6 snRNA, and 16 for U1 snRNA (Table 20).

For some of these RNA genes, homogeneous multigene families that occur in tandem arrays are again under-represented owing to the restriction enzymes used in constructing the BAC libraries and, in some instances, the decision to delay the sequencing of BAC clones with low complexity fingerprints indicative of tandemly repeated DNA. The U2 RNA genes are located at the RNU2 locus, a tandem array of 10–20 copies of nearly identical 6.1-kb units at 17q21–q22 (refs 270–272). Similarly, the U3 snoRNA genes (included in the aggregate count of C/D snoRNAs in Table 20) are clustered at the RNU3 locus at 17p11.2, not in a tandem array, but in a complex inverted repeat structure of about 5–10 copies per haploid genome[273]. The U1 RNA genes are clustered with about 30 copies at the RNU1 locus at 1p36.1, but this cluster is thought to be loose and irregularly organized; no two U1 genes have been cloned on the same cosmid[271]. In the draft genome sequence, we see six copies of U2 RNA that meet our criteria for true genes, three of which appear to be in the expected position on chromosome 17. For U3, so far we see one true copy at the correct place on chromosome 17p11.2. For U1, we see 16 true genes, 6 of which are loosely clustered within 0.6 Mb at 1p36.1 and another 6 are elsewhere on chromosome 1. Again, these and other clusters will be a matter for the finishing process.

Table 20 Known non-coding RNA genes in the draft genome sequence

RNA gene*	Number expected†	Number found‡	Number of related genes§	Function
tRNA	1,310	497	324	Protein synthesis
SSU (18S) rRNA	150–200	0	40	Protein synthesis
5.8S rRNA	150–200	1	11	Protein synthesis
LSU (28S) rRNA	150–200	0	181	Protein synthesis
5S rRNA	200–300	4	520	Protein synthesis
U1	~30	16	134	Spliceosome component
U2	10–20	6	94	Spliceosome component
U4	??	4	87	Spliceosome component
U4atac	??	1	20	Component of minor (U11/U12) spliceosome
U5	??	1	31	Spliceosome component
U6	??	44	1,135	Spliceosome component
U6atac	??	4	32	Component of minor (U11/U12) spliceosome
U7	1	1	3	Histone mRNA 3' processing
U11	1	0	6	Component of minor (U11/U12) spliceosome
U12	1	1	0	Component of minor (U11/U12) spliceosome
SRP (7SL) RNA	4	3	773	Component of signal recognition particle (protein secretion)
RNAse P	1	1	2	tRNA 5' end processing
RNAse MRP	1	1	6	rRNA processing
Telomerase RNA	1	1	4	Template for addition of telomeres
hY1	1	1	353	Component of Ro RNP, function unknown
hY3	1	25	414	Component of Ro RNP, function unknown
hY4	1	3	115	Component of Ro RNP, function unknown
hY5 (4.5S RNA)	1	1	9	Component of Ro RNP, function unknown
Vault RNAs	3	3	1	Component of 13-MDa vault RNP, function unknown
7SK	1	1	330	Unknown
H19	1	1	2	Unknown
Xist	1	1	0	Initiation of X chromosome inactivation (dosage compensation)
Known C/D snoRNAs	81	69	558	Pre-rRNA processing or site-specific ribose methylation of rRNA
Known H/ACA snoRNAs	16	15	87	Pre-rRNA processing or site-specific pseudouridylation of rRNA

* Known ncRNA genes (or gene families, such as the C/D and H/ACA snoRNA families); reference sequences were extracted from GenBank and used to probe the draft genome sequence.
† Number of genes that were expected in the human genome, based on previous literature (note that earlier experimental techniques probably tend to overestimate copy number, by counting closely related pseudogenes).
‡ The copy number of 'true' full-length genes identified in the draft genome sequence.
§ The copy number of other significantly related copies (pseudogenes, fragments, paralogues) found. Except for the 497 true tRNA genes, all sequence similarities were identified by WashU BLASTN 2.0MP (W. Gish, unpublished; http://blast.wustl.edu), with parameters '-kap wordmask = seg B = 50000 W = 8' and the default +5/−4 DNA scoring matrix. True genes were operationally defined as BLAST hits with $\geq 95\%$ identity over $\geq 95\%$ of the length of the query. Related sequences were operationally defined as all other BLAST hits with P-values ≤ 0.001.

articles

Table 21 Characteristics of human genes

	Median	Mean	Sample (size)
Internal exon	122 bp	145 bp	RefSeq alignments to draft genome sequence, with confirmed intron boundaries (43,317 exons)
Exon number	7	8.8	RefSeq alignments to finished sequence (3,501 genes)
Introns	1,023 bp	3,365 bp	RefSeq alignments to finished sequence (27,238 introns)
3′ UTR	400 bp	770 bp	Confirmed by mRNA or EST on chromosome 22 (689)
5′ UTR	240 bp	300 bp	Confirmed by mRNA or EST on chromosome 22 (463)
Coding sequence	1,100 bp	1,340 bp	Selected RefSeq entries (1,804)
(CDS)	367 aa	447 aa	
Genomic extent	14 kb	27 kb	Selected RefSeq entries (1,804)

Median and mean values for a number of properties of human protein-coding genes. The 1,804 selected RefSeq entries were those that could be unambiguously aligned to finished sequence over their entire length.

Our observations also confirm the striking proliferation of ncRNA-derived pseudogenes (Table 20). There are hundreds or thousands of sequences in the draft genome sequence related to some of the ncRNA genes. The most prolific pseudogene counts generally come from RNA genes transcribed by RNA polymerase III promoters, including U6, the hY RNAs and SRP-RNA. These ncRNA pseudogenes presumably arise through reverse transcription. The frequency of such events gives insight into how ncRNA genes can evolve into SINE retroposons, such as the tRNA-derived SINEs found in many vertebrates and the SRP-RNA-derived Alu elements found in humans.

Protein-coding genes

Identifying the protein-coding genes in the human genome is one of the most important applications of the sequence data, but also one of the most difficult challenges. We describe below our efforts to create an initial human gene and protein index.

Exploring properties of known genes. Before attempting to identify new genes, we explored what could be learned by aligning the cDNA sequences of known genes to the draft genome sequence. Genomic alignments allow one to study exon–intron structure and local GC content, and are valuable for biomedical studies because they connect genes with the genetic and cytogenetic map, link them with regulatory sequences and facilitate the development of polymerase chain reaction (PCR) primers to amplify exons. Until now, genomic alignment was available for only about a quarter of known genes.

The 'known' genes studied were those in the RefSeq database[110], a manually curated collection designed to contain nonredundant representatives of most full-length human mRNA sequences in GenBank (RefSeq intentionally contains some alternative splice forms of the same genes). The version of RefSeq used contained 10,272 mRNAs.

The RefSeq genes were aligned with the draft genome sequence, using both the Spidey (S. Wheelan, personal communication) and Acembly (D. Thierry-Mieg and J. Thierry-Mieg, unpublished; http://www.acedb.org) computer programs. Because this sequence is incomplete and contains errors, not all genes could be fully aligned and some may have been incorrectly aligned. More than 92% of the RefSeq entries could be aligned at high stringency over at least part of their length, and 85% could be aligned over more than half of their length. Some genes (16%) had high stringency alignments to more than one location in the draft genome sequence owing, for example, to paralogues or pseudogenes. In such cases, we considered only the best match. In a few of these cases, the assignment may not be correct because the true matching region has not yet been sequenced. Three per cent of entries appeared to be alternative splice products of the same gene, on the basis of their alignment to the same location in the draft genome sequence. In all, we obtained at least partial genomic alignments for 9,212 distinct known genes and essentially complete alignment for 5,364 of them.

Previous efforts to study human gene structure[116,274,275] have been hampered by limited sample sizes and strong biases in favour of compact genes. Table 21 gives the mean and median values of some basic characteristics of gene structures. Some of the values may be

underestimates. In particular, the UTRs given in the RefSeq database are likely to be incomplete; they are considerably shorter, for example, than those derived from careful reconstructions on chromosome 22. Intron sizes were measured only for genes in finished genomic sequence, to mitigate the bias arising from the fact that

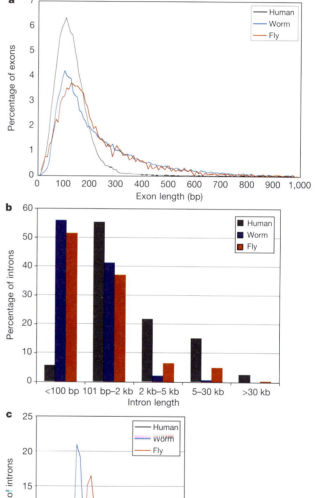

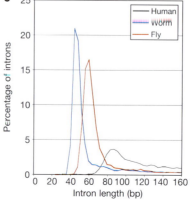

Figure 35 Size distributions of exons, introns and short introns, in sequenced genomes. **a**, Exons; **b**, introns; **c**, short introns (enlarged from **b**). Confirmed exons and introns for the human were taken from RefSeq alignments and for worm and fly from Acembly alignments of ESTs (J. and D. Thierry-Mieg and, for worm, Y. Kohara, unpublished).

long introns are more likely than short introns to be interrupted by gaps in the draft genome sequence. Nonetheless, there may be some residual bias against long genes and long introns.

There is considerable variation in overall gene size and intron size, with both distributions having very long tails. Many genes are over 100 kb long, the largest known example being the dystrophin gene (DMD) at 2.4 Mb. The variation in the size distribution of coding sequences and exons is less extreme, although there are still some remarkable outliers. The titin gene[276] has the longest currently known coding sequence at 80,780 bp; it also has the largest number of exons (178) and longest single exon (17,106 bp).

It is instructive to compare the properties of human genes with those from worm and fly. For all three organisms, the typical length of a coding sequence is similar (1,311 bp for worm, 1,497 bp for fly and 1,340 bp for human), and most internal exons fall within a common peak between 50 and 200 bp (Fig. 35a). However, the worm and fly exon distributions have a fatter tail, resulting in a larger mean size for internal exons (218 bp for worm versus 145 bp for human). The conservation of preferred exon size across all three species supports suggestions of a conserved exon-based component of the splicing machinery[277]. Intriguingly, the few extremely short human exons show an unusual base composition. In 42 detected

human exons of less than 19 bp, the nucleotide frequencies of A, G, T and C are 39, 33, 15 and 12%, respectively, showing a strong purine bias. Purine-rich sequences may enhance splicing[278,279], and it is possible that such sequences are required or strongly selected for to ensure correct splicing of very short exons. Previous studies have shown that short exons require intronic, but not exonic, splicing enhancers[280].

In contrast to the exons, the intron size distributions differ substantially among the three species (Fig. 35b, c). The worm and fly each have a reasonably tight distribution, with most introns near the preferred minimum intron length (47 bp for worm, 59 bp for fly) and an extended tail (overall average length of 267 bp for worm and 487 bp for fly). Intron size is much more variable in humans, with a peak at 87 bp but a very long tail resulting in a mean of more than 3,300 bp. The variation in intron size results in great variation in gene size.

The variation in gene size and intron size can partly be explained by the fact that GC-rich regions tend to be gene-dense with many compact genes, whereas AT-rich regions tend to be gene-poor with many sprawling genes containing large introns. The correlation of gene density with GC content is shown in Fig. 36a, b; the relative density increases more than tenfold as GC content increases from

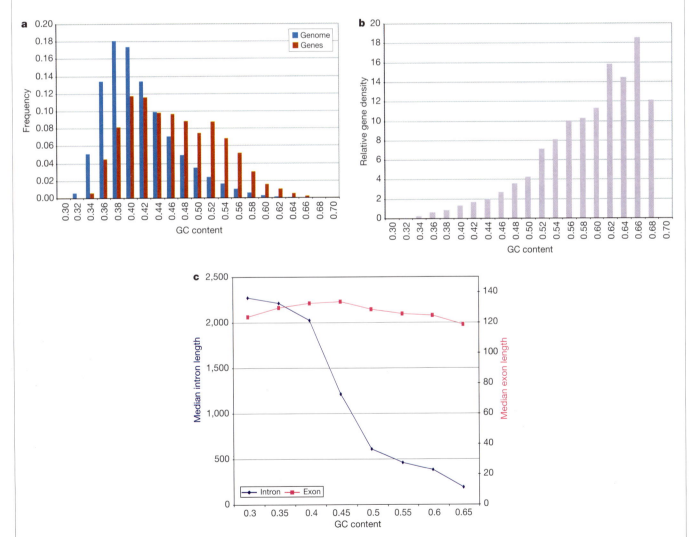

Figure 36 GC content. **a**, Distribution of GC content in genes and in the genome. For 9,315 known genes mapped to the draft genome sequence, the local GC content was calculated in a window covering either the whole alignment or 20,000 bp centred around the midpoint of the alignment, whichever was larger. Ns in the sequence were not counted. GC content for the genome was calculated for adjacent nonoverlapping 20,000-bp windows across the sequence. Both the gene and genome distributions have been normalized to sum to one. **b**, Gene density as a function of GC content, obtained by taking the ratio of the data in **a**. Values are less accurate at higher GC levels because the denominator is small. **c**, Dependence of mean exon and intron lengths on GC content. For exons and introns, the local GC content was derived from alignments to finished sequence only, and were calculated from windows covering the feature or 10,000 bp centred on the feature, whichever was larger.

30% to 50%. The correlation appears to be due primarily to intron size, which drops markedly with increasing GC content (Fig. 36c). In contrast, coding properties such as exon length (Fig. 36c) or exon number (data not shown) vary little. Intergenic distance is also probably lower in high-GC areas, although this is hard to prove directly until all genes have been identified.

The large number of confirmed human introns allows us to analyse variant splice sites, confirming and extending recent reports[281]. Intron positions were confirmed by applying a stringent criterion that EST or mRNA sequence show an exact match of 8 bp in the flanking exonic sequence on each side. Of 53,295 confirmed introns, 98.12% use the canonical dinucleotides GT at the 5′ splice site and AG at the 3′ site (GT–AG pattern). Another 0.76% use the related GC–AG. About 0.10% use AT–AC, which is a rare alternative pattern primarily recognized by the variant U12 splicing machinery[282]. The remaining 1% belong to 177 types, some of which undoubtedly reflect sequencing or alignment errors.

Finally, we looked at alternative splicing of human genes. Alternative splicing can allow many proteins to be produced from a single gene and can be used for complex gene regulation. It appears to be prevalent in humans, with lower estimates of about 35% of human genes being subject to alternative splicing[283–285]. These studies may have underestimated the prevalence of alternative splicing, because they examined only EST alignments covering only a portion of a gene.

To investigate the prevalence of alternative splicing, we analysed reconstructed mRNA transcripts covering the entire coding regions of genes on chromosome 22 (omitting small genes with coding regions of less than 240 bp). Potential transcripts identified by alignments of ESTs and cDNAs to genomic sequence were verified by human inspection. We found 642 transcripts, covering 245 genes (average of 2.6 distinct transcripts per gene). Two or more alternatively spliced transcripts were found for 145 (59%) of these genes. A similar analysis for the gene-rich chromosome 19 gave 1,859 transcripts, corresponding to 544 genes (average 3.2 distinct transcripts per gene). Because we are sampling only a subset of all transcripts, the true extent of alternative splicing is likely to be greater. These figures are considerably higher than those for worm, in which analysis reveals alternative splicing for 22% of genes for which ESTs have been found, with an average of 1.34 (12,816/9,516) splice variants per gene. (The apparently higher extent of alternative splicing seen in human than in worm was not an artefact resulting from much deeper coverage of human genes by ESTs and mRNAs. Although there are many times more ESTs available for human than worm, these ESTs tend to have shorter average length (because many were the product of early sequencing efforts) and many match no human genes. We calculated the actual coverage per bp used in the analysis of the human and worm genes; the coverage is only modestly higher (about 50%) for the human, with a strong bias towards 3′ UTRs which tend to show much less alternative splicing. We also repeated the analysis using equal coverage for the two organisms and confirmed that higher levels of alternative splicing were still seen in human.)

Seventy per cent of alternative splice forms found in the genes on chromosomes 19 and 22 affect the coding sequence, rather than merely changing the 3′ or 5′ UTR. (This estimate may be affected by the incomplete representation of UTRs in the RefSeq database and in the transcripts studied.) Alternative splicing of the terminal exon was seen for 20% of 6,105 mRNAs that were aligned to the draft genome sequence and correspond to confirmed 3′ EST clusters. In addition to alternative splicing, we found evidence of the terminal exon employing alternative polyadenylation sites (separated by >100 bp) in 24% of cases.

Towards a complete index of human genes. We next focused on creating an initial index of human genes and proteins. This index is quite incomplete, owing to the difficulty of gene identification in human DNA and the imperfect state of the draft genome sequence.

Nonetheless, it is valuable for experimental studies and provides important insights into the nature of human genes and proteins.

The challenge of identifying genes from genomic sequence varies greatly among organisms. Gene identification is almost trivial in bacteria and yeast, because the absence of introns in bacteria and their paucity in yeast means that most genes can be readily recognized by *ab initio* analysis as unusually long ORFs. It is not as simple, but still relatively straightforward, to identify genes in animals with small genomes and small introns, such as worm and fly. A major factor is the high signal-to-noise ratio—coding sequences comprise a large proportion of the genome and a large proportion of each gene (about 50% for worm and fly), and exons are relatively large.

Gene identification is more difficult in human DNA. The signal-to-noise ratio is lower: coding sequences comprise only a few per cent of the genome and an average of about 5% of each gene; internal exons are smaller than in worms; and genes appear to have more alternative splicing. The challenge is underscored by the work on human chromosomes 21 and 22. Even with the availability of finished sequence and intensive experimental work, the gene content remains uncertain, with upper and lower estimates differing by as much as 30%. The initial report of the finished sequence of chromosome 22 (ref. 94) identified 247 previously known genes, 298 predicted genes confirmed by sequence homology or ESTs and 325 *ab initio* predictions without additional support. Many of the confirmed predictions represented partial genes. In the past year, 440 additional exons (10%) have been added to existing gene annotations by the chromosome 22 annotation group, although the number of confirmed genes has increased by only 17 and some previously identified gene predictions have been merged[286].

Before discussing the gene predictions for the human genome, it is useful to consider background issues, including previous estimates of the number of human genes, lessons learned from worms and flies and the representativeness of currently 'known' human genes.

Previous estimates of human gene number. Although direct enumeration of human genes is only now becoming possible with the advent of the draft genome sequence, there have been many attempts in the past quarter of a century to estimate the number of genes indirectly. Early estimates based on reassociation kinetics estimated the mRNA complexity of typical vertebrate tissues to be 10,000–20,000, and were extrapolated to suggest around 40,000 for the entire genome[287]. In the mid-1980s, Gilbert suggested that there might be about 100,000 genes, based on the approximate ratio of the size of a typical gene ($\sim 3 \times 10^4$ bp) to the size of the genome (3×10^9 bp). Although this was intended only as a back-of-the-envelope estimate, the pleasing roundness of the figure seems to have led to it being widely quoted and adopted in many textbooks. (W. Gilbert, personal communication; ref. 288). An estimate of 70,000–80,000 genes was made by extrapolating from the number of CpG islands and the frequency of their association with known genes[129].

As human sequence information has accumulated, it has been possible to derive estimates on the basis of sampling techniques[289]. Such studies have sought to extrapolate from various types of data, including ESTs, mRNAs from known genes, cross-species genome comparisons and analysis of finished chromosomes. Estimates based on ESTs[290] have varied widely, from 35,000 (ref. 130) to 120,000 genes[291]. Some of the discrepancy lies in differing estimates of the amount of contaminating genomic sequence in the EST collection and the extent to which multiple distinct ESTs correspond to a single gene. The most rigorous analyses[130] exclude as spurious any ESTs that appear only once in the data set and carefully calibrate sensitivity and specificity. Such calculations consistently produce low estimates, in the region of 35,000.

Comparison of whole-genome shotgun sequence from the pufferfish *T. nigroviridis* with the human genome[292] can be used to estimate the density of exons (detected as conserved sequences between fish and human). These analyses also suggest around 30,000 human genes.

Extrapolations have also been made from the gene counts for chromosomes 21 and 22 (refs 93, 94), adjusted for differences in gene densities on these chromosomes, as inferred from EST mapping. These estimates are between 30,500 and 35,500, depending on the precise assumptions used[286].

Insights from invertebrates. The worm and fly genomes contain a large proportion of novel genes (around 50% of worm genes and 30% of fly genes), in the sense of showing no significant similarity to organisms outside their phylum[293–295]. Such genes may have been present in the original eukaryotic ancestor, but were subsequently lost from the lineages of the other eukaryotes for which sequence is available; they may be rapidly diverging genes, so that it is difficult to recognize homologues solely on the basis of sequence; they may represent true innovations developed within the lineage; or they may represent acquisitions by horizontal transfer. Whatever their origin, these genes tend to have different biological properties from highly conserved genes. In particular, they tend to have low expression levels as assayed both by direct studies and by a paucity of corresponding ESTs, and are less likely to produce a visible phenotype in loss-of-function genetic experiments[294,296].

Gene prediction. Current gene prediction methods employ combinations of three basic approaches: direct evidence of transcription provided by ESTs or mRNAs[297–299]; indirect evidence based on sequence similarity to previously identified genes and proteins[300,301]; and *ab initio* recognition of groups of exons on the basis of hidden Markov models (HMMs) that combine statistical information about splice sites, coding bias and exon and intron lengths (for example, Genscan[275], Genie[302,303] and FGENES[304]).

The first approach relies on direct experimental data, but is subject to artefacts arising from contaminating ESTs derived from unspliced mRNAs, genomic DNA contamination and nongenic transcription (for example, from the promoter of a transposable element). The first two problems can be mitigated by comparing transcripts with the genomic sequence and using only those that show clear evidence of splicing. This solution, however, tends to discard evidence from genes with long terminal exons or single exons. The second approach tends correctly to identify gene-derived sequences, although some of these may be pseudogenes. However, it obviously cannot identify truly novel genes that have no sequence similarity to known genes. The third approach would suffice alone if one could accurately define the features used by cells for gene recognition, but our current understanding is insufficient to do so. The sensitivity and specificity of *ab initio* predictions are greatly affected by the signal-to-noise ratio. Such methods are more accurate in the fly and worm than in human. In fly, *ab initio* methods can correctly predict around 90% of individual exons and can correctly predict all coding exons of a gene in about 40% of cases[303]. For human, the comparable figures are only about 70% and 20%, respectively[94,305]. These estimates may be optimistic, owing to the design of the tests used.

In any collection of gene predictions, we can expect to see various errors. Some gene predictions may represent partial genes, because of inability to detect some portions of a gene (incomplete sensitivity) or to connect all the components of a gene (fragmentation); some may be gene fusions; and others may be spurious predictions (incomplete specificity) resulting from chance matches or pseudogenes.

Creating an initial gene index. We set out to create an initial integrated gene index (IGI) and an associated integrated protein index (IPI) for the human genome. We describe the results obtained from a version of the draft genome sequence based on the sequence data available in July 2000, to allow time for detailed analysis of the gene and protein content. The additional sequence data that has since become available will affect the results quantitatively, but are unlikely to change the conclusions qualitatively.

We began with predictions produced by the Ensembl system[306]. Ensembl starts with *ab initio* predictions produced by Genscan[275] and then attempts to confirm them by virtue of similarity to proteins, mRNAs, ESTs and protein motifs (contained in the Pfam database[307]) from any organism. In particular, it confirms introns if they are bridged by matches and exons if they are flanked by confirmed introns. It then attempts to extend protein matches using the GeneWise computer program[308]. Because it requires confirmatory evidence to support each gene component, it frequently produces partial gene predictions. In addition, when there is evidence of alternative splicing, it reports multiple overlapping transcripts. In total, Ensembl produced 35,500 gene predictions with 44,860 transcripts.

To reduce fragmentation, we next merged Ensembl-based gene predictions with overlapping gene predictions from another program, Genie[302]. Genie starts with mRNA or EST matches and employs an HMM to extend these matches by using *ab initio* statistical approaches. To avoid fragmentation, it attempts to link information from 5′ and 3′ ESTs from the same cDNA clone and thereby to produce a complete coding sequence from an initial ATG to a stop codon. As a result, it may generate complete genes more accurately than Ensembl in cases where there is extensive EST support. (Genie also generates potential alternative transcripts, but we used only the longest transcript in each group.) We merged 15,437 Ensembl predictions into 9,526 clusters, and the longest transcript in each cluster (from either Genie or Ensembl) was taken as the representative.

Next, we merged these results with known genes contained in the RefSeq (version of 29 September 2000), SWISSPROT (release 39.6 of 30 August 2000) and TrEMBL databases (TrEMBL release 14.17 of 1 October 2000, TrEMBL_new of 1 October 2000). Incorporating these sequences gave rise to overlapping sequences because of alternative splice forms and partial sequences. To construct a nonredundant set, we selected the longest sequence from each overlapping set by using direct protein comparison and by mapping the gene predictions back onto the genome to construct the overlapping sets. This may occasionally remove some close paralogues in the event that the correct genomic location has not yet been sequenced, but this number is expected to be small.

Finally, we searched the set to eliminate any genes derived from contaminating bacterial sequences, recognized by virtue of near identity to known bacterial plasmids, transposons and chromosomal genes. Although most instances of such contamination had been removed in the assembly process, a few cases had slipped through and were removed at this stage.

The process resulted in version 1 of the IGI (IGI.1). The composition of the corresponding IPI.1 protein set, obtained by translating IGI.1, is given in Table 22. There are 31,778 protein predictions, with 14,882 from known genes, 4,057 predictions from Ensembl merged with Genie and 12,839 predictions from Ensembl alone. The average lengths are 469 amino acids for the known proteins, 443 amino acids for protein predictions from the Ensembl–Genie merge, and 187 amino acids for those from Ensembl alone. (The smaller average size for the predictions from Ensembl alone reflects its tendency to predict partial genes where there is supporting evidence for only part of the gene; the remainder of the gene will often not be predicted at all, rather than included as part of another prediction. Accordingly, the smaller size cannot be used to estimate the rate of fragmentation in such predictions.)

The set corresponds to fewer than 31,000 actual genes, because some genes are fragmented into more than one partial prediction and some predictions may be spurious or correspond to pseudogenes. As discussed below, our best estimate is that IGI.1 includes about 24,500 true genes.

Evaluation of IGI/IPI. We used several approaches to evaluate the sensitivity, specificity and fragmentation of the IGI/IPI set.

Comparison with 'new' known genes. One approach was to examine

articles

Table 22 Properties of the IGI/IPI human protein set

Source	Number	Average length (amino acids)	Matches to nonhuman proteins	Matches to RIKEN mouse cDNA set	Matches to RIKEN mouse cDNA set but not to nonhuman proteins
RefSeq/SwissProt/TrEMBL	14,882	469	12,708 (85%)	11,599 (78%)	776 (36%)
Ensembl–Genie	4,057	443	2,989 (74%)	3,016 (74%)	498 (47%)
Ensembl	12,839	187	81,126 (63%)	7,372 (57%)	1,449 (31%)
Total	31,778	352	23,813 (75%)	219,873 (69%)	2,723 (34%)

The matches to nonhuman proteins were obtained by using Smith-Waterman sequence alignment with an E-value threshold of 10^{-3} and the matches to the RIKEN mouse cDNAs by using TBLASTN with an E-value threshold of 10^{-6}. The last column shows that a significant number of the IGI members that do not have nonhuman protein matches do match sequences in the RIKEN mouse cDNA set, suggesting that both the IGI and the RIKEN sets contain a significant number of novel proteins.

newly discovered genes arising from independent work that were not used in our gene prediction effort. We identified 31 such genes: 22 recent entries to RefSeq and 9 from the Sanger Centre's gene identification program on chromosome X. Of these, 28 were contained in the draft genome sequence and 19 were represented in the IGI/IPI. This suggests that the gene prediction process has a sensitivity of about 68% (19/28) for the detection of novel genes in the draft genome sequence and that the current IGI contains about 61% (19/31) of novel genes in the human genome. On average, 79% of each gene was detected. The extent of fragmentation could also be estimated: 14 of the genes corresponded to a single prediction in the IGI/IPI, three genes corresponded to two predictions, one gene to three predictions and one gene to four predictions. This corresponds to a fragmentation rate of about 1.4 gene predictions per true gene.

Comparison with RIKEN mouse cDNAs. In a less direct but larger-scale approach, we compared the IGI gene set to a set of mouse cDNAs sequenced by the Genome Exploration Group of the RIKEN Genomic Sciences Center[309]. This set of 15,294 cDNAs, subjected to full-insert sequencing, was enriched for novel genes by selecting cDNAs with novel 3′ ends from a collection of nearly one million ESTs from diverse tissues and developmental timepoints. We determined the proportion of the RIKEN cDNAs that showed sequence similarity to the draft genome sequence and the proportion that showed sequence similarity to the IGI/IPI. Around 81% of the genes in the RIKEN mouse set showed sequence similarity to the human genome sequence, whereas 69% showed sequence similarity to the IGI/IPI. This suggests a sensitivity of 85% (69/81). This is higher than the sensitivity estimate above, perhaps because some of the matches may be due to paralogues rather than orthologues. It is consistent with the IGI/IPI representing a substantial fraction of the human proteome.

Conversely, 69% (22,013/31,898) of the IGI matches the RIKEN cDNA set. Table 22 shows the breakdown of these matches among the different components of the IGI. This is lower than the proportion of matches among known proteins, although this is expected because known proteins tend to be more highly conserved (see above) and because the predictions are on average shorter than known proteins. Table 22 also shows the numbers of matches to the RIKEN cDNAs among IGI members that do not match known proteins. The results indicate that both the IGI and the RIKEN set contain a significant number of genes that are novel in the sense of not having known protein homologues.

Comparison with genes on chromosome 22. We also compared the IGI/IPI with the gene annotations on chromosome 22, to assess the proportion of gene predictions corresponding to pseudogenes and to estimate the rate of overprediction. We compared 477 IGI gene predictions to 539 confirmed genes and 133 pseudogenes on chromosome 22 (with the immunoglobulin lambda locus excluded owing to its highly atypical gene structure). Of these, 43 hit 36 annotated pseudogenes. This suggests that 9% of the IGI predictions may correspond to pseudogenes and also suggests a fragmentation rate of 1.2 gene predictions per gene. Of the remaining hits, 63 did not overlap with any current annotations. This would suggest a rate of spurious predictions of about 13% (63/477), although the true rate is likely to be much lower because many of these may correspond to

unannotated portions of existing gene predictions or to currently unannotated genes (of which there are estimated to be about 100 on this chromosome[94]).

Chromosomal distribution. Finally, we examined the chromosomal distribution of the IGI gene set. The average density of gene predictions is 11.1 per Mb across the genome, with the extremes being chromosome 19 at 26.8 per Mb and chromosome Y at 6.4 per Mb. It is likely that a significant number of the predictions on chromosome Y are pseudogenes (this chromosome is known to be rich in pseudogenes) and thus that the density for chromosome Y is an overestimate. The density of both genes and Alus on chromosome 19 is much higher than expected, even accounting for the high GC content of the chromosome; this supports the idea that Alu density is more closely correlated with gene density than with GC content itself.

Summary. We are clearly still some way from having a complete set of human genes. The current IGI contains significant numbers of partial genes, fragmented and fused genes, pseudogenes and spurious predictions, and it also lacks significant numbers of true genes. This reflects the current state of gene prediction methods in vertebrates even in finished sequence, as well as the additional challenges related to the current state of the draft genome sequence. Nonetheless, the gene predictions provide a valuable starting point for a wide range of biological studies and will be rapidly refined in the coming year.

The analysis above allows us to estimate the number of distinct genes in the IGI, as well as the number of genes in the human genome. The IGI set contains about 15,000 known genes and about 17,000 gene predictions. Assuming that the gene predictions are subject to a rate of overprediction (spurious predictions and pseudogenes) of 20% and a rate of fragmentation of 1.4, the IGI would be estimated to contain about 24,500 actual human genes. Assuming that the gene predictions contain about 60% of previously unknown human genes, the total number of genes in the human genome would be estimated to be about 31,000. This is consistent with most recent estimates based on sampling, which suggest a gene number of 30,000–35,000. If there are 30,000–35,000 genes, with an average coding length of about 1,400 bp and average genomic extent of about 30 kb, then about 1.5% of the human genome would consist of coding sequence and one-third of the genome would be transcribed in genes.

The IGI/IPI was constructed primarily on the basis of gene predictions from Ensembl. However, we also generated an expanded set (IGI+) by including additional predictions from two other gene prediction programs, Genie and GenomeScan (C. Burge, personal communication). These predictions were not included in the core IGI set, because of the concern that each additional set will provide diminishing returns in identifying true genes while contributing its own false positives (increased sensitivity at the expense of specificity). Genie produced an additional 2,837 gene predictions not overlapping the IGI, and GenomeScan produced 6,534 such gene predictions. If all of these gene predictions were included in the IGI, the number of the 31 new 'known' genes (see above) contained in the IGI would rise from 19 to 24. This would amount to an increase of about 26% in sensitivity, at the expense of increasing the number of predicted genes (excluding knowns) by 55%. Allowing a higher overprediction rate of 30% for gene predictions in this expanded set, the analysis above suggests that IGI+ set contains about 28,000 true genes and yields an estimate of about 32,000 human genes. We are investigating ways to filter the expanded set, to produce an IGI with the advantage of the increased sensitivity resulting from combining multiple gene prediction programs without the corresponding loss of specificity. Meanwhile, the IGI+ set can be used by researchers searching for genes that cannot be found in the IGI.

Some classes of genes may have been missed by all of the gene-finding methods. Genes could be missed if they are expressed at low levels or in rare tissues (being absent or very under-represented in EST and mRNA databases) and have sequences that evolve rapidly (being hard to detect by protein homology and genome comparison). Both the worm and fly gene sets contain a substantial number of such genes[293,294]. Single-exon genes encoding small proteins may also have been missed, because EST evidence that supports them cannot be distinguished from genomic contamination in the EST dataset and because homology may be hard to detect for small proteins[310].

The human thus appears to have only about twice as many genes as worm or fly. However, human genes differ in important respects from those in worm and fly. They are spread out over much larger regions of genomic DNA, and they are used to construct more alternative transcripts. This may result in perhaps five times as many primary protein products in the human as in the worm or fly.

The predicted gene and protein sets described here are clearly far from final. Nonetheless, they provide a valuable starting point for experimental and computational research. The predictions will improve progressively as the sequence is finished, as further confirmatory evidence becomes available (particularly from other vertebrate genome sequences, such as those of mouse and *T. nigroviridis*), and as computational methods improve. We intend to create and release updated versions of the IGI and IPI regularly, until they converge to a final accurate list of every human gene. The gene predictions will be linked to RefSeq, HUGO and SWISSPROT identifiers where available, and tracking identifiers between versions will be included, so that individual genes under study can be traced forwards as the human sequence is completed.

Comparative proteome analysis

Knowledge of the human proteome will provide unprecedented opportunities for studies of human gene function. Often clues will be provided by sequence similarity with proteins of known function in model organisms. Such initial observations must then be followed up by detailed studies to establish the actual function of these molecules in humans.

For example, 35 proteins are known to be involved in the vacuolar protein-sorting machinery in yeast. Human genes encoding homologues can be found in the draft human sequence for 34 of these yeast proteins, but precise relationships are not always clear. In nine cases there appears to be a single clear human orthologue (a gene that arose as a consequence of speciation); in 12 cases there are matches to a family of human paralogues (genes that arose owing to intra-genome duplication); and in 13 cases there are matches to specific protein domains[311–314]. Hundreds of similar stories emerge from the draft sequence, but each merits a detailed interpretation in context. To treat these subjects properly, there will be many following studies, the first of which appear in accompanying papers[315–323].

Here, we aim to take a more global perspective on the content of the human proteome by comparing it with the proteomes of yeast, worm, fly and mustard weed. Such comparisons shed useful light on the commonalities and differences among these eukaryotes[294,324,325]. The analysis is necessarily preliminary, because of the imperfect nature of the human sequence, uncertainties in the gene and protein sets for all of the multicellular organisms considered and our incomplete knowledge of protein structures. Nonetheless, some general patterns emerge. These include insights into fundamental

mechanisms that create functional diversity, including invention of protein domains, expansion of protein and domain families, evolution of new protein architectures and horizontal transfer of genes. Other mechanisms, such as alternative splicing, post-translational modification and complex regulatory networks, are also crucial in generating diversity but are much harder to discern from the primary sequence. We will not attempt to consider the effects of alternative splicing on proteins; we will consider only a single splice form from each gene in the various organisms, even when multiple splice forms are known.

Functional and evolutionary classification. We began by classifying the human proteome on the basis of functional categories and evolutionary conservation. We used the InterPro annotation protocol to identify conserved biochemical and cellular processes. InterPro is a tool for combining sequence-pattern information from four databases. The first two databases (PRINTS[326] and Prosite[327]) primarily contain information about motifs corresponding to specific family subtypes, such as type II receptor tyrosine kinases (RTK-II) in particular or tyrosine kinases in general. The second two databases (Pfam[307] and Prosite Profile[327]) contain information (in the form of profiles or HMMs) about families of structural domains—for example, protein kinase domains. InterPro integrates the motif and domain assignments into a hierarchical classification system; so a protein might be classified at the most detailed level as being an RTK-II, at a more general level as being a kinase specific for tyrosine, and at a still more general level as being a protein kinase. The complete hierarchy of InterPro entries is described at http://www.ebi.ac.uk/interpro/. We collapsed the InterPro entries into 12 broad categories, each reflecting a set of cellular functions.

The InterPro families are partly the product of human judgement and reflect the current state of biological and evolutionary knowledge. The system is a valuable way to gain insight into large collections of proteins, but not all proteins can be classified at present. The proportions of the yeast, worm, fly and mustard weed protein sets that are assigned to at least one InterPro family is, for each organism, about 50% (Table 23; refs 307, 326, 327).

About 40% of the predicted human proteins in the IPI could be assigned to InterPro entries and functional categories. On the basis of these assignments, we could compare organisms according to the number of proteins in each category (Fig. 37). Compared with the two invertebrates, humans appear to have many proteins involved in cytoskeleton, defence and immunity, and transcription and translation. These expansions are clearly related to aspects of vertebrate physiology. Humans also have many more proteins that are classified as falling into more than one functional category (426 in human versus 80 in worm and 57 in fly, data not shown). Interestingly, 32% of these are transmembrane receptors.

We obtained further insight into the evolutionary conservation of proteins by comparing each sequence to the complete nonredundant database of protein sequences maintained at NCBI, using the BLASTP computer program[328] and then breaking down the matches according to organismal taxonomy (Fig. 38). Overall, 74% of the proteins had significant matches to known proteins.

Such classifications are based on the presence of clearly detectable homologues in existing databases. Many of these genes have surely evolved from genes that were present in common ancestors but have since diverged substantially. Indeed, one can detect more distant relationships by using sensitive computer programs that can recognize weakly conserved features. Using PSI-BLAST, we can recognize probable nonvertebrate homologues for about 45% of the 'vertebrate-specific' set. Nonetheless, the classification is useful for gaining insights into the commonalities and differences among the proteomes of different organisms.

Probable horizontal transfer. An interesting category is a set of 223 proteins that have significant similarity to proteins from bacteria, but no comparable similarity to proteins from yeast, worm, fly and

Table 23 Properties of genome and proteome in essentially completed eukaryotic proteomes

	Human	Fly	Worm	Yeast	Mustard weed
Number of identified genes	~32,000*	13,338	18,266	6,144	25,706
% with InterPro matches	51	56	50	50	52
Number of annotated domain families	1,262	1,035	1,014	851	1,010
Number of InterPro entries per gene	0.53	0.84	0.63	0.6	0.62
Number of distinct domain architectures	1,695	1,036	1,018	310	–
Percentage of 1-1-1-1	1.40	4.20	3.10	9.20	–
% Signal sequences	20	20	24	11	–
% Transmembrane proteins	20	25	28	15	–
% Repeat-containing	10	11	9	5	–
% Coiled-coil	11	13	10	9	–

The numbers of distinct architectures were calculated using SMART[339] and the percentages of repeat-containing proteins were estimated using Prospero[452] and a P-value threshold of 10^{-5}. The protein sets used in the analysis were taken from http://www.ebi.ac.uk/proteome/ for yeast, worm and fly. The proteins from mustard weed were taken from the TAIR website (http://www.arabidopsis.org/) on 5 September 2000. The protein set was searched against the InterPro database (http://www.ebi.ac.uk/interpro/) using the InterProscan software. Comparison of protein sequences with the InterPro database allows prediction of protein families, domain and repeat families and sequence motifs. The searches used Pfam release 5.2[307], Prints release 26.1[326], Prosite release 16[327] and Prosite preliminary profiles. InterPro analysis results are available as Supplementary Information. The fraction of 1-1-1-1 is the percentage of the genome that falls into orthologous groups composed of only one member each in human, fly, worm and yeast.
* The gene number for the human is still uncertain (see text). Table is based on 31,778 known genes and gene predictions.

mustard weed, or indeed from any other (nonvertebrate) eukaryote. These sequences should not represent bacterial contamination in the draft human sequence, because we filtered the sequence to eliminate sequences that were essentially identical to known bacterial plasmid, transposon or chromosomal DNA (such as the host strains for the large-insert clones). To investigate whether these were genuine human sequences, we designed PCR primers for 35 of these genes and confirmed that most could be readily detected directly in human genomic DNA (Table 24). Orthologues of many of these genes have also been detected in other vertebrates (Table 24).

A more detailed computational analysis indicated that at least 113 of these genes are widespread among bacteria, but, among eukaryotes, appear to be present only in vertebrates. It is possible that the genes encoding these proteins were present in both early prokaryotes and eukaryotes, but were lost in each of the lineages of yeast, worm, fly, mustard weed and, possibly, from other nonvertebrate eukaryote lineages. A more parsimonious explanation is that these genes entered the vertebrate (or prevertebrate) lineage by horizontal transfer from bacteria. Many of these genes contain introns, which presumably were acquired after the putative horizontal transfer event. Similar observations indicating probable lineage-specific horizontal gene transfers, as well as intron insertion in the acquired genes, have been made in the worm genome[329].

We cannot formally exclude the possibility that gene transfer occurred in the opposite direction—that is, that the genes were invented in the vertebrate lineage and then transferred to bacteria. However, we consider this less likely. Under this scenario, the broad distribution of these genes among bacteria would require extensive horizontal dissemination after their initial acquisition. In addition, the functional repertoire of these genes, which largely encode intracellular enzymes (Table 24), is uncharacteristic of vertebrate-specific evolutionary innovations (which appear to be primarily extracellular proteins; see below).

We did not identify a strongly preferred bacterial source for the putative horizontally transferred genes, indicating the likelihood of multiple independent gene transfers from different bacteria (Table 24). Notably, several of the probable recent acquisitions have established (or likely) roles in metabolism of xenobiotics or stress response. These include several hydrolases of different specificities, including epoxide hydrolase, and several dehydrogenases (Table 24). Of particular interest is the presence of two paralogues of monoamine oxidase (MAO), an enzyme of the mitochondrial outer membrane that is central in the metabolism of neuromediators and is a target of important psychiatric drugs[330–333]. This example shows that at least some of the genes thought to be horizontally transferred into the vertebrate lineage appear to be involved in important physiological functions and so probably have been fixed and maintained during evolution because

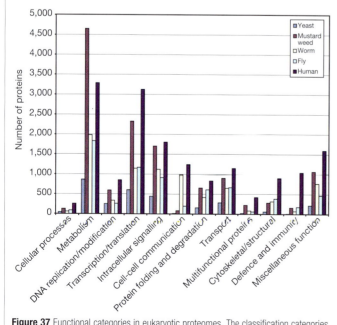

Figure 37 Functional categories in eukaryotic proteomes. The classification categories were derived from functional classification systems, including the top-level biological function category of the Gene Ontology project (GO; see http://www.geneontology.org).

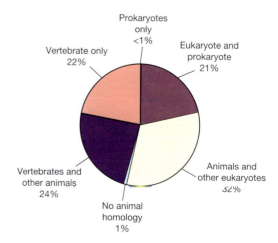

Figure 38 Distribution of the homologues of the predicted human proteins. For each protein, a homologue to a phylogenetic lineage was considered present if a search of the NCBI nonredundant protein sequence database, using the gapped BLASTP program, gave a random expectation (E) value of ≤ 0.001. Additional searches for probable homologues with lower sequence conservation were performed using the PSI-BLAST program, run for three iterations using the same cut-off for inclusion of sequences into the profile[328].

of the increased selective advantage(s) they provide.

Genes shared with fly, worm and yeast. IPI.1 contains apparent homologues of 61% of the fly proteome, 43% of the worm proteome and 46% of the yeast proteome. We next considered the groups of proteins containing likely orthologues and paralogues (genes that arose from intragenome duplication) in human, fly, worm and yeast.

Briefly, we performed all-against-all sequence comparison[334] for the combined protein sets of human, yeast, fly and worm. Pairs of sequences that were one another's best matches in their respective genomes were considered to be potential orthologues. These were then used to identify orthologous groups across three organisms[335]. Recent species-specific paralogues were defined by using the all-against-all sequence comparison to cluster the protein set for each organism. For each sequence found in an orthologous group, the recent paralogues were defined to be the largest species-specific cluster including it. The set of paralogues may be inflated by unrecognized splice variants and by fragmentation.

We identified 1,308 groups of proteins, each containing at least one predicted orthologue in each species and many containing additional paralogues. The 1,308 groups contained 3,129 human proteins, 1,445 fly proteins, 1,503 worm proteins and 1,441 yeast proteins. These 1,308 groups represent a conserved core of proteins that are mostly responsible for the basic 'housekeeping' functions of the cell, including metabolism, DNA replication and repair, and translation.

In 564 of the 1,308 groups, one orthologue (and no additional paralogues) could be unambiguously assigned for each of human, fly, worm and yeast. These groups will be referred to as 1-1-1-1 groups. More than half (305) of these groups could be assigned to the functional categories shown in Fig. 37. Within these functional categories, the numbers of groups containing single orthologues in each of the four proteomes was: 19 for cellular processes, 66 for metabolism, 31 for DNA replication and modification, 106 for transcription/translation, 13 for intracellular signalling, 24 for protein folding and degradation, 38 for transport, 5 for

Table 24 Probable vertebrate-specific acquisitions of bacterial genes

Human protein (accession)	Predicted function	Known orthologues in other vertebrates	Bacterial homologues		Human origin confirmed by PCR
			Range	Best hit	
AAG01853.1	Formiminotransferase cyclodeaminase	Pig, rat, chicken	*Thermotoga, Thermoplasma, Methylobacter*	*Thermotoga maritima*	Yes
CAB81772.1 AAB59448.1 AAA36608.1 AAC41747.1	Na/glucose cotransporter	Rodents, ungulates	Most bacteria	*Vibrio parahaemolyticus*	Yes (CAB81772, AAC41747.1) NT* (AAB59448.1, AAA36608.1)
BAA1143.21	Epoxide hydrolase (α/β-hydrolase)	Mouse, *Danio*, fugu fish	Most bacteria	*Pseudomonas aeruginosa*	Yes
CAB59628.1	Protein-methionine-*S*-oxide reductase	Cow	Most bacteria	*Synechocystis sp.*	Yes
BAA91273.1	Hypertension-associated protein SA/ acetate-CoA ligase	Mouse, rat, cow	Most bacteria	*Bacillus halodurans*	NT*
CAA75608.1	Glucose-6-phosphate transporter/ glycogen storage disease type 1b protein	Mouse, rat	Most bacteria	*Chlamydophila pneumoniae*	Yes
AAA59548.1 AAB27229.1	Monoamine oxidase	Cow, rat, salmon	Most bacteria	*Mycobacterium tuberculosis*	Yes
AAF12736.1 AAA51565.1	Acyl-CoA dehydrogenase, mitochondrial protein	Mouse, rat, pig	Most bacteria	*P. aeruginosa*	Yes
IGI_M1_ctg19153_147	Aldose-1-epimerase	Pig (also found in plants)	*Streptomyces, Bacillus*	*Streptomyces coelicolor*	Yes
BAA92632.1	Predicted carboxylase (C-terminal domain, N-terminal domain unique)	None	*Streptomyces, Rhizobium, Bacillus*	*S. coelicolor*	Yes
BAA34458.1	Uncharacterized protein	None	Gamma-proteobacteria	*Escherichia coli*	Yes
AAF24044.1	Uncharacterized protein	None	Most bacteria	*T. maritima*	Yes
BAA34458.1	β-Lactamase superfamily hydrolase	None	Most bacteria	*Synechocystis sp.*	Yes
BAA91839.1	Oxidoreductase (Rossmann fold) fused to a six-transmembrane protein	None (several human paralogues of both parts)	*Actinomycetes, Leptospira*; more distant homologues in other bacteria	*S. coelicolor*	Yes
BAA92073.1	Oxidoreductase (Rossmann fold)	None	*Synechocystis, Pseudomonas*	*Synechocystis sp.*	Yes
BAA92133.1	α/β-hydrolase	None	*Rickettsia*; more distant homologues in other bacteria	*Rickettsia prowazekii*	Yes
BAA91174.1	ADP-ribosylglycohydrolase	None	*Streptomyces, Aquifex, Archaeoglobus* (archaeon), *E. coli*	*S. coelicolor*	Yes
AAA60043.1	Thymidine phosporylase/endothelial cell growth factor	None	Most bacteria	*Bacillus stearothermophilus*	Yes
BAA86552.1	Ribosomal protein S6-glutamic acid ligase	None	Most bacteria and archaea	*Haemophilus influenzae*	Yes
IGI_M1_ctg12741_7	Ribosomal protein S6-glutamic acid ligase (paralogue of the above)	None	Most bacteria and archaea	*H. influenzae*	Yes
IGI_M1_ctg13238_61	Hydratase	None	*Synechocystis, Sphingomonas*	*Synechocystis sp.*	Yes
IGI_M1_ctg13305_116	Homologue of histone macro-2A C-terminal domain, predicted phosphatase	None (several human paralogues, RNA viruses)	*Thermotoga, Alcaligenes, E. coli*, more distant homologues in other bacteria	*T. maritima*	Yes
IGI_M1_ctg14420_10	Sugar transporter	None	Most bacteria	*Synechocystis sp.*	Yes
IGI_M1_ctg16010_18	Predicted metal-binding protein	None	Most bacteria	*Borrelia burgdorferi*	Yes
IGI_M1_ctg16227_58	Pseudouridine synthase	None	Most bacteria	*Zymomonas mobilis*	Yes
IGI_M1_ctg25107_24	Surfactin synthetase domain	None	Gram-positive bacteria, *Actinomycetes, Cyanobacteria*	*Bacillus subtilis*	Yes

* NT, not tested.
Representative genes confirmed by PCR to be present in the human genome. The similarity to a bacterial homologue was considered to be 'significantly' greater than that to eukaryotic homologues if the difference in alignment scores returned by BLASTP was greater than 30 bits (~9 orders of magnitude in terms of *E*-value). A complete, classified and annotated list of probable vertebrate-specific horizontal gene transfers detected in this analysis is available as Supplementary Information. cDNA sequences for each protein were searched, using the SSAHA algorithm, against the draft genome sequence. Primers were designed and PCR was performed using three human genomic samples and a random BAC clone. The predicted genes were considered to be present in the human genome if a band of the expected size was found in all three human samples but not in the control clone.

multifunctional proteins and 3 for cytoskeletal/structural. No such groups were found for defence and immunity or cell–cell communication.

The 1-1-1-1 groups probably represent key functions that have not undergone duplication and elaboration in the various lineages. They include many anabolic enzymes responsible for such functions as respiratory chain and nucleotide biosynthesis. In contrast, there are few catabolic enzymes. As anabolic pathways branch less frequently than catabolic pathways, this indicates that alternative routes and displacements are more frequent in catabolic reactions. If proteins from the single-celled yeast are excluded from the analysis, there are 1,195 1-1-1 groups. The additional groups include many examples of more complex signalling proteins, such as receptor-type and src-like tyrosine kinases, likely to have arisen early in the metazoan lineage. The fact that this set comprises only a small proportion of the proteome of each of the animals indicates that, apart from a modest conserved core, there has been extensive elaboration and innovation within the protein complement.

Most proteins do not show simple 1-1-1 orthologous relationships across the three animals. To illustrate this, we investigated the nuclear hormone receptor family. In the human proteome, this family consists of 60 different 'classical' members, each with a zinc finger and a ligand-binding domain. In comparison, the fly proteome has 19 and the worm proteome has 220. As shown in Fig. 39, few simple orthologous relationships can be derived among these homologues. And, where potential subgroups of orthologues and

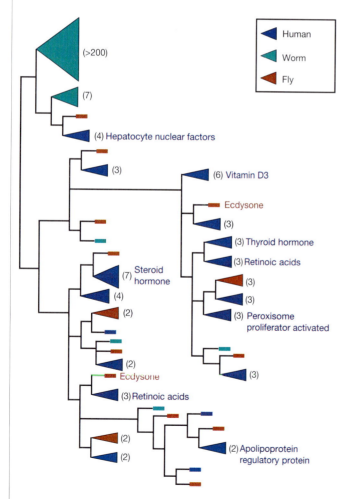

Figure 39 Simplified cladogram (relationship tree) of the 'many-to-many' relationships of classical nuclear receptors. Triangles indicate expansion within one lineage; bars represent single members. Numbers in parentheses indicate the number of paralogues in each group.

paralogues could be identified, it was apparent that the functions of the subgroup members could differ significantly. For example, the fly receptor for the fly-specific hormone ecdysone and the human retinoic acid receptors cluster together on the basis of sequence similarity. Such examples underscore that the assignment of functional similarity on the basis of sequence similarities among these three organisms is not trivial in most cases.

New vertebrate domains and proteins. We then explored how the proteome of vertebrates (as represented by the human) differs from those of the other species considered. The 1,262 InterPro families were scanned to identify those that contain only vertebrate proteins. Only 94 (7%) of the families were 'vertebrate-specific'. These represent 70 protein families and 24 domain families. Only one of the 94 families represents enzymes, which is consistent with the ancient origins of most enzymes[336]. The single vertebrate-specific enzyme family identified was the pancreatic or eosinophil-associated ribonucleases. These enzymes evolved rapidly, possibly to combat vertebrate pathogens[337].

The relatively small proportion of vertebrate-specific multicopy families suggests that few new protein domains have been invented in the vertebrate lineage, and that most protein domains trace at least as far back as a common animal ancestor. This conclusion must be tempered by the fact that the InterPro classification system is incomplete; additional vertebrate-specific families undoubtedly exist that have not yet been recognized in the InterPro system.

The 94 vertebrate-specific families appear to reflect important physiological differences between vertebrates and other eukaryotes. Defence and immunity proteins (23 families) and proteins that function in the nervous system (17 families) are particularly enriched in this set. These data indicate the recent emergence or rapid divergence of these proteins.

Representative human proteins were previously known for nearly all of the vertebrate-specific families. This was not surprising, given the anthropocentrism of biological research. However, the analysis did identify the first mammalian proteins belonging to two of these families. Both of these families were originally defined in fish. The first is the family of polar fish antifreeze III proteins. We found a human sialic acid synthase containing a domain homologous to polar fish antifreeze III protein (BAA91818.1). This finding suggests that fish created the antifreeze function by adaptation of this domain. We also found a human protein (CAB60269.1) homologous to the ependymin found in teleost fish. Ependymins are major glycoproteins of fish brains that have been claimed to be involved in long-term memory formation[338]. The function of the mammalian ependymin homologue will need to be elucidated.

New architectures from old domains. Whereas there appears to be only modest invention at the level of new vertebrate protein domains, there appears to be substantial innovation in the creation of new vertebrate proteins. This innovation is evident at the level of domain architecture, defined as the linear arrangement of domains within a polypeptide. New architectures can be created by shuffling, adding or deleting domains, resulting in new proteins from old parts.

We quantified the number of distinct protein architectures found in yeast, worm, fly and human by using the SMART annotation resource[339] (Fig. 40). The human proteome set contained 1.8 times as many protein architectures as worm or fly and 5.8 times as many as yeast. This difference is most prominent in the recent evolution of novel extracellular and transmembrane architectures in the human lineage. Human extracellular proteins show the greatest innovation: the human has 2.3 times as many extracellular architectures as fly and 2.0 times as many as worm. The larger number of human architectures does not simply reflect differences in the number of domains known in these organisms; the result remains qualitatively the same even if the number of architectures in each organism is normalized by dividing by the total number of domains (not shown). (We also checked that the larger number of human

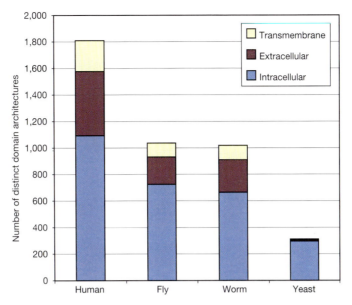

Figure 40 Number of distinct domain architectures in the four eukaryotic genomes, predicted using SMART[339]. The number of architectures is split into three cellular environments: intracellular, extracellular and membrane-associated. The increase in architectures for the human, relative to the other lineages, is seen when these numbers are normalized with respect to the numbers of domains predicted in each phylum. To avoid artefactual results from the relatively low detection rate for some repeat types, tandem occurrences of tetratricopeptide, armadillo, EF-hand, leucine-rich, WD40 or ankyrin repeats or C2H2-type zinc fingers were treated as single occurrences.

architectures could not be an artefact resulting from erroneous gene predictions. Three-quarters of the architectures can be found in known genes, which already yields an increase of about 50% over worm and fly. We expect the final number of human architectures to grow as the complete gene set is identified.)

A related measure of proteome complexity can be obtained by considering an individual domain and counting the number of different domain types with which it co-occurs. For example, the trypsin-like serine protease domain (number 12 in Fig. 41) co-occurs with 18 domain types in human (including proteins involved in the mammalian complement system, blood coagulation, and fibrinolytic and related systems). By contrast, the trypsin-like serine protease domain occurs with only eight other domains in fly, five in worm and one in yeast. Similar results for 27 common domains are shown in Fig. 41. In general, there are more different co-occurring domains in the human proteome than in the other proteomes.

One mechanism by which architectures evolve is through the fusion of additional domains, often at one or both ends of the proteins. Such 'domain accretion'[340] is seen in many human proteins when compared with proteins from other eukaryotes. The effect is illustrated by several chromatin-associated proteins (Fig. 42). In these examples, the domain architectures of human proteins differ from those found in yeast, worm and fly proteins only by the addition of domains at their termini.

Among chromatin-associated proteins and transcription factors, a significant proportion of domain architectures is shared between the vertebrate and fly, but not with worm (Fig. 43a). The trend was even more prominent in architectures of proteins involved in another key cellular process, programmed cell death (Fig. 43b). These examples might seem to bear upon the unresolved issue of the evolutionary branching order of worms, flies and humans, suggesting that worms branched off first. However, there were other cases in which worms and humans shared architectures not present in fly. A global analysis of shared architectures could not conclusively distinguish between the two models, given the possibility of lineage-specific loss of architectures. Comparison of protein architectures may help to resolve the evolutionary issue, but it will require more detailed analyses of many protein families.

New physiology from old proteins. An important aspect of

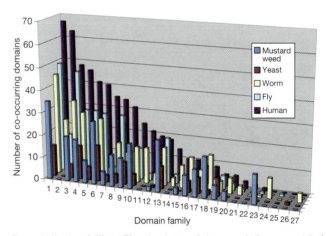

Figure 41 Number of different Pfam domain types that co-occur in the same protein, for each of the 10 most common domain families in each of the five eukaryotic proteomes. Because some common domain families are shared, there are 27 families rather than 50. The data are ranked according to decreasing numbers of human co-occurring Pfam domains. The domain families are: (1) eukaryotic protein kinase [IPR000719]; (2) immunoglobulin domain [IPR003006]; (3) ankyrin repeat [IPR002110]; (4) RING finger [IPR001841]; (5) C2H2-type zinc finger [IPR000822]; (6) ATP/GTP-binding P-loop [IPR001687]; (7) reverse transcriptase (RNA-dependent DNA polymerase) [IPR000477]; (8) leucine-rich repeat [IPR001611]; (9) G-proteinβ WD-40 repeats [IPR001680]; (10) RNA-binding region RNP-1 (RNA recognition motif) [IPR000504]; (11) C-type lectin domain [IPR001304]; (12) serine proteases, trypsin family [IPR001254]; (13) helicase C-terminal domain [IPR001650]; (14) collagen triple helix repeat [IPR000087]; (15) rhodopsin-like GPCR superfamily [IPR000276]; (16) esterase/lipase/thioesterase [IPR000379]; (17) Myb DNA-binding domain [IPR001005]; (18) F-box domain [IPR001810]; (19) ATP-binding transport protein, 2nd P-loop motif [IPR001051]; (20) homeobox domain [IPR001356]; (21) C4-type steroid receptor zinc finger [IPR001628]; (22) sugar transporter [IPR001066]; (23) PPR repeats [IPR002885]; (24) seven-helix G-protein-coupled receptor, worm (probably olfactory) family [IPR000168]; (25) cytochrome P450 enzyme [IPR001128]; (26) fungal transcriptional regulatory protein, N terminus [IPR001138]; (27) domain of unknown function DUF38 [IPR002900].

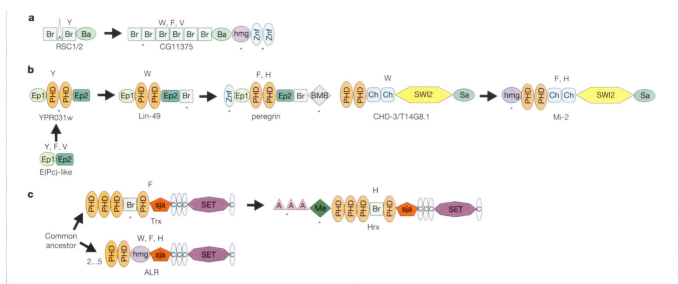

Figure 42 Examples of domain accretion in chromatin proteins. Domain accretion in various lineages before the animal divergence, in the apparent coelomate lineage and the vertebrate lineage are shown using schematic representations of domain architectures (not to scale). Asterisks, mobile domains that have participated in the accretion. Species in which a domain architecture has been identified are indicated above the diagram (Y, yeast; W, worm; F, fly; V, vertebrate). Protein names are below the diagrams. The domains are SET, a chromatin protein methyltransferase domain; SWI2, a superfamily II helicase/ATPase domain; Sa, sant domain; Br, bromo domain; Ch, chromodomain; C, a cysteine triad motif associated with the Msl-2 and SET domains; A, AT hook motif; EP1/EP2, enhancer of polycomb domains 1 and 2; Znf, zinc finger; sja, SET-JOR-associated domain (L. Aravind, unpublished); Me, DNA methylase/Hrx-associated DNA binding zinc finger; Ba, bromo-associated homology motif. **a–c**, Different examples of accretion.

vertebrate innovation lies in the expansion of protein families. Table 25 shows the most prevalent protein domains and protein families in humans, together with their relative ranks in the other species. About 60% of families are more numerous in the human than in any of the other four organisms. This shows that gene duplication has been a major evolutionary force during vertebrate evolution. A comparison of relative expansions in human versus fly is shown in Fig. 44.

Many of the families that are expanded in human relative to fly and worm are involved in distinctive aspects of vertebrate physiology. An example is the family of immunoglobulin (IG) domains, first identified in antibodies thirty years ago. Classic (as opposed to divergent) IG domains are completely absent from the yeast and mustard weed proteomes and, although prokaryotic homologues exist, they have probably been transferred horizontally from metazoans[341]. Most IG superfamily proteins in invertebrates are cell-surface proteins. In vertebrates, the IG repertoire includes immune functions such as those of antibodies, MHC proteins, antibody receptors and many lymphocyte cell-surface proteins. The large expansion of IG domains in vertebrates shows the versatility of a single family in evoking rapid and effective response to infection.

Two prominent families are involved in the control of development. The human genome contains 30 fibroblast growth factors (FGFs), as opposed to two FGFs each in the fly and worm. It contains 42 transforming growth factor-βs (TGFβs) compared with nine and six in the fly and worm, respectively. These growth factors are involved in organogenesis, such as that of the liver and the lung. A fly FGF protein, branchless, is involved in developing respiratory organs (tracheae) in embryos[342]. Thus, developmental triggers of morphogenesis in vertebrates have evolved from related but simpler systems in invertebrates[343].

Another example is the family of intermediate filament proteins, with 127 family members. This expansion is almost entirely due to 111 keratins, which are chordate-specific intermediate filament proteins that form filaments in epithelia. The large number of human keratins suggests multiple cellular structural support roles for the many specialized epithelia of vertebrates.

Finally, the olfactory receptor genes comprise a huge gene family of about 1,000 genes and pseudogenes[344,345]. The number of olfactory receptors testifies to the importance of the sense of smell in vertebrates. A total of 906 olfactory receptor genes and pseudogenes could be identified in the draft genome sequence, two-thirds of which were not previously annotated. About 80% are found in about two dozen clusters ranging from 6 to 138 genes and encompassing about 30 Mb (~1%) of the human genome. Despite the importance of smell among our vertebrate ancestors, hominids

a Conserved domain architectures in chromatin proteins

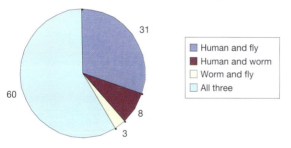

b Conserved domain architectures in apoptotic proteins

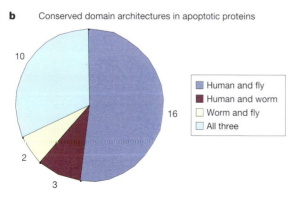

Figure 43 Conservation of architectures between animal species. The pie charts illustrate the shared domain architectures of apparent orthologues that are conserved in at least two of the three sequenced animal genomes. If an architecture was detected in fungi or plants, as well as two of the animal lineages, it was omitted as ancient and its absence in the third animal lineage attributed to gene loss. **a**, Chromatin-associated proteins. **b**, Components of the programmed cell death system.

appear to have considerably less interest in this sense. About 60% of the olfactory receptors in the draft genome sequence have disrupted ORFs and appear to be pseudogenes, consistent with recent reports[344,346] suggesting massive functional gene loss in the last 10 Myr[347,348]. Interestingly, there appears to be a much higher proportion of intact genes among class I than class II olfactory receptors, suggesting functional importance.

Vertebrates are not unique in employing gene family expansion. For many domain types, expansions appear to have occurred independently in each of the major eukaryotic lineages. A good example is the classical C2H2 family of zinc finger domains, which have expanded independently in the yeast, worm, fly and human lineages (Fig. 45). These independent expansions have resulted in numerous C2H2 zinc finger domain-containing proteins that are specific to each lineage. In flies, the important components of the C2H2 zinc finger expansion are architectures in which it is combined with the POZ domain and the C4DM domain (a metal-binding domain found only in fly). In humans, the most prevalent expansions are combinations of the C2H2 zinc finger with POZ (independent of the one in insects) and the vertebrate-specific KRAB and SCAN domains.

The homeodomain is similarly expanded in all animals and is present in both architectures that are conserved and lineage-specific architectures (Fig. 45). This indicates that the ancestral animal probably encoded a significant number of homeodomain proteins, but subsequent evolution involved multiple, independent expansions and domain shuffling after lineages diverged. Thus, the most prevalent transcription factor families are different in worm, fly and human (Fig. 45). This has major biological implications because transcription factors are critical in animal development and differ-

entiation. The emergence of major variations in the developmental body plans that accompanied the early radiation of the animals[349] could have been driven by lineage-specific proliferation of such transcription factors. Beyond these large expansions of protein families, protein components of particular functional systems such as the cell death signalling system show a general increase in diversity and numbers in the vertebrates relative to other animals. For example, there are greater numbers of and more novel architectures in cell death regulatory proteins such as BCL-2, TNFR and NFκB from vertebrates.

Conclusion. Five lines of evidence point to an increase in the complexity of the proteome from the single-celled yeast to the multicellular invertebrates and to vertebrates such as the human. Specifically, the human contains greater numbers of genes, domain and protein families, paralogues, multidomain proteins with multiple functions, and domain architectures. According to these measures, the relatively greater complexity of the human proteome is a consequence not simply of its larger size, but also of large-scale protein innovation.

An important question is the extent to which the greater phenotypic complexity of vertebrates can be explained simply by two- or threefold increases in proteome complexity. The real explanation may lie in combinatorial amplification of these modest differences, by mechanisms that include alternative splicing, post-translational modification and cellular regulatory networks. The potential numbers of different proteins and protein–protein interactions are vast, and their actual numbers cannot readily be discerned from the genome sequence. Elucidating such system-level properties presents one of the great challenges for modern biology.

Table 25 The most populous InterPro families in the human proteome and other species

InterPro ID	Human		Fly		Worm		Yeast		Mustard weed		
	No. of genes	Rank	No. of genes	Rank	No. of genes	Rank	No. of genes	Rank	No. of genes	Rank	
IPR003006	765	(1)	140	(9)	64	(34)	0	(na)	0	(na)	Immunoglobulin domain
PR000822	706	(2)	357	(1)	151	(10)	48	(7)	115	(20)	C2H2 zinc finger
IPR000719	575	(3)	319	(2)	437	(2)	121	(1)	1049	(1)	Eukaryotic protein kinase
IPR000276	569	(4)	97	(14)	358	(3)	0	(na)	16	(84)	Rhodopsin-like GPCR superfamily
IPR001687	433	(5)	198	(4)	183	(7)	97	(2)	331	(5)	P-loop motif
IPR000477	350	(6)	10	(65)	50	(41)	6	(36)	80	(35)	Reverse transcriptase (RNA-dependent DNA polymerase)
IPR000504	300	(7)	157	(6)	96	(21)	54	(6)	255	(8)	rrm domain
IPR001680	277	(8)	162	(5)	102	(19)	91	(3)	210	(10)	G-protein β WD-40 repeats
IPR002110	276	(9)	105	(13)	107	(17)	19	(23)	120	(18)	Ankyrin repeat
IPR001356	267	(10)	148	(7)	109	(15)	9	(33)	118	(19)	Homeobox domain
IPR001849	252	(11)	77	(22)	71	(31)	27	(17)	27	(73)	PH domain
IPR002048	242	(12)	111	(12)	81	(25)	15	(27)	167	(12)	EF-hand family
IPR000561	222	(13)	81	(20)	113	(14)	0	(na)	17	(83)	EGF-like domain
IPR001452	215	(14)	72	(23)	62	(35)	25	(18)	3	(97)	SH3 domain
IPR001841	210	(15)	114	(11)	126	(12)	35	(12)	379	(4)	RING finger
IPR001611	188	(16)	115	(10)	54	(38)	7	(35)	392	(2)	Leucine-rich repeat
IPR001909	171	(17)	0	(na)	0	(na)	0	(na)	0	(na)	KRAB box
IPR001777	165	(18)	63	(27)	51	(40)	2	(40)	4	(96)	Fibronectin type III domain
IPR001478	162	(19)	70	(24)	66	(33)	2	(40)	15	(85)	PDZ domain
IPR001650	155	(20)	87	(17)	78	(27)	79	(4)	148	(13)	Helicase C-terminal domain
IPR001440	150	(21)	86	(18)	46	(43)	36	(11)	125	(17)	TPR repeat
IPR002216	133	(22)	65	(26)	99	(20)	2	(40)	31	(69)	Ion transport protein
IPR001092	131	(23)	84	(19)	41	(46)	7	(35)	106	(24)	Helix–loop–helix DNA-binding domain
IPR000008	123	(24)	43	(34)	36	(49)	9	(33)	82	(34)	C2 domain
IPR001664	119	(25)	4	(71)	22	(63)	1	(41)	2	(98)	SH2 domain
IPR001254	118	(26)	210	(3)	12	(73)	1	(41)	15	(85)	Serine protease, trypsin family
IPR002126	114	(27)	19	(56)	16	(69)	0	(na)	0	(na)	Cadherin domain
IPR000210	113	(28)	78	(21)	117	(13)	1	(41)	54	(50)	BTB/POZ domain
IPR000387	112	(29)	35	(40)	108	(16)	12	(30)	21	(79)	Tyrosine-specific protein phosphatase and dual specificity protein phosphatase family
IPR000087	106	(30)	18	(57)	169	(9)	0	(na)	5	(95)	Collagen triple helix repeat
IPR000379	94	(31)	141	(8)	134	(11)	40	(10)	194	(11)	Esterase/lipase/thioesterase
IPR000910	89	(32)	38	(38)	18	(67)	8	(34)	18	(82)	HMG1/2 (high mobility group) box
IPR000130	87	(33)	56	(29)	92	(22)	8	(34)	12	(88)	Neutral zinc metallopeptidase
IPR001965	84	(34)	37	(39)	24	(61)	16	(26)	71	(39)	PHD-finger
IPR000636	83	(35)	32	(43)	24	(61)	1	(41)	14	(86)	Cation channels (non-ligand gated)
IPR001781	81	(36)	38	(38)	36	(49)	4	(38)	8	(92)	LIM domain
IPR002035	81	(36)	8	(67)	45	(44)	3	(39)	17	(83)	VWA domain
IPR001715	80	(37)	33	(42)	30	(55)	3	(39)	18	(82)	Calponin homology domain
IPR000198	77	(38)	20	(55)	20	(65)	10	(32)	9	(91)	RhoGAP domain

Forty most populous Interpro families found in the human proteome compared with equivalent numbers from other species. na, not applicable (used when there are no proteins in an organism in that family).

Segmental history of the human genome

In bacteria, genomic segments often convey important information about function: genes located close to one another often encode proteins in a common pathway and are regulated in a common operon. In mammals, genes found close to each other only rarely have common functions, but they are still interesting because they have a common history. In fact, the study of genomic segments can shed light on biological events as long as 500 Myr ago and as recently as 20,000 years ago.

Conserved segments between human and mouse

Humans and mice shared a common ancestor about 100 Myr ago. Despite the 200 Myr of evolutionary distance between the species, a significant fraction of genes show synteny between the two, being preserved within conserved segments. Genes tightly linked in one mammalian species tend to be linked in others. In fact, conserved segments have been observed in even more distant species: humans show conserved segments with fish[350,351] and even with invertebrates such as fly and worm[352]. In general, the likelihood that a syntenic relationship will be disrupted correlates with the physical distance between the loci and the evolutionary distance between the species.

Studying conserved segments between human and mouse has several uses. First, conservation of gene order has been used to identify likely orthologues between the species, particularly when investigating disease phenotypes. Second, the study of conserved segments among genomes helps us to deduce evolutionary ancestry.

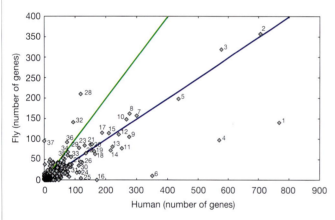

Figure 44 Relative expansions of protein families between human and fly. These data have not been normalized for proteomic size differences. Blue line, equality between normalized family sizes in the two organisms. Green line, equality between unnormalized family sizes. Numbered InterPro entries: (1) immunoglobulin domain [IPR003006]; (2) zinc finger, C2H2 type [IPR000822]; (3) eukaryotic protein kinase [IPR000719]; (4) rhodopsin-like GPCR superfamily [IPR000276]; (5) ATP/GTP-binding site motif A (P-loop) [IPR001687]; (6) reverse transcriptase (RNA-dependent DNA polymerase) [IPR000477]; (7) RNA-binding region RNP-1 (RNA recognition motif) [IPR000504]; (8) G-proteinβ WD-40 repeats [IPR001680]; (9) ankyrin repeat [IPR002110]; (10) homeobox domain [IPR001356]; (11) PH domain [IPR001849]; (12) EF-hand family [IPR002048]; (13) EGF-like domain [IPR000561]; (14) Src homology 3 (SH3) domain [IPR001452]; (15) RING finger [IPR001841]; (16) KRAB box [IPR001909]; (17) leucine-rich repeat [IPR001611]; (18) fibronectin type III domain [IPR001777]; (19) PDZ domain (also known as DHR or GLGF) [IPR001478]; (20) TPR repeat [IPR001440]; (21) helicase C-terminal domain [IPR001650]; (22) ion transport protein [IPR002216]; (23) helix–loop–helix DNA-binding domain [IPR001092]; (24) cadherin domain [IPR002126]; (25) intermediate filament proteins [IPR001664]; (26) C2 domain [IPR000008]; (27) Src homology 2 (SH2) domain [IPR000980]; (28) serine proteases, trypsin family [IPR001254]; (29) BTB/POZ domain [IPR000210]; (30) tyrosine-specific protein phosphatase and dual specificity protein phosphatase family [IPR000387]; (31) collagen triple helix repeat [IPR000087]; (32) esterase/lipase/thioesterase [IPR000379]; (33) neutral zinc metallopeptidases, zinc-binding region [IPR000130]; (34) ATP-binding transport protein, 2nd P-loop motif [IPR001051]; (35) ABC transporters family [IPR001617]; (36) cytochrome P450 enzyme [IPR001128]; (37) insect cuticle protein [IPR000618].

And third, detailed comparative maps may assist in the assembly of the mouse sequence, using the human sequence as a scaffold.

Two types of linkage conservation are commonly described[353]. 'Conserved synteny' indicates that at least two genes that reside on a common chromosome in one species are also located on a common chromosome in the other species. Syntenic loci are said to lie in a 'conserved segment' when not only the chromosomal position but the linear order of the loci has been preserved, without interruption by other chromosomal rearrangements.

An initial survey of homologous loci in human and mouse[354] suggested that the total number of conserved segments would be about 180. Subsequent estimates based on increasingly detailed comparative maps have remained close to this projection[353,355,356] (http://www.informatics.jax.org). The distribution of segment lengths has corresponded reasonably well to the truncated negative exponential curve predicted by the random breakage model[357].

The availability of a draft human genome sequence allows the first global human–mouse comparison in which human physical distances can be measured in Mb, rather than cM or orthologous gene counts. We identified likely orthologues by reciprocal comparison of the human and mouse mRNAs in the LocusLink database, using megaBLAST. For each orthologous pair, we mapped the location of the human gene in the draft genome sequence and then checked the location of the mouse gene in the Mouse Genome Informatics database (http://www.informatics.jax.org). Using a conservative threshold, we identified 3,920 orthologous pairs in which the human gene could be mapped on the draft genome sequence with high confidence. Of these, 2,998 corresponding mouse genes had a known position in the mouse genome. We then searched for definitive conserved segments, defined as human regions containing orthologues of at least two genes from the same mouse chromosome region (< 15 cM) without interruption by segments from other chromosomes.

We identified 183 definitive conserved segments (Fig. 46). The average segment length was 15.4 Mb, with the largest segment being 90.5 Mb and the smallest 24 kb. There were also 141 'singletons', segments that contained only a single locus; these are not counted in the statistics. Although some of these could be short conserved segments, they could also reflect incorrect choices of orthologues or problems with the human or mouse maps. Because of this conservative approach, the observed number of definitive segments is likely be lower than the correct total. One piece of evidence for this conclusion comes from a more detailed analysis on human chromosome 7 (ref. 358), which identified 20 conserved segments, of which three were singletons. Our analysis revealed only 13 definitive segments on this chromosome, with nine singletons.

The frequency of observing a particular gene count in a conserved segment is plotted on a logarithmic scale in Fig. 47. If chromosomal breaks occur in a random fashion (as has been proposed) and differences in gene density are ignored, a roughly straight line should result. There is a clear excess for $n = 1$, suggesting that 50% or more of the singletons are indeed artefactual. Thus, we estimate that true number of conserved segments is around 190–230, in good agreement with the original Nadeau–Taylor prediction[354].

Figure 48 shows a plot of the frequency of lengths of conserved segments, where the x-axis scale is shown in Mb. As before, there is a fair amount of scatter in the data for the larger segments (where the numbers are small), but the trend appears to be consistent with a random breakage model.

We attempted to ascertain whether the breakpoint regions have any special characteristics. This analysis was complicated by imprecision in the positioning of these breaks, which will tend to blur any relationships. With 2,998 orthologues, the average interval within which a break is known to have occurred is about 1.1 Mb. We compared the aggregate features of these breakpoint intervals with the genome as a whole. The mean gene density was lower in breakpoint regions than in the conserved segments (13.8 versus

18.6 per Mb). This suggests that breakpoints may be more likely to occur or to undergo fixation in gene-poor intervals than in gene-rich intervals. The occurrence of breakpoints may be promoted by homologous recombination among repeated sequences[359]. When the sequence of the mouse genome is finished, this analysis can be revisited more precisely.

A number of examples of extended conserved segments and syntenies are apparent in Fig. 46. As has been noted, almost all human genes on chromosome 17 are found on mouse chromosome 11, with two members of the placental lactogen family from mouse 13 inserted. Apart from two singleton loci, human chromosome 20 appears to be entirely orthologous to mouse chromosome 2, apparently in a single segment. The largest apparently contiguous conserved segment in the human genome is on chromosome 4, including roughly 90.5 Mb of human DNA that is orthologous to mouse chromosome 5. This analysis also allows us to infer the likely location of thousands of mouse genes for which the human orthologue has been located in the draft genome sequence but the mouse locus has not yet been mapped.

With about 200 conserved segments between mouse and human and about 100 Myr of evolution from their common ancestor[360], we obtain an estimated rate of about 1.0 chromosomal rearrangement being fixed per Myr. However, there is good evidence that the rate of chromosomal rearrangement (like the rate of nucleotide substitutions; see above) differs between the two species. Among mammals, rodents may show unusually rapid chromosome alteration. By comparison, very few rearrangements have been observed among primates, and studies of a broader array of mammalian orders, including cats, cows, sheep and pigs, suggest an average rate of chromosome alteration of only about 0.2 rearrangements per Myr

in these lineages[361]. Additional evidence that rodents are outliers comes from a recent analysis of synteny between the human and zebrafish genomes. From a study of 523 orthologues, it was possible to project 418 conserved segments[350]. Assuming 400 Myr since a common vertebrate ancestor of zebrafish and humans[362], we obtain an estimate of 0.52 rearrangements per Myr. Recent estimates of rearrangement rates in plants have suggested bimodality, with some pairs showing rates of 0.15–0.41 rearrangements per Myr, and others showing higher rates of 1.1–1.3 rearrangements per Myr[363]. With additional detailed genome maps of multiple species, it should be possible to determine whether this particular molecular clock is truly operating at a different rate in various branches of the evolutionary tree, and whether variations in that rate are bimodal or continuous. It should also be possible to reconstruct the karyotypes of common ancestors.

Ancient duplicated segments in the human genome

Another approach to genomic history is to study segmental duplications within the human genome. Earlier, we discussed examples of recent duplications of genomic segments to pericentromeric and subtelomeric regions. Most of these events appear to be evolutionary dead-ends resulting in nonfunctional pseudogenes; however, segmental duplication is also an important mode of evolutionary innovation: a duplication permits one copy of each gene to drift and potentially to acquire a new function.

Segmental duplications can occur through unequal crossing over to create gene families in specific chromosomal regions. This mechanism can create both small families, such as the five related genes of the β-globin cluster on chromosome 11, and large ones, such as the olfactory receptor gene clusters, which together contain nearly 1,000 genes and pseudogenes.

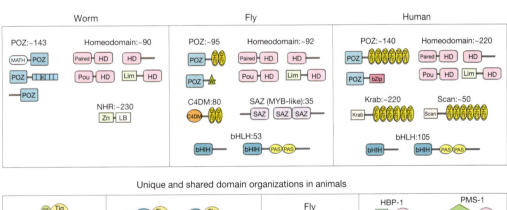

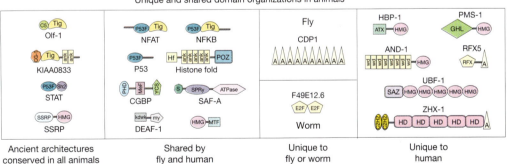

Figure 45 Lineage-specific expansions of domains and architectures of transcription factors. Top, specific families of transcription factors that have been expanded in each of the proteomes. Approximate numbers of domains identified in each of the (nearly) complete proteomes representing the lineages are shown next to the domains, and some of the most common architectures are shown. Some are shared by different animal lineages; others are lineage-specific. Bottom, samples of architectures from transcription factors that are shared by all animals (ancient architectures), shared by fly and human and unique to each lineage. Domains: K, kelch; HD, homeodomain; Zn, zinc-binding domain; LB, ligand-binding domain; C4DM, novel Zn cluster with four cysteines, probably involved in protein–protein interactions (L. Aravind, unpublished); MATH, meprin-associated TRAF

domain; CG-1, novel domain in KIAA0909-like transcription factors (L. Aravind, unpublished); MTF, myelin transcription factor domain; SAZ, specialized Myb-like helix-turn-helix (HTH) domain found in Stonewall, ADF-1 and Zeste (L. Aravind, unpublished); A, AT-hook motif; E2F, winged HTH DNA-binding domain; GHL, gyraseB-histidine kinase-MutL ATPase domain; ATX, ATaXin domain; RFX, RFX winged HTH DNA binding domain; My, MYND domain; KDWK, KDWK DNA-binding domain; POZ, Pox zinc finger domain; S, SAP domain; P53F, P53 fold domain; HF, histone fold; ANK, ankyrin repeat; TIG, transcription factor Ig domain; SSRP, structure-specific recognition protein domain; C5, 5-cysteine metal binding domain; C2H2, classic zinc finger domain; WD, WD40 repeats.

The most extreme mechanism is whole-genome duplication (WGD), through a polyploidization event in which a diploid organism becomes tetraploid. Such events are classified as autopolyploidy or allopolyploidy, depending on whether they involve hybridization between members of the same species or different species. Polyploidization is common in the plant kingdom, with many known examples among wild and domesticated crop species. Alfalfa (*Medicago sativa*) is a naturally occurring autotetraploid[364], and *Nicotiana tabacum*, some species of cotton (*Gossypium*) and several of the common brassicas are allotetraploids containing pairs of 'homeologous' chromosome pairs.

In principle, WGD provides the raw material for great bursts of innovation by allowing the duplication and divergence of entire pathways. Ohno[365] suggested that WGD has played a key role in evolution. There is evidence for an ancient WGD event in the ancestry of yeast and several independent such events in the ancestry of mustard weed[366–369]. Such ancient WGD events can be hard to detect because only a minority of the duplicated loci may be retained, with the result that the genes in duplicated segments cannot be aligned in a one-to-one correspondence but rather require many gaps. In addition, duplicated segments may be subsequently rearranged. For example, the ancient duplication in the yeast genome appears to have been followed by loss of more than 90% of the newly duplicated genes[366].

One of the most controversial hypotheses about vertebrate evolution is the proposal that two WGD events occurred early in the vertebrate lineage, around the time of jawed fishes some 500 Myr ago. Some authors[370–373] have seen support for this theory in the fact that many human genes occur in sets of four homologues—most notably the four extensive HOX gene clusters on chromosomes 2, 7, 12 and 17, whose duplication dates to around the correct time. However, other authors have disputed this interpretation[374], suggesting that these cases may reflect unrelated duplications of specific regions rather than successive WGD.

We analysed the draft genome sequence for evidence that might bear on this question. The analysis provides many interesting observations, but no convincing evidence of ancient WGD. We looked for evidence of pairs of chromosomal regions containing many homologous genes. Although we found many pairs containing a few homologous genes, the human genome does not appear to contain any pairs of regions where the density of duplicated genes approaches the densities seen in yeast or mustard weed[366–369].

We also examined human proteins in the IPI for which the orthologues among fly or worm proteins occur in the ratios 2:1:1, 3:1:1, 4:1:1 and so on (Fig. 49). The number of such families falls smoothly, with no peak at four and some instances of five or more homologues. Although this does not rule out two rounds of WGD followed by extensive gene loss and some unrelated gene duplication, it provides no support for the theory. More probatively, if two successive rounds of genome duplication occurred, phylogenetic analysis of the proteins having 4:1:1 ratios between human, fly and worm would be expected to show more trees with the topology (A,B)(C,D) for the human sequences than (A,(B,(C,D)))[375]. However, of 57 sets studied carefully, only 24% of the trees constructed from the 4:1:1 set have the former topology; this is not significantly different from what would be expected under the hypothesis of random sequential duplication of individual loci.

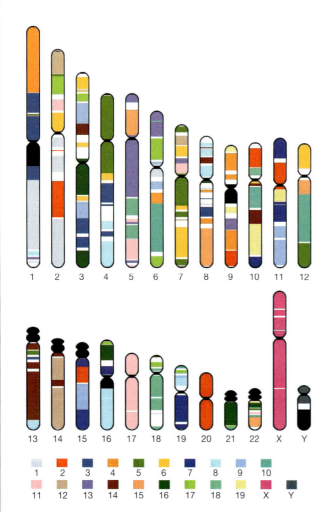

Figure 46 Conserved segments in the human and mouse genome. Human chromosomes, with segments containing at least two genes whose order is conserved in the mouse genome as colour blocks. Each colour corresponds to a particular mouse chromosome. Centromeres, subcentromeric heterochromatin of chromosomes 1, 9 and 16, and the repetitive short arms of 13, 14, 15, 21 and 22 are in black.

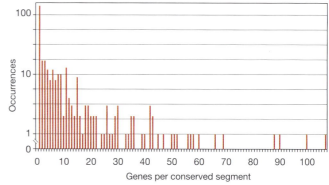

Figure 47 Distribution of number of genes per conserved segment between human and mouse genomes.

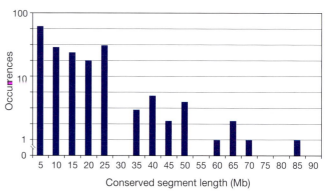

Figure 48 Distribution of lengths (in 5-Mb bins) of conserved segments between human and mouse genomes, omitting singletons.

We also searched for sets of four chromosomes where there are multiple genes with homologues on each of the four. The strongest example was chromosomes 2, 7, 12 and 17, containing the HOX clusters as well as additional genes. These four chromosomes appear to have an excess of quadruplicated genes. The genes are not all clustered in a single region; this may reflect intrachromosomal rearrangement since the duplication of these genes, or it may indicate that they result from several independent events. Of the genes with homologues on chromosomes 2, 12 and 17, many of those missing on chromosome 7 are clustered on chromosome 3, suggesting a translocation. Several additional examples of groups of four chromosomes were found, although they were connected by fewer homologous genes.

Although the analyses are sensitive to the imperfect quality of the gene predictions, our results so far are insufficient to settle whether two rounds of WGD occurred around 500 Myr ago. It may be possible to resolve the issue by systematically estimating the time of each of the many gene duplication events on the basis of sequence divergence, although this is beyond the scope of this report. Another approach to determining whether a widespread duplication occurred at a particular time in vertebrate evolution would be to sequence the genomes of organisms whose lineages diverged from vertebrates at appropriate times, such as amphioxus.

Recent history from human polymorphism

The recent history of genomic segments can be probed by studying the properties of SNPs segregating in the current human population. The sequence information generated in the course of this project has yielded a huge collection of SNPs. These SNPs were extracted in two ways: by comparing overlapping large-insert clones derived from distinct haplotypes (either different individuals or different chromosomes within an individual) and by comparing random reads from whole-genome shotgun libraries derived from multiple individuals. The analysis confirms an average heterozygosity rate in the human population of about 1 in 1,300 bp (ref. 97).

More than 1.42 million SNPs have been assembled into a genome-wide map and are analysed in detail in an accompanying paper[97]. SNP density is also displayed across the genome in Fig. 9. The SNPs have an average spacing of 1.9 kb and 63% of 5-kb intervals contain a SNP. These polymorphisms are of immediate utility for medical genetic studies. Whereas investigators studying a gene previously had to expend considerable effort to discover polymorphisms across the region of interest, the current collection now provides then with about 15 SNPs for gene loci of average size.

The density of SNPs (adjusted for ascertainment—that is, polymorphisms per base screened) varies considerably across the genome[97] and sheds light on the unique properties and history of each genomic region. The average heterozygosity at a locus will tend to increase in proportion to the local mutation rate and the 'age' of the locus (which can be defined as the average number of generations since the most recent common ancestor of two randomly chosen copies in the population). For example, positive selection can cause a locus to be unusually 'young' and balancing selection can cause it to be unusually 'old'. An extreme example is the HLA region, in which a high SNP density is observed, reflecting the fact that diverse HLA haplotypes have been maintained for many millions of years by balancing selection and greatly predate the origin of the human species.

SNPs can also be used to study linkage disequilibrium in the human genome[376]. Linkage disequilibrium refers to the persistence of ancestral haplotypes—that is, genomic segments carrying particular combinations of alleles descended from a common ancestor. It can provide a powerful tool for mapping disease genes[377,378] and for probing population history[379–381]. There has been considerably controversy concerning the typical distance over which linkage disequilibrium extends in the human genome[382–387]. With the collection of SNPs now available, it should be possible to resolve this important issue.

Applications to medicine and biology

In most research papers, the authors can only speculate about future applications of the work. Because the genome sequence has been released on a daily basis over the past four years, however, we can already cite many direct applications. We focus on a handful of applications chosen primarily from medical research.

Disease genes

A key application of human genome research has been the ability to find disease genes of unknown biochemical function by positional cloning[388]. This method involves mapping the chromosomal region containing the gene by linkage analysis in affected families and then scouring the region to find the gene itself. Positional cloning is powerful, but it has also been extremely tedious. When the approach was first proposed in the early 1980s[9], a researcher wishing to perform positional cloning had to generate genetic markers to trace inheritance; perform chromosomal walking to obtain genomic DNA covering the region; and analyse a region of around 1 Mb by either direct sequencing or indirect gene identification methods. The first two barriers were eliminated with the development in the mid-1990s of comprehensive genetic and physical maps of the human chromosomes, under the auspices of the Human Genome Project. The remaining barrier, however, has continued to be formidable.

All that is changing with the availability of the human draft genome sequence. The human genomic sequence in public databases allows rapid identification *in silico* of candidate genes, followed by mutation screening of relevant candidates, aided by information on gene structure. For a mendelian disorder, a gene search can now often be carried out in a matter of months with only a modestly sized team.

At least 30 disease genes[55,389–422] (Table 26) have been positionally cloned in research efforts that depended directly on the publicly available genome sequence. As most of the human sequence has only arrived in the past twelve months, it is likely that many similar discoveries are not yet published. In addition, there are many cases in which the genome sequence played a supporting role, such as providing candidate microsatellite markers for finer genetic linkage analysis.

The genome sequence has also helped to reveal the mechanisms leading to some common chromosomal deletion syndromes. In several instances, recurrent deletions have been found to result from homologous recombination and unequal crossing over between large, nearly identical intrachromosomal duplications. Examples

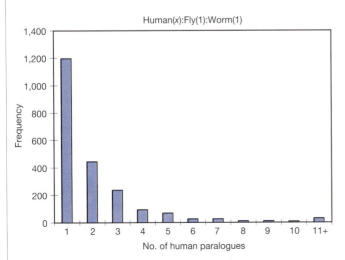

Figure 49 Number of human paralogues of genes having single orthologues in worm and fly.

include the DiGeorge/velocardiofacial syndrome region on chromosome 22 (ref. 238) and the Williams–Beuren syndrome recurrent deletion on chromosome 7 (ref. 239).

The availability of the genome sequence also allows rapid identification of paralogues of disease genes, which is valuable for two reasons. First, mutations in a paralogous gene may give rise to a related genetic disease. A good example, discovered through use of the genome sequence, is achromatopsia (complete colour blindness). The CNGA3 gene, encoding the α-subunit of the cone photoreceptor cyclic GMP-gated channel, had been shown to harbour mutations in some families with achromatopsia. Computational searching of the genome sequences revealed the paralogous gene encoding the corresponding β-subunit, CNGB3 (which had not been apparent from EST databases). The CNGB3 gene was rapidly shown to be the cause of achromatopsia in other families[407,408]. Another example is provided by the presenilin-1 and presenilin-2 genes, in which mutations can cause early-onset Alzheimer's disease[423,424]. Second, the paralogue may provide an opportunity for therapeutic intervention, as exemplified by attempts to reactivate the fetally expressed haemoglobin genes in individuals with sickle cell disease or β-thalassaemia, caused by mutations in the β-globin gene[425].

We undertook a systematic search for paralogues of 971 known human disease genes with entries in both the Online Mendelian Inheritance in Man (OMIM) database (http://www.ncbi.nlm.nih.gov/Omim/) and either the SwissProt or TrEMBL protein databases. We identified 286 potential paralogues (with the requirement of a match of at least 50 amino acids with identity greater than 70% but less than 90% if on the same chromosome, and less than 95% if on a different chromosome). Although this analysis may have identified some pseudogenes, 89% of the matches showed homology over more than one exon in the new target sequence, suggesting that many are functional. This analysis shows the potential for rapid identification of disease gene paralogues in silico.

Drug targets

Over the past century, the pharmaceutical industry has largely depended upon a limited set of drug targets to develop new therapies. A recent compendium[426,427] lists 483 drug targets as accounting for virtually all drugs on the market. Knowing the complete set of human genes and proteins will greatly expand the search for suitable drug targets. Although only a minority of human genes may be drug targets, it has been predicted that the number will exceed several thousand, and this prospect has led to a massive expansion of genomic research in pharmaceutical research and development. A few examples will illustrate the point.

(1) The neurotransmitter serotonin (5-HT) mediates rapid excitatory responses through ligand-gated channels. The previously identified 5-HT$_{3A}$ receptor gene produces functional receptors, but with a much smaller conductance than observed in vivo. Cross-hybridization experiments and analysis of ESTs failed to reveal any other homologues of the known receptor. Recently, however, by searching the human draft genome sequence at low stringency, a putative homologue was identified within a PAC clone from the long arm of chromosome 11 (ref. 428). The homologue was shown to be expressed in the amygdala, caudate and hippocampus, and a full-length cDNA was subsequently obtained. The gene, which codes for a serotonin receptor, was named 5-HT$_{3B}$. When assembled in a heterodimer with 5-HT$_{3A}$, it was shown to account for the large-conductance neuronal serotonin channel. Given the central role of the serotonin pathway in mood disorders and schizophrenia, the discovery of a major new therapeutic target is of considerable interest.

(2) The contractile and inflammatory actions of the cysteinyl leukotrienes, formerly known as the slow reacting substance of anaphylaxis (SRS-A), are mediated through specific receptors. The second such receptor, CysLT$_2$, was identified using the combination of a rat EST and the human genome sequence. This led to the cloning of a gene with 38% amino-acid identity to the only other receptor that had previously been identified[429]. This new receptor, which shows high-affinity binding to several leukotrienes, maps to a region of chromosome 13 that is linked to atopic asthma. The gene is expressed in airway smooth muscles and in the heart. As the leukotriene pathway has been a significant target for the development of drugs against asthma, the discovery of a new receptor has obvious and important consequences.

(3) Abundant deposition of β-amyloid in senile plaques is the hallmark of Alzheimer's disease. β-Amyloid is generated by proteolytic processing of the amyloid precursor protein (APP). One of the enzymes involved is the β-site APP-cleaving enzyme (BACE), which is a transmembrane aspartyl protease. Computational searching of the public human draft genome sequence recently identified a new sequence homologous to BACE, encoding a protein now named BACE2[430,431]. BACE2, which has 52% amino-acid sequence identity to BACE, contains two active protease sites and maps to the obligatory Down's syndrome region of chromosome 21, as does APP. This raises the question of whether the extra copies of both BACE2 and APP may contribute to accelerated deposition of β-amyloid in the brains of Down's syndrome patients. The development of antagonists to BACE and BACE2 represents a promising approach to preventing Alzheimer's disease.

Given these examples, we undertook a systematic effort to identify paralogues of the classic drug target proteins in the draft genome sequence. The target list[427] was used to identify 603 entries in the SwissProt database with unique accession numbers. These were then searched against the current genome sequence database, using the requirement that a match should have 70–100% identity to at least 50 amino acids. Matches to named proteins were ignored, as we assumed that these represented known homologues.

We found 18 putative novel paralogues (Table 27), including apparent dopamine receptors, purinergic receptors and insulin-like growth factor receptors. In six cases, the novel paralogue matches at least one EST, adding confidence that this search process can identify novel functional genes. For the remaining 12 putative paralogues without an EST match, all have long ORFs and all but

Table 26 Disease genes positionally cloned using the draft genome sequence

Locus	Disorder	Reference(s)
BRCA2	Breast cancer susceptibility	55
AIRE	Autoimmune polyglandular syndrome type 1 (APS1 or APECED)	389
PEX1	Peroxisome biogenesis disorder	390, 391
PDS	Pendred syndrome	392
XLP	X-linked lymphoproliferative disease	393
DFNA5	Nonsyndromic deafness	394
ATP2A2	Darier's disease	395
SEDL	X-linked spondyloepiphyseal dysplasia tarda	396
WISP3	Progressive pseudorheumatoid dysplasia	397
CCM1	Cerebral cavernous malformations	398, 399
COL11A2/DFNA13	Nonsyndromic deafness	400
LGMD 2G	Limb-girdle muscular dystrophy	401
EVC	Ellis-Van Creveld syndrome, Weyer's acrodental dysostosis	402
ACTN4	Familial focal segmental glomerulosclerosis	403
SCN1A	Generalized epilepsy with febrile seizures plus type 2	404
AASS	Familial hyperlysinaemia	405
NDRG1	Hereditary motor and sensory neuropathy-Lom	406
CNGB3	Total colour-blindness	407, 408
MUL	Mulibrey nanism	409
USH1C	Usher type 1C	410, 411
MYH9	May-Hegglin anomaly	412, 413
PRKAR1A	Carney's complex	414
MYH9	Nonsyndromic hereditary deafness DFNA17	415
SCA10	Spinocerebellar ataxia type 10	416
OPA1	Optic atrophy	417
XLCSNB	X-linked congenital stationary night blindness	418
FGF23	Hypophosphataemic rickets	419
GAN	Giant axonal neuropathy	420
AAAS	Triple-A syndrome	421
HSPG2	Schwartz-Jampel syndrome	422

one show similarity spanning multiple exons separated by introns, so these are not processed pseudogenes. They are likely to represent interesting new candidate drug targets.

Basic biology

Although the examples above reflect medical applications, there are also many similar applications to basic physiology and cell biology. To cite one satisfying example, the publicly available sequence was used to solve a mystery that had vexed investigators for several decades: the molecular basis of bitter taste[432]. Humans and other animals are polymorphic for response to certain bitter tastes. Recently, investigators mapped this trait in both humans and mice and then searched the relevant region of the human draft genome sequence for G-protein coupled receptors. These studies led, in quick succession, to the discovery of a new family of such proteins, the demonstration that they are expressed almost exclusively in taste buds, and the experimental confirmation that the receptors in cultured cells respond to specific bitter substances[433–435].

The next steps

Considerable progress has been made in human sequencing, but much remains to be done to produce a finished sequence. Even more work will be required to extract the full information contained in the sequence. Many of the key next steps are already underway.

Finishing the human sequence

The human sequence will serve as a foundation for biomedical research in the years ahead, and it is thus crucial that the remaining gaps be filled and ambiguities be resolved as quickly as possible. This will involve a three-step program.

The first stage involves producing finished sequence from clones spanning the current physical map, which covers more than 96% of the euchromatic regions of the genome. About 1 Gb of finished sequence is already completed. Almost all of the remaining clones are already sequenced to at least draft coverage, and the rest have been selected for sequencing. All clones are expected to reach 'full shotgun' coverage (8–10-fold redundancy) by about mid-2001 and finished form (99.99% accuracy) not long thereafter, using established and increasingly automated protocols.

The next stage will be to screen additional libraries to close gaps between clone contigs. Directed probing of additional large-insert clone libraries should close many of the remaining gaps. Unclosed gaps will be sized by FISH techniques or other methods. Two chromosomes, 22 and 21, have already been assembled in this 'essentially complete' form in this manner[93,94], and chromosomes 20, Y, 19, 14 and 7 are likely to reach this status in the next few

months. All chromosomes should be essentially completed by 2003, if not sooner.

Finally, techniques must be developed to close recalcitrant gaps. Several hundred such gaps in the euchromatic sequence will probably remain in the genome after exhaustive screening of existing large-insert libraries. New methodologies will be needed to recover sequence from these segments, and to define biological reasons for their lack of representation in standard libraries. Ideally, it would be desirable to obtain complete sequence from all heterochromatic regions, such as centromeres and ribosomal gene clusters, although most of this sequence will consist of highly polymorphic tandem repeats containing few protein-coding genes.

Developing the IGI and IPI

The draft genome sequence has provided an initial look at the human gene content, but many ambiguities remain. A high priority will be to refine the IGI and IPI to the point where they accurately reflect every gene and every alternatively spliced form. Several steps are needed to reach this ambitious goal.

Finishing the human sequence will assist in this effort, but the experiences gained on chromosomes 21 and 22 show that sequence alone is not enough to allow complete gene identification. One powerful approach is cross-species sequence comparison with related organisms at suitable evolutionary distances. The sequence coverage from the pufferfish *T. nigroviridis* has already proven valuable in identifying potential exons[292]; this work is expected to continue from its current state of onefold coverage to reach at least fivefold coverage later this year. The genome sequence of the laboratory mouse will provide a particularly powerful tool for exon identification, as sequence similarity is expected to identify 95–97% of the exons, as well as a significant number of regulatory domains[436–438]. A public-private consortium is speeding this effort, by producing freely accessible whole-genome shotgun coverage that can be readily used for cross-species comparison[439]. More than onefold coverage from the C57BL/6J strain has already been completed and threefold is expected within the next few months. In the slightly longer term, a program is under way to produce a finished sequence of the laboratory mouse.

Another important step is to obtain a comprehensive collection of full-length human cDNAs, both as sequences and as actual clones. The Mammalian Gene Collection project has been underway for a year[18] and expects to produce 10,000–15,000 human full-length cDNAs over the coming year, which will be available without restrictions on use. The Genome Exploration Group of the RIKEN Genomic Sciences Center is similarly developing a collection of cDNA clones from mouse[309], which is a valuable complement

Table 27 New paralogues of common drug targets identified by searching the draft human genome sequence

Drug target	Drug target			Novel match IGI number	Chromosome containing paralogue	Per cent identity	dbEST GenBank accession
	HGM symbol	SwissProt accession	Chromosome				
Aquaporin 7	AQP7	O14520	9	IGI_M1_ctg15869_11	2	92.3	AW593324
Arachidonate 12-lipoxygenase	ALOX12	P18054	17	IGI_M1_ctg17216_23	17	70.1	
Calcitonin	CALCA	P01258	11	IGI_M1_ctg14138_20	12	93.6	
Calcium channel, voltage-dependent, γ-subunit	CACNG2	Q9Y698	22	IGI_M1_ctg17137_10	19	70.7	
DNA polymerase-δ, small subunit	POLD2	P49005	7	IGI_M1_ctg12903_29	5	86.8	
Dopamine receptor, D1-α	DRD1	P21728	5	IGI_M1_ctg25203_33	16	70.7	
Dopamine receptor, D1-β	DRD5	P21918	4	IGI_M1_ctg17190_14	1	88.0	AI148329
Eukaryotic translation elongation factor, 1δ	EEF1D	P29692	2	IGI_M1_ctg16401_37	17	77.6	BE719683
FKBP, tacrolimus binding protein, FK506 binding protein	FKBP1B	Q16645	2	IGI_M1_ctg14291_56	6	79.4	
Glutamic acid decarboxylase	GAD1	Q99259	2	IGI_M1_ctg12341_103	18	70.5	
Glycine receptor, α1	GLRA1	P23415	5	IGI_M1_ctg16547_14	X	85.5	
Heparan N-deacetylase/N-sulphotransferase	NDST1	P52848	5	IGI_M1_ctg13263_18	4	81.5	
Insulin-like growth factor-1 receptor	IGF1R	P08069	15	IGI_M1_ctg18444_3	19	71.8	
Na,K-ATPase, α-subunit	ATP1A1	P05023	1	IGI_M1_ctg14877_54	1	83	
Purinergic receptor 7 (P2X), ligand-gated ion channel	P2RX7	Q99572	12	IGI_M1_ctg15140_15	12	80.3	H84353
Tubulin, ε-chain	TUBE*	Q9UJT0	6	IGI_M1_ctg13826_4	5	78.5	AA970498
Tubulin, χ-chain	TUBG1	P23258	17	IGI_M1_ctg12599_5	7	84.0	
Voltage-gated potassium channel, KV3.3	KCNC3	Q14003	19	IGI_M1_ctg13492_5	12	80.1	H49142

* HGM symbol unknown.

because of the availability of tissues from all developmental time points. A challenge will be to define the gene-specific patterns of alternative splicing, which may affect half of human genes. Existing collections of ESTs and cDNAs may allow identification of the most abundant of these isoforms, but systematic exploration of this problem may require exhaustive analysis of cDNA libraries from multiple tissues or perhaps high-throughput reverse transcription–PCR studies. Deep understanding of gene function will probably require knowledge of the structure, tissue distribution and abundance of these alternative forms.

Large-scale identification of regulatory regions

The one-dimensional script of the human genome, shared by essentially all cells in all tissues, contains sufficient information to provide for differentiation of hundreds of different cell types, and the ability to respond to a vast array of internal and external influences. Much of this plasticity results from the carefully orchestrated symphony of transcriptional regulation. Although much has been learned about the cis-acting regulatory motifs of some specific genes, the regulatory signals for most genes remain uncharacterized. Comparative genomics of multiple vertebrates offers the best hope for large-scale identification of such regulatory sites[440]. Previous studies of sequence alignment of regulatory domains of orthologous genes in multiple species has shown a remarkable correlation between sequence conservation, dubbed 'phylogenetic footprints'[441], and the presence of binding motifs for transcription factors. This approach could be particularly powerful if combined with expression array technologies that identify cohorts of genes that are coordinately regulated, implicating a common set of cis-acting regulatory sequences[442–445]. It will also be of considerable interest to study epigenetic modifications such as cytosine methylation on a genome-wide scale, and to determine their biological consequences[446,447]. Towards this end, a pilot Human Epigenome Project has been launched[448,449].

Sequencing of additional large genomes

More generally, comparative genomics allows biologists to peruse evolution's laboratory notebook—to identify conserved functional features and recognize new innovations in specific lineages. Determination of the genome sequence of many organisms is very desirable. Already, projects are underway to sequence the genomes of the mouse, rat, zebrafish and the pufferfishes *T. nigroviridis* and *Takifugu rubripes*. Plans are also under consideration for sequencing additional primates and other organisms that will help define key developments along the vertebrate and nonvertebrate lineages.

To realize the full promise of comparative genomics, however, it needs to become simple and inexpensive to sequence the genome of any organism. Sequencing costs have dropped 100-fold over the last 10 years, corresponding to a roughly twofold decrease every 18 months. This rate is similar to 'Moore's law' concerning improvements in semiconductor manufacture. In both sequencing and semiconductors, such improvement does not happen automatically, but requires aggressive technological innovation fuelled by major investment. Improvements are needed to move current dideoxy sequencing to smaller volumes and more rapid sequencing times, based upon advances such as microchannel technology. More revolutionary methods, such as mass spectrometry, single-molecule sequencing and nanopore approaches[76], have not yet been fully developed, but hold great promise and deserve strong encouragement.

Completing the catalogue of human variation

The human draft genome sequence has already allowed the identification of more than 1.4 million SNPs, comprising a substantial proportion of all common human variation. This program should be extended to obtain a nearly complete catalogue of common variants and to identify the common ancestral haplotypes present in the population. In principle, these genetic tools should make it possible to perform association studies and linkage disequilibrium studies[376] to identify the genes that confer even relatively modest risk

for common diseases. Launching such an intense era of human molecular epidemiology will also require major advances in the cost efficiency of genotyping technology, in the collection of carefully phenotyped patient cohorts and in statistical methods for relating large-scale SNP data to disease phenotype.

From sequence to function

The scientific program outlined above focuses on how the genome sequence can be mined for biological information. In addition, the sequence will serve as a foundation for a broad range of functional genomic tools to help biologists to probe function in a more systematic manner. These will need to include improved techniques and databases for the global analysis of: RNA and protein expression, protein localization, protein–protein interactions and chemical inhibition of pathways. New computational techniques will be needed to use such information to model cellular circuitry. A full discussion of these important directions is beyond the scope of this paper.

Concluding thoughts

The Human Genome Project is but the latest increment in a remarkable scientific program whose origins stretch back a hundred years to the rediscovery of Mendel's laws and whose end is nowhere in sight. In a sense, it provides a capstone for efforts in the past century to discover genetic information and a foundation for efforts in the coming century to understand it.

We find it humbling to gaze upon the human sequence now coming into focus. In principle, the string of genetic bits holds long-sought secrets of human development, physiology and medicine. In practice, our ability to transform such information into understanding remains woefully inadequate. This paper simply records some initial observations and attempts to frame issues for future study. Fulfilling the true promise of the Human Genome Project will be the work of tens of thousands of scientists around the world, in both academia and industry. It is for this reason that our highest priority has been to ensure that genome data are available rapidly, freely and without restriction.

The scientific work will have profound long-term consequences for medicine, leading to the elucidation of the underlying molecular mechanisms of disease and thereby facilitating the design in many cases of rational diagnostics and therapeutics targeted at those mechanisms. But the science is only part of the challenge. We must also involve society at large in the work ahead. We must set realistic expectations that the most important benefits will not be reaped overnight. Moreover, understanding and wisdom will be required to ensure that these benefits are implemented broadly and equitably. To that end, serious attention must be paid to the many ethical, legal and social implications (ELSI) raised by the accelerated pace of genetic discovery. This paper has focused on the scientific achievements of the human genome sequencing efforts. This is not the place to engage in a lengthy discussion of the ELSI issues, which have also been a major research focus of the Human Genome Project, but these issues are of comparable importance and could appropriately fill a paper of equal length.

Finally, it is has not escaped our notice that the more we learn about the human genome, the more there is to explore.

"We shall not cease from exploration. And the end of all our exploring will be to arrive where we started, and know the place for the first time."—T. S. Eliot[450] □

Received 7 December 2000; accepted 9 January 2001.

1. Correns, C. Untersuchungen über die Xenien bei Zea mays. *Berichte der Deutsche Botanische Gesellschaft* **17**, 410–418 (1899).
2. De Vries, H. Sur la loie de disjonction des hybrides. *Comptes Rendue Hebdemodaires, Acad. Sci. Paris* **130**, 845–847 (1900).
3. von Tschermack, E. Uber Künstliche Kreuzung bei Pisum sativum. *Berichte der Deutsche Botanische Gesellschaft* **18**, 232–239. (1900).
4. Sanger, F. *et al.* Nucleotide sequence of bacteriophage Φ X174 DNA. *Nature* **265**, 687–695 (1977).
5. Sanger, F. *et al.* The nucleotide sequence of bacteriophage ΦX174. *J Mol Biol* **125**, 225–246 (1978).

6. Sanger, F., Coulson, A. R., Hong, G. F., Hill, D. F. & Petersen, G. B. Nucleotide-sequence of bacteriophage Lambda DNA. *J. Mol. Biol.* **162**, 729–773 (1982).

7. Fiers, W. *et al.* Complete nucleotide sequence of SV40 DNA. *Nature* **273**, 113–120 (1978).

8. Anderson, S. *et al.* Sequence and organization of the human mitochondrial genome. *Nature* **290**, 457–465 (1981).

9. Botstein, D., White, R. L., Skolnick, M. & Davis, R. W. Construction of a genetic linkage map in man using restriction fragment length polymorphisms. *Am. J. Hum. Genet.* **32**, 314–331 (1980).

10. Olson, M. V. *et al.* Random-clone strategy for genomic restriction mapping in yeast. *Proc. Natl Acad. Sci. USA* **83**, 7826–7830 (1986).

11. Coulson, A., Sulston, J., Brenner, S. & Karn, J. Toward a physical map of the genome of the nematode *Caenorhabditis elegans. Proc. Natl Acad. Sci. USA* **83**, 7821–7825 (1986).

12. Putney, S. D., Herlihy, W. C. & Schimmel, P. A new troponin T and cDNA clones for 13 different muscle proteins, found by shotgun sequencing. *Nature* **302**, 718–721 (1983).

13. Milner, R. J. & Sutcliffe, J. G. Gene expression in rat brain. *Nucleic Acids Res.* **11**, 5497–5520 (1983).

14. Adams, M. D. *et al.* Complementary DNA sequencing: expressed sequence tags and human genome project. *Science* **252**, 1651–1656 (1991).

15. Adams, M. D. *et al.* Initial assessment of human gene diversity and expression patterns based upon 83 million nucleotides of cDNA sequence. *Nature* **377**, 3–174 (1995).

16. Okubo, K. *et al.* Large scale cDNA sequencing for analysis of quantitative and qualitative aspects of gene expression. *Nature Genet.* **2**, 173–179 (1992).

17. Hillier, L. D. *et al.* Generation and analysis of 280,000 human expressed sequence tags. *Genome Res.* **6**, 807–828 (1996).

18. Strausberg, R. L., Feingold, E. A., Klausner, R. D. & Collins, F. S. The mammalian gene collection. *Science* **286**, 455–457 (1999).

19. Berry, R. *et al.* Gene-based sequence-tagged-sites (STSs) as the basis for a human gene map. *Nature Genet.* **10**, 415–423 (1995).

20. Houlgatte, R. *et al.* The Genexpress Index: a resource for gene discovery and the genic map of the human genome. *Genome Res.* **5**, 272–304 (1995).

21. Sinsheimer, R. L. The Santa Cruz Workshop—May 1985. *Genomics* **5**, 954–956 (1989).

22. Palca, J. Human genome—Department of Energy on the map. *Nature* **321**, 371 (1986).

23. National Research Council *Mapping and Sequencing the Human Genome* (National Academy Press, Washington DC, 1988).

24. Bishop, J. E. & Waldholz, M. *Genome* (Simon and Schuster, New York, 1990).

25. Kevles, D. J. & Hood, L. (eds) *The Code of Codes: Scientific and Social Issues in the Human Genome Project* (Harvard Univ. Press, Cambridge, Massachusetts, 1992).

26. Cook-Deegan, R. *The Gene Wars: Science, Politics, and the Human Genome* (W. W. Norton & Co., New York, London, 1994).

27. Donis-Keller, H. *et al.* A genetic linkage map of the human genome. *Cell* **51**, 319–337 (1987).

28. Gyapay, G. *et al.* The 1993–94 Genethon human genetic linkage map. *Nature Genet.* **7**, 246–339 (1994).

29. Hudson, T. J. *et al.* An STS-based map of the human genome. *Science* **270**, 1945–1954 (1995).

30. Dietrich, W. F. *et al.* A comprehensive genetic map of the mouse genome. *Nature* **380**, 149–152 (1996).

31. Nusbaum, C. *et al.* A YAC-based physical map of the mouse genome. *Nature Genet.* **22**, 388–393 (1999).

32. Oliver, S. G. *et al.* The complete DNA sequence of yeast chromosome III. *Nature* **357**, 38–46 (1992).

33. Wilson, R. *et al.* 2.2 Mb of contiguous nucleotide sequence from chromosome III of *C. elegans. Nature* **368**, 32–38 (1994).

34. Chen, E. Y. *et al.* The human growth hormone locus: nucleotide sequence, biology, and evolution. *Genomics* **4**, 479–497 (1989).

35. McCombie, W. R. *et al.* Expressed genes, Alu repeats and polymorphisms in cosmids sequenced from chromosome 4p16.3. *Nature Genet.* **1**, 348–353 (1992).

36. Martin-Gallardo, A. *et al.* Automated DNA sequencing and analysis of 106 kilobases from human chromosome 19q13.3. *Nature Genet.* **1**, 34–39 (1992).

37. Edwards, A. *et al.* Automated DNA sequencing of the human HPRT locus. *Genomics* **6**, 593–608 (1990).

38. Marshall, E. A strategy for sequencing the genome 5 years early. *Science* **267**, 783–784 (1995).

39. Project to sequence human genome moves on to the starting blocks. *Nature* **375**, 93–94 (1995).

40. Shizuya, H. *et al.* Cloning and stable maintenance of 300-kilobase-pair fragments of human DNA in *Escherichia coli* using an F-factor-based vector. *Proc. Natl Acad. Sci. USA* **89**, 8794–8797 (1992).

41. Burke, D. T., Carle, G. F. & Olson, M. V. Cloning of large segments of exogenous DNA into yeast by means of artificial chromosome vectors. *Science* **236**, 806–812 (1987).

42. Marshall, E. A second private genome project. *Science* **281**, 1121 (1998).

43. Marshall, E. NIH to produce a 'working draft' of the genome by 2001. *Science* **281**, 1774–1775 (1998).

44. Pennisi, E. Academic sequencers challenge Celera in a sprint to the finish. *Science* **283**, 1822–1823 (1999).

45. Bouck, J., Miller, W., Gorrell, J. H., Muzny, D. & Gibbs, R. A. Analysis of the quality and utility of random shotgun sequencing at low redundancies. *Genome Res.* **8**, 1074–1084 (1998).

46. Collins, F. S. *et al.* New goals for the U. S. Human Genome Project: 1998–2003. *Science* **282**, 682–689 (1998).

47. Sanger, F. & Coulson, A. R. A rapid method for determining sequences in DNA by primed synthesis with DNA polymerase. *J. Mol. Biol.* **94**, 441–448 (1975).

48. Maxam, A. M. & Gilbert, W. A new method for sequencing DNA. *Proc. Natl Acad. Sci. USA* **74**, 560–564 (1977).

49. Anderson, S. Shotgun DNA sequencing using cloned DNase I-generated fragments. *Nucleic Acids Res.* **9**, 3015–3027 (1981).

50. Gardner, R. C. *et al.* The complete nucleotide sequence of an infectious clone of cauliflower mosaic virus by M13mp7 shotgun sequencing. *Nucleic Acids Res.* **9**, 2871–2888 (1981).

51. Deininger, P. L. Random subcloning of sonicated DNA: application to shotgun DNA sequence analysis. *Anal. Biochem.* **129**, 216–223 (1983).

52. Chissoe, S. L. *et al.* Sequence and analysis of the human ABL gene, the BCR gene, and regions involved in the Philadelphia chromosomal translocation. *Genomics* **27**, 67–82 (1995).

53. Rowen, L., Koop, B. F. & Hood, L. The complete 685-kilobase DNA sequence of the human beta T cell receptor locus. *Science* **272**, 1755–1762 (1996).

54. Koop, B. F. *et al.* Organization, structure, and function of 95 kb of DNA spanning the murine T-cell receptor C alpha/C delta region. *Genomics* **13**, 1209–1230 (1992).

55. Wooster, R. *et al.* Identification of the breast cancer susceptibility gene BRCA2. *Nature* **378**, 789–792 (1995).

56. Fleischmann, R. D. *et al.* Whole-genome random sequencing and assembly of *Haemophilus influenzae* Rd. *Science* **269**, 496–512 (1995).

57. Lander, E. S. & Waterman, M. S. Genomic mapping by fingerprinting random clones: a mathematical analysis. *Genomics* **2**, 231–239 (1988).

58. Weber, J. L. & Myers, E. W. Human whole-genome shotgun sequencing. *Genome Res.* **7**, 401–409 (1997).

59. Green, P. Against a whole-genome shotgun. *Genome Res.* **7**, 410–417 (1997).

60. Venter, J. C. *et al.* Shotgun sequencing of the human genome. *Science* **280**, 1540–1542 (1998).

61. Venter, J. C. *et al.* The sequence of the human genome. *Science* **291**, 1304–1351 (2001).

62. Smith, L. M. *et al.* Fluorescence detection in automated DNA sequence analysis. *Nature* **321**, 674–679 (1986).

63. Ju, J. Y., Ruan, C. C., Fuller, C. W., Glazer, A. N. & Mathies, R. A. Fluorescence energy-transfer dye-labeled primers for DNA sequencing and analysis. *Proc. Natl Acad. Sci. USA* **92**, 4347–4351 (1995).

64. Lee, L. G. *et al.* New energy transfer dyes for DNA sequencing. *Nucleic Acids Res.* **25**, 2816–2822 (1997).

65. Rosenblum, B. B. *et al.* New dye-labeled terminators for improved DNA sequencing patterns. *Nucleic Acids Res.* **25**, 4500–4504 (1997).

66. Metzker, M. L., Lu, J. & Gibbs, R. A. Electrophoretically uniform fluorescent dyes for automated DNA sequencing. *Science* **271**, 1420–1422 (1996).

67. Prober, J. M. *et al.* A system for rapid DNA sequencing with fluorescent chain-terminating dideoxynucleotides. *Science* **238**, 336–341 (1987).

68. Reeve, M. A. & Fuller, C. W. A novel thermostable polymerase for DNA sequencing. *Nature* **376**, 796–797 (1995).

69. Tabor, S. & Richardson, C. C. Selective inactivation of the exonuclease activity of bacteriophage T7 DNA polymerase by in vitro mutagenesis. *J. Biol. Chem.* **264**, 6447–6458 (1989).

70. Tabor, S. & Richardson, C. C. DNA sequence analysis with a modified bacteriophage T7 DNA polymerase—effect of pyrophosphorolysis and metal ions. *J. Biol. Chem.* **265**, 8322–8328 (1990).

71. Murray, V. Improved double-stranded DNA sequencing using the linear polymerase chain reaction. *Nucleic Acids Res.* **17**, 8889 (1989).

72. Guttman, A., Cohen, A. S., Heiger, D. N. & Karger, B. L. Analytical and micropreparative ultrahigh resolution of oligonucleotides by polyacrylamide-gel high-performance capillary electrophoresis. *Anal. Chem.* **62**, 137–141 (1990).

73. Luckey, J. A. *et al.* High-speed DNA sequencing by capillary electrophoresis. *Nucleic Acids Res.* **18**, 4417–4421 (1990).

74. Swerdlow, H., Wu, S., Harke, H. & Dovichi, N. J. Capillary gel-electrophoresis for DNA sequencing—laser-induced fluorescence detection with the sheath flow cuvette. *J. Chromatogr.* **516**, 61–67 (1990).

75. Meldrum, D. Automation for genomics, part one: preparation for sequencing. *Genome Res.* **10**, 1081–1092 (2000).

76. Meldrum, D. Automation for genomics, part two: sequencers, microarrays, and future trends. *Genome Res.* **10**, 1288–1303 (2000).

77. Ewing, B. & Green, P. Base-calling of automated sequencer traces using phred. II. Error probabilities. *Genome Res.* **8**, 186–194 (1998).

78. Ewing, B., Hillier, L., Wendl, M. C. & Green, P. Base-calling of automated sequencer traces using phred. I. Accuracy assessment. *Genome Res.* **8**, 175–185 (1998).

79. Bentley, D. R. Genomic sequence information should be released immediately and freely in the public domain. *Science* **274**, 533–534 (1996).

80. Guyer, M. Statement on the rapid release of genomic DNA sequence. *Genome Res.* **8**, 413 (1998).

81. Dietrich, W. *et al.* A genetic map of the mouse suitable for typing intraspecific crosses. *Genetics* **131**, 423–447 (1992).

82. Kim, U. J. *et al.* Construction and characterization of a human bacterial artificial chromosome library. *Genomics* **34**, 213–218 (1996).

83. Osoegawa, K. *et al.* Bacterial artificial chromosome libraries for mouse sequencing and functional analysis. *Genome Res.* **10**, 116–128 (2000).

84. Marra, M. A. *et al.* High throughput fingerprint analysis of large-insert clones. *Genome Res.* **7**, 1072–1084 (1997).

85. Marra, M. *et al.* A map for sequence analysis of the *Arabidopsis thaliana* genome. *Nature Genet.* **22**, 265–270 (1999).

86. The International Human Genome Mapping Consortium. A physical map of the human genome. *Nature* **409**, 934–941 (2001).

87. Zhao, S. *et al.* Human BAC ends quality assessment and sequence analyses. *Genomics* **63**, 321–332 (2000).

88. Mahairas, G. G. *et al.* Sequence-tagged connectors: A sequence approach to mapping and scanning the human genome. *Proc. Natl Acad. Sci. USA* **96**, 9739–9744 (1999).

89. Tilford, C. A. *et al.* A physical map of the human Y chromosome. *Nature* **409**, 943–945 (2001).

90. Bentley, D. R. *et al.* The physical maps for sequencing human chromosomes 1, 6, 9, 10, 13, 20 and X. *Nature* **409**, 942–943 (2001).

91. Montgomery, K. T. *et al.* A high-resolution map of human chromosome 12. *Nature* **409**, 945–946 (2001).

92. Brüls, T. *et al.* A physical map of human chromosome 14. *Nature* **409**, 947–948 (2001).

93. Hattori, M. *et al.* The DNA sequence of human chromosome 21. *Nature* **405**, 311–319 (2000).

94. Dunham, I. *et al.* The DNA sequence of human chromosome 22. *Nature* **402**, 489–495 (1999).

95. Cox, D. *et al.* Radiation hybrid map of the human genome. *Science* (in the press).

96. Osoegawa, K. *et al.* An improved approach for construction of bacterial artificial chromosome libraries. *Genomics* **52**, 1–8 (1998).

97. The International SNP Map Working Group. A map of human genome sequence variation containing 1.42 million single nucleotide polymorphisms. *Nature* **409**, 928–933 (2001).

98. Collins, F. S., Brooks, L. D. & Chakravarti, A. A DNA polymorphism discovery resource for research on human genetic variation. *Genome Res.* **8**, 1229–1231 (1998).

99. Stewart, E. A. *et al.* An STS-based radiation hybrid map of the human genome. *Genome Res.* **7**, 422–433 (1997).

100. Deloukas, P. *et al.* A physical map of 30,000 human genes. *Science* **282**, 744–746 (1998).

101. Dib, C. *et al.* A comprehensive genetic map of the human genome based on 5,264 microsatellites.

Nature **380,** 152–154 (1996).
102. Broman, K. W., Murray, J. C., Sheffield, V. C., White, R. L. & Weber, J. L. Comprehensive human genetic maps: individual and sex-specific variation in recombination. *Am. J. Hum. Genet.* **63,** 861–869 (1998).
103. The BAC Resource Consortium. Integration of cytogenetic landmarks into the draft sequence of the human genome. *Nature* **409,** 953–958 (2001).
104. Kent, W. J. & Haussler, D. GigAssembler: an algorithm for the initial assembly of the human working draft . Technical Report UCSC-CRL-00-17 (Univ. California at Santa Cruz, Santa Cruz, California, 2001).
105. Morton, N. E. Parameters of the human genome. *Proc. Natl Acad. Sci. USA* **88,** 7474–7476 (1991).
106. Podugolnikova, O. A. & Blumina, M. G. Heterochromatic regions on chromosomes 1, 9, 16, and Y in children with some disturbances occurring during embryo development. *Hum. Genet.* **63,** 183–188 (1983).
107. Lundgren, R., Berger, R. & Kristoffersson, U. Constitutive heterochromatin C-band polymorphism in prostatic cancer. *Cancer Genet. Cytogenet.* **51,** 57–62 (1991).
108. Lee, C., Wevrick, R., Fisher, R. B., Ferguson-Smith, M. A. & Lin, C. C. Human centromeric DNAs. *Hum. Genet.* **100,** 291–304 (1997).
109. Riethman, H. C. *et al.* Integration of telomere sequences with the draft human genome sequence. *Nature* **409,** 953–958 (2001).
110. Pruit, K. D. & Maglott, D. R. RefSeq and LocusLink: NCBI gene-centered resources. *Nucleic Acids Res.* **29,** 137–140 (2001).
111. Wolfsberg, T. G., McEntyre, J. & Schuler, G. D. Guide to the draft human genome. *Nature* **409,** 824–826 (2001).
112. Hurst, L. D. & Eyre-Walker, A. Evolutionary genomics: reading the bands. *Bioessays* **22,** 105–107 (2000).
113. Saccone, S. *et al.* Correlations between isochores and chromosomal bands in the human genome. *Proc. Natl Acad. Sci. USA* **90,** 11929–11933 (1993).
114. Zoubak, S., Clay, O. & Bernardi, G. The gene distribution of the human genome. *Gene* **174,** 95–102 (1996).
115. Gardiner, K. Base composition and gene distribution: critical patterns in mammalian genome organization. *Trends Genet.* **12,** 519–524 (1996).
116. Duret, L., Mouchiroud, D. & Gautier, C. Statistical analysis of vertebrate sequences reveals that long genes are scarce in GC-rich isochores. *J. Mol. Evol.* **40,** 308–317 (1995).
117. Saccone, S., De Sario, A., Della Valle, G. & Bernardi, G. The highest gene concentrations in the human genome are in telomeric bands of metaphase chromosomes. *Proc. Natl Acad. Sci. USA* **89,** 4913–4917 (1992).
118. Bernardi, G. *et al.* The mosaic genome of warm-blooded vertebrates. *Science* **228,** 953–958 (1985).
119. Bernardi, G. Isochores and the evolutionary genomics of vertebrates. *Gene* **241,** 3–17 (2000).
120. Fickett, J. W., Torney, D. C. & Wolf, D. R. Base compositional structure of genomes. *Genomics* **13,** 1056–1064 (1992).
121. Churchill, G. A. Stochastic models for heterogeneous DNA sequences. *Bull. Math. Biol.* **51,** 79–94 (1989).
122. Bird, A., Taggart, M., Frommer, M., Miller, O. J. & Macleod, D. A fraction of the mouse genome that is derived from islands of nonmethylated, CpG-rich DNA. *Cell* **40,** 91–99 (1985).
123. Bird, A. P. CpG islands as gene markers in the vertebrate nucleus. *Trends Genet.* **3,** 342–347 (1987).
124. Chan, M. F., Liang, G. & Jones, P. A. Relationship between transcription and DNA methylation. *Curr. Top. Microbiol. Immunol.* **249,** 75–86 (2000).
125. Holliday, R. & Pugh, J. E. DNA modification mechanisms and gene activity during development. *Science* **187,** 226–232 (1975).
126. Larsen, F., Gundersen, G., Lopez, R. & Prydz, H. CpG islands as gene markers in the human genome. *Genomics* **13,** 1095–1107 (1992).
127. Tazi, J. & Bird, A. Alternative chromatin structure at CpG islands. *Cell* **60,** 909–920 (1990).
128. Gardiner-Garden, M. & Frommer, M. CpG islands in vertebrate genomes. *J. Mol. Biol.* **196,** 261–282 (1987).
129. Antequera, F. & Bird, A. Number of CpG islands and genes in human and mouse. *Proc. Natl Acad. Sci. USA* **90,** 11995–11999 (1993).
130. Ewing, B. & Green, P. Analysis of expressed sequence tags indicates 35,000 human genes. *Nature Genet.* **25,** 232–234 (2000).
131. Yu, A. Comparison of human genetic and sequence-based physical maps. *Nature* **409,** 951–953 (2001).
132. Kaback, D. B., Guacci, V., Barber, D. & Mahon, J. W. Chromosome size-dependent control of meiotic recombination. *Science* **256,** 228–232 (1992).
133. Riles, L. *et al.* Physical maps of the 6 smallest chromosomes of *Saccharomyces cerevisiae* at a resolution of 2.6-kilobase pairs. *Genetics* **134,** 81–150 (1993).
134. Lynn, A. *et al.* Patterns of meiotic recombination on the long arm of human chromosome 21. *Genome Res.* **10,** 1319–1332 (2000).
135. Laurie, D. A. & Hulten, M. A. Further studies on bivalent chiasma frequency in human males with normal karyotypes. *Ann. Hum. Genet.* **49,** 189–201 (1985).
136. Roeder, G. S. Meiotic chromosomes: it takes two to tango. *Genes Dev.* **11,** 2600–2621 (1997).
137. Wu, T.-C. & Lichten, M. Meiosis-induced double-strand break sites determined by yeast chromatin structure. *Science* **263,** 515–518 (1994).
138. Gerton, J. L. *et al.* Global mapping of meiotic recombination hotspots and coldspots in the yeast *Saccharomyces cerevisiae. Proc. Natl Acad. Sci. USA* **97,** 11383–11390 (2000).
139. Li, W. -H. *Molecular Evolution* (Sinauer, Sunderland, Massachusetts, 1997).
140. Gregory, T. R. & Hebert, P. D. The modulation of DNA content: proximate causes and ultimate consequences. *Genome Res.* **9,** 317–324 (1999).
141. Hartl, D. L. Molecular melodies in high and low C. *Nature Rev. Genet.* **1,** 145–149 (2000).
142. Smit, A. F. Interspersed repeats and other mementos of transposable elements in mammalian genomes. *Curr. Opin. Genet. Dev.* **9,** 657–663 (1999).
143. Prak, E. L. & Haig, H. K. Jr Mobile elements and the human genome. *Nature Rev. Genet.* **1,** 134–144 (2000).
144. Okada, N., Hamada, M., Ogiwara, I. & Ohshima, K. SINEs and LINEs share common 3′ sequences: a review. *Gene* **205,** 229–243 (1997).
145. Esnault, C., Maestre, J. & Heidmann, T. Human LINE retrotransposons generate processed pseudogenes. *Nature Genet.* **24,** 363–367 (2000).
146. Wei, W. *et al.* Human L1 retrotransposition: cis-preference vs. trans-complementation. *Mol. Cell. Biol.* **21,** 1429–1439 (2001)
147. Malik, H. S., Henikoff, S. & Eickbush, T. H. Poised for contagion: evolutionary origins of the infectious abilities of invertebrate retroviruses. *Genome Res.* **10,** 1307–1318 (2000).
148. Smit, A. F. The origin of interspersed repeats in the human genome. *Curr. Opin. Genet. Dev.* **6,** 743–748 (1996).
149. Clark, J. B. & Tidwell, M. G. A phylogenetic perspective on P transposable element evolution in Drosophila. *Proc. Natl Acad. Sci. USA* **94,** 11428–11433 (1997).
150. Haring, E., Hagemann, S. & Pinsker, W. Ancient and recent horizontal invasions of Drosophilids by P elements. *J. Mol. Evol.* **51,** 577–586 (2000).
151. Koga, A. *et al.* Evidence for recent invasion of the medaka fish genome by the Tol2 transposable element. *Genetics* **155,** 273–281 (2000).
152. Robertson, H. M. & Lampe, D. J. Recent horizontal transfer of a mariner transposable element among and between Diptera and Neuroptera. *Mol. Biol. Evol.* **12,** 850–862 (1995).
153. Simmons, G. M. Horizontal transfer of hobo transposable elements within the *Drosophila melanogaster* species complex: evidence from DNA sequencing. *Mol. Biol. Evol.* **9,** 1050–1060 (1992).
154. Malik, H. S., Burke, W. D. & Eickbush, T. H. The age and evolution of non-LTR retrotransposable elements. *Mol. Biol. Evol.* **16,** 793–805 (1999).
155. Kordis, D. & Gubensek, F. Bov-B long interspersed repeated DNA (LINE) sequences are present in *Vipera ammodytes* phospholipase A2 genes and in genomes of Viperidae snakes. *Eur. J. Biochem.* **246,** 772–779 (1997).
156. Jurka, J. Repbase update: a database and an electronic journal of repetitive elements. *Trends Genet.* **16,** 418–420 (2000).
157. Sarich, V. M. & Wilson, A. C. Generation time and genome evolution in primates. *Science* **179,** 1144–1147 (1973).
158. Smit, A. F., Toth, G., Riggs, A. D., & Jurka, J. Ancestral, mammalian-wide subfamilies of LINE-1 repetitive sequences. *J. Mol. Biol.* **246,** 401–417 (1995).
159. Lim, J. K. & Simmons, M. J. Gross chromosome rearrangements mediated by transposable elements in *Drosophila melanogaster. Bioessays* **16,** 269–275 (1994).
160. Caceres, M., Ranz, J. M., Barbadilla, A., Long, M. & Ruiz, A. Generation of a widespread Drosophila inversion by a transposable element. *Science* **285,** 415–418 (1999).
161. Gray, Y. H. It takes two transposons to tango: transposable-element-mediated chromosomal rearrangements. *Trends Genet.* **16,** 461–468 (2000).
162. Zhang, J. & Peterson, T. Genome rearrangements by nonlinear transposons in maize. *Genetics* **153,** 1403–1410 (1999).
163. Smit, A. F. Identification of a new, abundant superfamily of mammalian LTR-transposons. *Nucleic Acids Res.* **21,** 1863–1872 (1993).
164. Cordonnier, A., Casella, J. F. & Heidmann, T. Isolation of novel human endogenous retrovirus-like elements with foamy virus-related pol sequence. *J. Virol.* **69,** 5890–5897 (1995).
165. Medstrand, P. & Mager, D. L. Human-specific integrations of the HERV-K endogenous retrovirus family. *J. Virol.* **72,** 9782–9787 (1998).
166. Myers, E. W. *et al.* A whole-genome assembly of Drosophila. *Science* **287,** 2196–2204 (2000).
167. Petrov, D. A., Lozovskaya, E. R. & Hartl, D. L. High intrinsic rate of DNA loss in *Drosophila. Nature* **384,** 346–349 (1996).
168. Li, W. H., Ellsworth, D. L., Krushkal, J., Chang, B. H. & Hewett-Emmett, D. Rates of nucleotide substitution in primates and rodents and the generation-time effect hypothesis. *Mol. Phylogenet. Evol.* **5,** 182–187 (1996).
169. Goodman, M. *et al.* Toward a phylogenetic classification of primates based on DNA evidence complemented by fossil evidence. *Mol. Phylogenet. Evol.* **9,** 585–598 (1998).
170. Kazazian, H. H. Jr & Moran, J. V. The impact of L1 retrotransposons on the human genome. *Nature Genet.* **19,** 19–24 (1998).
171. Malik, H. S. & Eickbush, T. H. NeSL-1, an ancient lineage of site-specific non-LTR retrotransposons from *Caenorhabditis elegans. Genetics* **154,** 193–203 (2000).
172. Casavant, N. C. *et al.* The end of the LINE?: lack of recent L1 activity in a group of South American rodents. *Genetics* **154,** 1809–1817 (2000).
173. Meunier-Rotival, M., Soriano, P., Cuny, G., Strauss, F. & Bernardi, G. Sequence organization and genomic distribution of the major family of interspersed repeats of mouse DNA. *Proc. Natl Acad. Sci. USA* **79,** 355–359 (1982).
174. Soriano, P., Meunier-Rotival, M. & Bernardi, G. The distribution of interspersed repeats is nonuniform and conserved in the mouse and human genomes. *Proc. Natl Acad. Sci. USA* **80,** 1816–1820 (1983).
175. Goldman, M. A., Holmquist, G. P., Gray, M. C., Caston, L. A. & Nag, A. Replication timing of genes and middle repetitive sequences. *Science* **224,** 686–692 (1984).
176. Manuelidis, L. & Ward, D. C. Chromosomal and nuclear distribution of the *Hin*dIII 1.9-kb human DNA repeat segment. *Chromosoma* **91,** 28–38 (1984).
177. Feng, Q., Moran, J. V., Kazazian, H. H. Jr & Boeke, J. D. Human L1 retrotransposon encodes a conserved endonuclease required for retrotransposition. *Cell* **87,** 905–916 (1996).
178. Jurka, J. Sequence patterns indicate an enzymatic involvement in integration of mammalian retroposons. *Proc. Natl Acad. Sci. USA* **94,** 1872–1877 (1997).
179. Arcot, S. S. *et al.* High-resolution cartography of recently integrated human chromosome 19-specific Alu fossils. *J. Mol. Biol.* **281,** 843–856 (1998).
180. Schmid, C. W. Does SINE evolution preclude Alu function? *Nucleic Acids Res.* **26,** 4541–4550 (1998).
181. Chu, W. M., Ballard, R., Carpick, B. W., Williams, B. R. & Schmid, C. W. Potential Alu function: regulation of the activity of double-stranded RNA-activated kinase PKR. *Mol. Cell. Biol.* **18,** 58–68 (1998).
182. Li, T., Spearow, J., Rubin, C. M. & Schmid, C. W. Physiological stresses increase mouse short interspersed element (SINE) RNA expression in vivo. *Gene* **239,** 367–372 (1999).
183. Liu, W. M., Chu, W. M., Choudary, P. V. & Schmid, C. W. Cell stress and translational inhibitors transiently increase the abundance of mammalian SINE transcripts. *Nucleic Acids Res.* **23,** 1758–1765 (1995).
184. Filipski, J. Correlation between molecular clock ticking, codon usage fidelity of DNA repair, chromosome banding and chromatin compactness in germline cells. *FEBS Lett.* **217,** 184–186 (1987).
185. Sueoka, N. Directional mutation pressure and neutral molecular evolution. *Proc. Natl Acad. Sci.*

USA **85,** 2653–2657 (1988).
186. Wolfe, K. H., Sharp, P. M. & Li, W. H. Mutation rates differ among regions of the mammalian genome. *Nature* **337,** 283–285 (1989).
187. Bains, W. Local sequence dependence of rate of base replacement in mammals. *Mutat. Res.* **267,** 43–54 (1992).
188. Mathews, C. K. & Ji, J. DNA precursor asymmetries, replication fidelity, and variable genome evolution. *Bioessays* **14,** 295–301 (1992).
189. Holmquist, G. P. & Filipski, J. Organization of mutations along the genome: a prime determinant of genome evolution. *Trends Ecol. Evol.* **9,** 65–68 (1994).
190. Eyre-Walker, A. Evidence of selection on silent site base composition in mammals: potential implications for the evolution of isochores and junk DNA. *Genetics* **152,** 675–683 (1999).
191. The International SNP Map Working Group. An SNP map of the human genome generated by reduced representation shotgun sequencing. *Nature* **407,** 513–516 (2000).
192. Bohossian, H. B., Skaletsky, H. & Page, D. C. Unexpectedly similar rates of nucleotide substitution found in male and female hominids. *Nature* **406,** 622–625 (2000).
193. Skowronski, J., Fanning, T. G. & Singer, M. F. Unit-length LINE-1 transcripts in human teratocarcinoma cells. *Mol. Cell. Biol.* **8,** 1385–1397 (1988).
194. Boissinot, S., Chevret, P. & Furano, A. V. L1 (LINE-1) retrotransposon evolution and amplification in recent human history. *Mol. Biol. Evol.* **17,** 915–928 (2000).
195. Moran, J. V. Human L1 retrotransposition: insights and peculiarities learned from a cultured cell retrotransposition assay. *Genetica* **107,** 39–51 (1999).
196. Kazazian, H. H. Jr *et al.* Haemophilia A resulting from *de novo* insertion of L1 sequences represents a novel mechanism for mutation in man. *Nature* **332,** 164–166 (1988).
197. Sheen, F.-m. *et al.* Reading between the LINEs: Human genomic variation introduced by LINE-1 retrotransposition. *Genome Res.* **10,** 1496–1508 (2000).
198. Dombroski, B. A., Mathias, S. L., Nanthakumar, E., Scott, A. F. & Kazazian, H. H. Jr Isolation of an active human transposable element. *Science* **254,** 1805–1808 (1991).
199. Holmes, S. E., Dombroski, B. A., Krebs, C. M., Boehm, C. D. & Kazazian, H. H. Jr A new retrotransposable human L1 element from the LRE2 locus on chromosome 1q produces a chimaeric insertion. *Nature Genet.* **7,** 143–148 (1994).
200. Sassaman, D. M. *et al.* Many human L1 elements are capable of retrotransposition. *Nature Genet.* **16,** 37–43 (1997).
201. Dombroski, B. A., Scott, A. F. & Kazazian, H. H. Jr Two additional potential retrotransposons isolated from a human L1 subfamily that contains an active retrotransposable element. *Proc. Natl Acad. Sci. USA* **90,** 6513–6517 (1993).
202. Kimberland, M. L. *et al.* Full-length human L1 insertions retain the capacity for high frequency retrotransposition in cultured cells. *Hum. Mol. Genet.* **8,** 1557–1560 (1999).
203. Moran, J. V. *et al.* High frequency retrotransposition in cultured mammalian cells. *Cell* **87,** 917–927 (1996).
204. Moran, J. V., DeBerardinis, R. J. & Kazazian, H. H. Jr Exon shuffling by L1 retrotransposition. *Science* **283,** 1530–1534 (1999).
205. Pickeral, O. K., Makalowski, W., Boguski, M. S. & Boeke, J. D. Frequent human genomic DNA transduction driven by LINE-1 retrotransposition. *Genome Res.* **10,** 411–415 (2000).
206. Miki, Y. *et al.* Disruption of the APC gene by a retrotransposal insertion of L1 sequence in a colon cancer. *Cancer Res.* **52,** 643–645 (1992).
207. Branciforte, D. & Martin, S. L. Developmental and cell type specificity of LINE-1 expression in mouse testis: implications for transposition. *Mol. Cell. Biol.* **14,** 2584–2592 (1994).
208. Trelogan, S. A. & Martin, S. L. Tightly regulated, developmentally specific expression of the first open reading frame from LINE-1 during mouse embryogenesis. *Proc. Natl Acad. Sci. USA* **92,** 1520–1524 (1995).
209. Jurka, J. & Kapitonov, V. V. Sectorial mutagenesis by transposable elements. *Genetica* **107,** 239–248 (1999).
210. Fraser, M. J., Ciszczon, T., Elick, T. & Bauser, C. Precise excision of TTAA-specific lepidopteran transposons piggyBac (IFP2) and tagalong (TFP3) from the baculovirus genome in cell lines from two species of Lepidoptera. *Insect Mol. Biol.* **5,** 141–151 (1996).
211. Brosius, J. Genomes were forged by massive bombardments with retroelements and retrosequences. *Genetica* **107,** 209–238 (1999).
212. Kruglyak, S., Durrett, R. T., Schug, M. D. & Aquadro, C. F. Equilibrium distribution of microsatellite repeat length resulting from a balance between slippage events and point mutations. *Proc. Natl Acad. Sci. USA* **95,** 10774–10778 (1998).
213. Toth, G., Gaspari, Z. & Jurka, J. Microsatellites in different eukaryotic genomes: survey and analysis. *Genome Res.* **10,** 967–981 (2000).
214. Ellegren, H. Heterogeneous mutation processes in human microsatellite DNA sequences. *Nature Genet.* **24,** 400–402 (2000).
215. Ji, Y., Eichler, E. E., Schwartz, S. & Nicholls, R. D. Structure of chromosomal duplicons and their role in mediating human genomic disorders. *Genome Res.* **10,** 597–610 (2000).
216. Eichler, E. E. Masquerading repeats: paralogous pitfalls of the human genome. *Genome Res.* **8,** 758–762 (1998).
217. Mazzarella, R. & D. Schlessinger, D. Pathological consequences of sequence duplications in the human genome. *Genome Res.* **8,** 1007–1021 (1998).
218. Eichler, E. E. *et al.* Interchromosomal duplications of the adrenoleukodystrophy locus: a phenomenon of pericentromeric plasticity. *Hum. Mol. Genet.* **6,** 991–1002 (1997).
219. Horvath, J. E., Schwartz, S. & Eichler, E. E. The mosaic structure of human pericentromeric DNA: a strategy for characterizing complex regions of the human genome. *Genome Res.* **10,** 839–852 (2000).
220. Brand-Arpon, V. *et al.* A genomic region encompassing a cluster of olfactory receptor genes and a myosin light chain kinase (MYLK) gene is duplicated on human chromosome regions 3q13-q21 and 3p13. *Genomics* **56,** 98–110 (1999).
221. Arnold, N., Wienberg, J., Ermert, K. & Zachau, H. G. Comparative mapping of DNA probes derived from the V kappa immunoglobulin gene regions on human and great ape chromosomes by fluorescence in situ hybridization. *Genomics* **26,** 147–150 (1995).
222. Eichler, E. E. *et al.* Duplication of a gene-rich cluster between 16p11.1 and Xq28: a novel pericentromeric-directed mechanism for paralogous genome evolution. *Hum. Mol. Genet.* **5,** 899–912 (1996).
223. Potier, M. *et al.* Two sequence-ready contigs spanning the two copies of a 200-kb duplication on human 21q: partial sequence and polymorphisms. *Genomics* **51,** 417–426 (1998).
224. Regnier, V. *et al.* Emergence and scattering of multiple neurofibromatosis (NF1)-related sequences during hominoid evolution suggest a process of pericentromeric interchromosomal transposition. *Hum. Mol. Genet.* **6,** 9–16 (1997).
225. Ritchie, R. J., Mattei, M. G. & Lalande, M. A large polymorphic repeat in the pericentromeric region of human chromosome 15q contains three partial gene duplications. *Hum. Mol. Genet.* **7,** 1253–1260 (1998).
226. Trask, B. J. *et al.* Members of the olfactory receptor gene family are contained in large blocks of DNA duplicated polymorphically near the ends of human chromosomes. *Hum. Mol. Genet.* **7,** 13–26 (1998).
227. Trask, B. J. *et al.* Large multi-chromosomal duplications encompass many members of the olfactory receptor gene family in the human genome. *Hum. Mol. Genet.* **7,** 2007–2020 (1998).
228. van Deutekom, J. C. *et al.* Identification of the first gene (FRG1) from the FSHD region on human chromosome 4q35. *Hum. Mol. Genet.* **5,** 581–590 (1996).
229. Zachau, H. G. The immunoglobulin kappa locus—or—what has been learned from looking closely at one-tenth of a percent of the human genome. *Gene* **135,** 167–173 (1993).
230. Zimonjic, D. B., Kelley, M. J., Rubin, J. S., Aaronson, S. A. & Popescu, N. C. Fluorescence in situ hybridization analysis of keratinocyte growth factor gene amplification and dispersion in evolution of great apes and humans. *Proc. Natl Acad. Sci. USA* **94,** 11461–11465 (1997).
231. van Geel, M. *et al.* The FSHD region on human chromosome 4q35 contains potential coding regions among pseudogenes and a high density of repeat elements. *Genomics* **61,** 55–65 (1999).
232. Horvath, J. E. *et al.* Molecular structure and evolution of an alpha satellite/non-alpha satellite junction at 16p11. *Hum. Mol. Genet.* **9,** 113–123 (2000).
233. Guy, J. *et al.* Genomic sequence and transcriptional profile of the boundary between pericentromeric satellites and genes on human chromosome arm 10q. *Hum. Mol. Genet.* **9,** 2029–2042 (2000).
234. Reiter, L. T., Murakami, T., Koeuth, T., Gibbs, R. A. & Lupski, J. R. The human COX10 gene is disrupted during homologous recombination between the 24 kb proximal and distal CMT1A-REPs. *Hum. Mol. Genet.* **6,** 1595–1603 (1997).
235. Amos-Landgraf, J. M. *et al.* Chromosome breakage in the Prader-Willi and Angelman syndromes involves recombination between large, transcribed repeats at proximal and distal breakpoints. *Am. J. Hum. Genet.* **65,** 370–386 (1999).
236. Christian, S. L., Fantes, J. A., Mewborn, S. K., Huang, B. & Ledbetter, D. H. Large genomic duplicons map to sites of instability in the Prader-Willi/Angelman syndrome chromosome region (15q11-q13). *Hum. Mol. Genet.* **8,** 1025–1037 (1999).
237. Edelmann, L., Pandita, R. K. & Morrow, B. E. Low-copy repeats mediate the common 3-Mb deletion in patients with velo-cardio-facial syndrome. *Am. J. Hum. Genet.* **64,** 1076–1086 (1999).
238. Shaikh, T. H. *et al.* Chromosome 22-specific low copy repeats and the 22q11.2 deletion syndrome: genomic organization and deletion endpoint analysis. *Hum. Mol. Genet.* **9,** 489–501 (2000).
239. Francke, U. Williams-Beuren syndrome: genes and mechanisms. *Hum. Mol. Genet.* **8,** 1947–1954 (1999).
240. Peoples, R. *et al.* A physical map, including a BAC/PAC clone contig, of the Williams-Beuren syndrome-deletion region at 7q11.23. *Am. J. Hum. Genet.* **66,** 47–68 (2000).
241. Eichler, E. E., Archidiacono, N. & Rocchi, M. CAGGG repeats and the pericentromeric duplication of the hominoid genome. *Genome Res.* **9,** 1048–1058 (1999).
242. O'Keefe, C. & Eichler, E. in *Comparative Genomics: Empirical and Analytical Approaches to Gene Order Dynamics, Map Alignment and the Evolution of Gene Families* (eds Sankoff, D. & Nadeau, J.) 29–46 (Kluwer Academic, Dordrecht, 2000).
243. Lander, E. S. The new genomics: Global views of biology. *Science* **274,** 536–539 (1996).
244. Eddy, S. R. Noncoding RNA genes. *Curr. Op. Genet. Dev.* **9,** 695–699 (1999).
245. Ban, N., Nissen, P., Hansen, J., Moore, P. B. & Steitz, T. A. The complete atomic structure of the large ribosomal subunit at 2.4 angstrom resolution. *Science* **289,** 905–920 (2000).
246. Nissen, P., Hansen, J., Ban, N., Moore, P. B. & Steitz, T. A. The structural basis of ribosome activity in peptide bond synthesis. *Science* **289,** 920–930 (2000).
247. Weinstein, L. B. & Steitz, J. A. Guided tours: from precursor snoRNA to functional snoRNP. *Curr. Opin. Cell Biol.* **11,** 378–384 (1999).
248. Bachellerie, J.-P. & Cavaille, J. in *Modification and Editing of RNA* (ed. Benne, H. G. a. R.) 255–272 (ASM, Washington DC, 1998).
249. Burge, C. & Sharp, P. A. Classification of introns: U2-type or U12-type. *Cell* **91,** 875–879 (1997).
250. Brown, C. J. *et al.* The Human Xist gene—analysis of a 17 kb inactive X-specific RNA that contains conserved repeats and is highly localized within the nucleus. *Cell* **71,** 527–542 (1992).
251. Kickhoefer, V. A., Vasu, S. K. & Rome, L. H. Vaults are the answer, what is the question? *Trends Cell Biol.* **6,** 174–178 (1996).
252. Hatlen, L. & Attardi, G. Proportion of the HeLa cell genome complementary to the transfer RNA and 5S RNA. *J. Mol. Biol.* **56,** 535–553 (1971).
253. Sprinzl, M., Horn, C., Brown, M., Ioudovitch, A. & Steinberg, S. Compilation of tRNA sequences and sequences of tRNA genes. *Nucleic Acids Res.* **26,** 148–153 (1998).
254. Long, E. O. & Dawid, I. B. Repeated genes in eukaryotes. *Annu. Rev. Biochem.* **49,** 727–764 (1980).
255. Crick, F. H. Codon–anticodon pairing: the wobble hypothesis. *J. Mol. Biol.* **19,** 548–555 (1966).
256. Guthrie, C. & Abelson, J. in *The Molecular Biology of the Yeast Saccharomyces: Metabolism and Gene Expression* (eds Strathern, J. & Broach J.) 487–528 (Cold Spring Harbor Laboratory Press, Cold Spring Harbor, New York, 1982).
257. Soll, D. & RajBhandary, U. (eds) *tRNA: Structure, Biosynthesis, and Function* (ASM, Washington DC, 1995).
258. Ikemura, T. Codon usage and tRNA content in unicellular and multicellular organisms. *Mol. Biol. Evol.* **2,** 13–34 (1985).
259. Bulmer, M. Coevolution of codon usage and transfer-RNA abundance. *Nature* **325,** 728–730 (1987).
260. Duret, L. tRNA gene number and codon usage in the *C. elegans* genome are co-adapted for optimal translation of highly expressed genes. *Trends Genet.* **16,** 287–289 (2000).
261. Sharp, P. M. & Matassi, G. Codon usage and genome evolution. *Curr. Opin. Genet. Dev.* **4,** 851–860 (1994).
262. Buckland, R. A. A primate transfer-RNA gene cluster and the evolution of human chromosome 1. *Cytogenet. Cell Genet.* **61,** 1–4 (1992).
263. Gonos, E. S. & Goddard, J. P. Human tRNA-Glu genes: their copy number and organization. *FEBS Lett.* **276,** 138–142 (1990).
264. Sylvester, J. E. *et al.* The human ribosomal RNA genes: structure and organization of the complete repeating unit. *Hum. Genet.* **73,** 193–198 (1986).

265. Sorensen, P. D. & Frederiksen, S. Characterization of human 5S ribosomal RNA genes. *Nucleic Acids Res.* **19**, 4147–4151 (1991).

266. Timofeeva, M. *et al.* [Organization of a 5S ribosomal RNA gene cluster in the human genome]. *Mol. Biol. (Mosk.)* **27**, 861–868 (1993).

267. Little, R. D. & Braaten, D. C. Genomic organization of human 5S rDNA and sequence of one tandem repeat. *Genomics* **4**, 376–383 (1989).

268. Maden, B. E. H. The numerous modified nucleotides in eukaryotic ribosomal RNA. *Prog. Nucleic Acid Res. Mol. Biol.* **39**, 241–303 (1990).

269. Tycowski, K. T., You, Z. H., Graham, P. J. & Steitz, J. A. Modification of U6 spliceosomal RNA is guided by other small RNAs. *Mol. Cell* **2**, 629–638 (1998).

270. Pavelitz, T., Liao, D. Q. & Weiner, A. M. Concerted evolution of the tandem array encoding primate U2 snRNA (the RNU2 locus) is accompanied by dramatic remodeling of the junctions with flanking chromosomal sequences. *EMBO J.* **18**, 3783–3792 (1999).

271. Lindgren, V., Ares, A., Weiner, A. M. & Francke, U. Human genes for U2 small nuclear RNA map to a major adenovirus 12 modification site on chromosome 17. *Nature* **314**, 115–116 (1985).

272. Van Arsdell, S. W. & Weiner, A. M. Human genes for U2 small nuclear RNA are tandemly repeated. *Mol. Cell. Biol.* **4**, 492–499 (1984).

273. Gao, L. I., Frey, M. R. & Matera, A. G. Human genes encoding U3 snRNA associate with coiled bodies in interphase cells and are clustered on chromosome 17p11. 2 in a complex inverted repeat structure. *Nucleic Acids Res.* **25**, 4740–4747 (1997).

274. Hawkins, J. D. A survey on intron and exon lengths. *Nucleic Acids Res.* **16**, 9893–9908 (1988).

275. Burge, C. & Karlin, S. Prediction of complete gene structures in human genomic DNA. *J. Mol. Biol.* **268**, 78–94 (1997).

276. Labeit, S. & Kolmerer, B. Titins: giant proteins in charge of muscle ultrastructure and elasticity. *Science* **270**, 293–296 (1995).

277. Sterner, D. A., Carlo, T. & Berget, S. M. Architectural limits on split genes. *Proc. Natl Acad. Sci. USA* **93**, 15081–15085 (1996).

278. Sun, Q., Mayeda, A., Hampson, R. K., Krainer, A. R. & Rottman, F. M. General splicing factor SF2/ASF promotes alternative splicing by binding to an exonic splicing enhancer. *Genes Dev.* **7**, 2598–2608 (1993).

279. Tanaka, K., Watakabe, A. & Shimura, Y. Polypurine sequences within a downstream exon function as a splicing enhancer. *Mol. Cell. Biol.* **14**, 1347–1354 (1994).

280. Carlo, T., Sterner, D. A. & Berget, S. M. An intron splicing enhancer containing a G-rich repeat facilitates inclusion of a vertebrate micro-exon. *RNA* **2**, 342–353 (1996).

281. Burset, M., Seledtsov, I. A. & Solovyev, V. V. Analysis of canonical and non-canonical splice sites in mammalian genomes. *Nucleic Acids Res.* **28**, 4364–4375 (2000).

282. Burge, C. B., Padgett, R. A. & Sharp, P. A. Evolutionary fates and origins of U12-type introns. *Mol. Cell* **2**, 773–785 (1998).

283. Mironov, A. A., Fickett, J. W. & Gelfand, M. S. Frequent alternative splicing of human genes. *Genome Res.* **9**, 1288–1293 (1999).

284. Hanke, J. *et al.* Alternative splicing of human genes: more the rule than the exception? *Trends Genet.* **15**, 389–390 (1999).

285. Brett, D. *et al.* EST comparison indicates 38% of human mRNAs contain possible alternative splice forms. *FEBS Lett.* **474**, 83–86 (2000).

286. Dunham, I. The gene guessing game. *Yeast* **17**, 218–224 (2000).

287. Lewin, B. *Gene Expression* (Wiley, New York, 1980).

288. Lewin, B. *Genes IV* 466–481 (Oxford Univ. Press, Oxford, 1990).

289. Smaglik, P. Researchers take a gamble on the human genome. *Nature* **405**, 264 (2000).

290. Fields, C., Adams, M. D., White, O. & Venter, J. C. How many genes in the human genome? *Nature Genet.* **7**, 345–346 (1994).

291. Liang, F. *et al.* Gene index analysis of the human genome estimates approximately 120,000 genes. *Nature Genet.* **25**, 239–240 (2000).

292. Roest Crollius, H. *et al.* Estimate of human gene number provided by genome-wide analysis using *Tetraodon nigroviridis* DNA sequence. *Nature Genet.* **25**, 235–238 (2000).

293. The C. elegans Sequencing Consortium. Genome sequence of the nematode *C. elegans*: A platform for investigating biology. *Science* **282**, 2012–2018 (1998).

294. Rubin, G. M. *et al.* Comparative genomics of the eukaryotes. *Science* **287**, 2204–2215 (2000).

295. Green, P. *et al.* Ancient conserved regions in new gene sequences and the protein databases. *Science* **259**, 1711–1716 (1993).

296. Fraser, A. G. *et al.* Functional genomic analysis of *C. elegans* chromosome I by systematic RNA interference. *Nature* **408**, 325–330 (2000).

297. Mott, R. EST_GENOME: a program to align spliced DNA sequences to unspliced genomic DNA. *Comput. Appl. Biosci.* **13**, 477–478 (1997).

298. Florea, L., Hartzell, G., Zhang, Z., Rubin, G. M. & Miller, W. A computer program for aligning a cDNA sequence with a genomic DNA sequence. *Genome Res.* **8**, 967–974 (1998).

299. Bailey, L. C. Jr, Searls, D. B. & Overton, G. C. Analysis of EST-driven gene annotation in human genomic sequence. *Genome Res.* **8**, 362–376 (1998).

300. Birney, E., Thompson, J. D. & Gibson, T. J. PairWise and SearchWise: finding the optimal alignment in a simultaneous comparison of a protein profile against all DNA translation frames. *Nucleic Acids Res.* **24**, 2730–2739 (1996).

301. Gelfand, M. S., Mironov, A. A. & Pevzner, P. A. Gene recognition via spliced sequence alignment. *Proc. Natl Acad. Sci. USA* **93**, 9061–9066 (1996).

302. Kulp, D., Haussler, D., Reese, M. G. & Eeckman, F. H. A generalized hidden Markov model for the recognition of human genes in DNA. *ISMB* **4**, 134–142 (1996).

303. Reese, M. G., Kulp, D., Tammana, H. & Haussler, D. Genie—gene finding in *Drosophila melanogaster*. *Genome Res.* **10**, 529–538 (2000).

304. Solovyev, V. & Salamov, A. The Gene-Finder computer tools for analysis of human and model organisms genome sequences. *ISMB* **5**, 294–302 (1997).

305. Guigo, R., Agarwal, P., Abril, J. F., Burset, M. & Fickett, J. W. An assessment of gene prediction accuracy in large DNA sequences. *Genome Res.* **10**, 1631–1642 (2000).

306. Hubbard, T. & Birney, E. Open annotation offers a democratic solution to genome sequencing. *Nature* **403**, 825 (2000).

307. Bateman, A. *et al.* The Pfam protein families database. *Nucleic Acids Res.* **28**, 263–266 (2000).

308. Birney, E. & Durbin, R. Using GeneWise in the Drosophila annotation experiment. *Genome Res.* **10**, 547–548 (2000).

309. The RIKEN Genome Exploration Research Group Phase II Team and the FANTOM Consortium. Functional annotation of a full-length mouse cDNA collection. *Nature* **409**, 685–690 (2001).

310. Basrai, M. A., Hieter, P. & Boeke, J. D. Small open reading frames: beautiful needles in the haystack. *Genome Res.* **7**, 768–771 (1997).

311. Janin, J. & Chothia, C. Domains in proteins: definitions, location, and structural principles. *Methods Enzymol.* **115**, 420–430 (1985).

312. Ponting, C. P., Schultz, J., Copley, R. R., Andrade, M. A. & Bork, P. Evolution of domain families. *Adv. Protein Chem.* **54**, 185–244 (2000).

313. Doolittle, R. F. The multiplicity of domains in proteins. *Annu. Rev. Biochem.* **64**, 287–314 (1995).

314. Bateman, A. & Birney, E. Searching databases to find protein domain organization. *Adv. Protein Chem.* **54**, 137–157 (2000).

315. Futreal, P. A. *et al.* Cancer and genomics. *Nature* **409**, 850–852 (2001).

316. Nestler, E. J. & Landsman, D. Learning about addiction from the human draft genome. *Nature* **409**, 834–835 (2001).

317. Tupler, R., Perini, G. & Green, M. R. Expressing the human genome. *Nature* **409**, 832–835 (2001).

318. Fahrer, A. M., Bazan, J. F., Papathanasiou, P., Nelms, K. A. & Goodnow, C. C. A genomic view of immunology. *Nature* **409**, 836–838 (2001).

319. Li, W. -H., Gu, Z., Wang, H. & Nekrutenko, A. Evolutionary analyses of the human genome. *Nature* **409**, 847–849 (2001).

320. Bock, J. B., Matern, H. T., Peden, A. A. & Scheller, R. H. A genomic perspective on membrane compartment organization. *Nature* **409**, 839–841 (2001).

321. Pollard, T. D. Genomics, the cytoskeleton and motility. *Nature* **409**, 842–843 (2001).

322. Murray, A. W. & Marks, D. Can sequencing shed light on cell cycling? *Nature* **409**, 844–846 (2001).

323. Clayton, J. D., Kyriacou, C. P. & Reppert, S. M. Keeping time with the human genome. *Nature* **409**, 829–831 (2001).

324. Chervitz, S. A. *et al.* Comparison of the complete protein sets of worm and yeast: orthology and divergence. *Science* **282**, 2022–2028 (1998).

325. Aravind, L. & Subramanian, G. Origin of multicellular eukaryotes—insights from proteome comparisons. *Curr. Opin. Genet. Dev.* **9**, 688–694 (1999).

326. Attwood, T. K. *et al.* PRINTS-S: the database formerly known as PRINTS. *Nucleic Acids Res.* **28**, 225–227 (2000).

327. Hofmann, K., Bucher, P., Falquet, L. & Bairoch, A. The PROSITE database, its status in 1999. *Nucleic Acids Res.* **27**, 215–219 (1999).

328. Altschul, S. F. *et al.* Gapped BLAST and PSI-BLAST: a new generation of protein database search programs. *Nucleic Acids Res.* **25**, 3389–3402 (1997).

329. Wolf, Y. I., Kondrashov, F. A. & Koonin, E. V. No footprints of primordial introns in a eukaryotic genome. *Trends Genet.* **16**, 333–334 (2000).

330. Brunner, H. G., Nelen, M., Breakefield, X. O., Ropers, H. H. & van Oost, B. B. A. Abnormal behavior associated with a point mutation in the structural gene for monoamine oxidase A. *Science* **262**, 578–580 (1993).

331. Cases, O. *et al.* Aggressive behavior and altered amounts of brain serotonin and norepinephrine in mice lacking MAOA. *Science* **268**, 1763–1766 (1995).

332. Brunner, H. G. *et al.* X-linked borderline mental retardation with prominent behavioral disturbance: phenotype, genetic localization, and evidence for disturbed monoamine metabolism. *Am. J. Hum. Genet.* **52**, 1032–1039 (1993).

333. Deckert, J. *et al.* Excess of high activity monoamine oxidase A gene promoter alleles in female patients with panic disorder. *Hum. Mol. Genet.* **8**, 621–624 (1999).

334. Smith, T. F. & Waterman, M. S. Identification of common molecular subsequences. *J. Mol. Biol.* **147**, 195–197 (1981).

335. Tatusov, R. L., Koonin, E. V. & Lipman, D. J. A genomic perspective on protein families. *Science* **278**, 631–637 (1997).

336. Ponting, C. P., Aravind, L., Schultz, J., Bork, P. & Koonin, E. V. Eukaryotic signalling domain homologues in archaea and bacteria. Ancient ancestry and horizontal gene transfer. *J. Mol. Biol.* **289**, 729–745 (1999).

337. Zhang, J., Dyer, K. D. & Rosenberg, H. F. Evolution of the rodent eosinophil-associated Rnase gene family by rapid gene sorting and positive selection. *Proc. Natl Acad. Sci. USA* **97**, 4701–4706 (2000).

338. Shashoua, V. E. Ependymin, a brain extracellular glycoprotein, and CNS plasticity. *Ann. NY Acad. Sci.* **627**, 94–114 (1991).

339. Schultz, J., Copley, R. R., Doerks, T., Ponting, C. P. & Bork, P. SMART: a web-based tool for the study of genetically mobile domains. *Nucleic Acids Res.* **28**, 231–234 (2000).

340. Koonin, E. V., Aravind, L. & Kondrashov, A. S. The impact of comparative genomics on our understanding of evolution. *Cell* **101**, 573–576 (2000).

341. Bateman, A., Eddy, S. R. & Chothia, C. Members of the immunoglobulin superfamily in bacteria. *Protein Sci.* **5**, 1939–1941 (1996).

342. Sutherland, D., Samakovlis, C. & Krasnow, M. A. Branchless encodes a Drosophila FGF homolog that controls tracheal cell migration and the pattern of branching. *Cell* **87**, 1091–1101 (1996).

343. Warburton, D. *et al.* The molecular basis of lung morphogenesis. *Mech. Dev.* **92**, 55–81 (2000).

344. Fuchs, T., Glusman, G., Horn-Saban, S., Lancet, D. & Pilpel, Y. The human olfactory subgenome: from sequence to structure to evolution. *Hum. Genet.* **108**, 1–13 (2001).

345. Glusman, G. *et al.* The olfactory receptor gene family: data mining, classification and nomenclature. *Mamm. Genome* **11**, 1016–1023 (2000).

346. Rouquier, S. *et al.* Distribution of olfactory receptor genes in the human genome. *Nature Genet.* **18**, 243–250 (1998).

347. Sharon, D. *et al.* Primate evolution of an olfactory receptor cluster: Diversification by gene conversion and recent emergence of a pseudogene. *Genomics* **61**, 24–36 (1999).

348. Gilad, Y. *et al.* Dichotomy of single-nucleotide polymorphism haplotypes in olfactory receptor genes and pseudogenes. *Nature Genet.* **26**, 221–224 (2000).

349. Gearhart, J. & Kirschner, M. *Cells, Embryos, and Evolution* (Blackwell Science, Malden, Massachusetts, 1997).

350. Barbazuk, W. B. *et al.* The syntenic relationship of the zebrafish and human genomes. *Genome Res.* **10**, 1351–1358 (2000).

351. McLysaght, A., Enright, A. J., Skrabanek, L. & Wolfe, K. H. Estimation of synteny conservation and genome compaction between pufferfish (Fugu) and human. *Yeast* **17**, 22–36 (2000).

352. Trachtulec, Z. *et al.* Linkage of TATA-binding protein and proteasome subunit C5 genes in mice and humans reveals synteny conserved between mammals and invertebrates. *Genomics* **44**, 1–7 (1997).

353. Nadeau, J. H. Maps of linkage and synteny homologies between mouse and man. *Trends Genet.* **5,** 82–86 (1989).
354. Nadeau, J. H. & Taylor, B. A. Lengths of chromosomal segments conserved since divergence of man and mouse. *Proc. Natl Acad. Sci. USA* **81,** 814–818 (1984).
355. Copeland, N. G. *et al.* A genetic linkage map of the mouse: current applications and future prospects. *Science* **262,** 57–66 (1993).
356. DeBry, R. W. & Seldin, M. F. Human/mouse homology relationships. *Genomics* **33,** 337–351 (1996).
357. Nadeau, J. H. & Sankoff, D. The lengths of undiscovered conserved segments in comparative maps. *Mamm. Genome* **9,** 491–495 (1998).
358. Thomas, J. W. *et al.* Comparative genome mapping in the sequence-based era: early experience with human chromosome 7. *Genome Res.* **10,** 624–633 (2000).
359. Pletcher, M. T. *et al.* Chromosome evolution: The junction of mammalian chromosomes in the formation of mouse chromosome 10. *Genome Res.* **10,** 1463–1467 (2000).
360. Novacek, M. J. Mammalian phylogeny: shaking the tree. *Nature* **356,** 121–125 (1992).
361. O'Brien, S. J. *et al.* Genome maps 10. Comparative genomics. Mammalian radiations. Wall chart. *Science* **286,** 463–478 (1999).
362. Romer, A. S. *Vertebrate Paleontology* (Univ. Chicago Press, Chicago and New York, 1966).
363. Paterson, A. H. *et al.* Toward a unified genetic map of higher plants, transcending the monocot-dicot divergence. *Nature Genet.* **14,** 380–382 (1996).
364. Jenczewski, E., Prosperi, J. M. & Ronfort, J. Differentiation between natural and cultivated populations of *Medicago sativa* (Leguminosae) from Spain: analysis with random amplified polymorphic DNA (RAPD) markers and comparison to allozymes. *Mol. Ecol.* **8,** 1317–1330 (1999).
365. Ohno, S. *Evolution by Gene Duplication* (George Allen and Unwin, London, 1970).
366. Wolfe, K. H. & Shields, D. C. Molecular evidence for an ancient duplication of the entire yeast genome. *Nature* **387,** 708–713 (1997).
367. Blanc, G., Barakat, A., Guyot, R., Cooke, R. & Delseny, M. Extensive duplication and reshuffling in the arabidopsis genome. *Plant Cell* **12,** 1093–1102 (2000).
368. Paterson, A. H. *et al.* Comparative genomics of plant chromosomes. *Plant Cell* **12,** 1523–1540 (2000).
369. Vision, T., Brown, D. & Tanksley, S. The origins of genome duplications in *Arabidopsis*. *Science* **290,** 2114–2117 (2000).
370. Sidow, A. & Bowman, B. H. Molecular phylogeny. *Curr. Opin. Genet. Dev.* **1,** 451–456 (1991).
371. Sidow, A. & Thomas, W. K. A molecular evolutionary framework for eukaryotic model organisms. *Curr. Biol.* **4,** 596–603 (1994).
372. Sidow, A. Gen(om)e duplications in the evolution of early vertebrates. *Curr. Opin. Genet. Dev.* **6,** 715–722 (1996).
373. Spring, J. Vertebrate evolution by interspecific hybridisation—are we polyploid? *FEBS Lett.* **400,** 2–8 (1997).
374. Skrabanek, L. & Wolfe, K. H. Eukaryote genome duplication—where's the evidence? *Curr. Opin. Genet. Dev.* **8,** 694–700 (1998).
375. Hughes, A. L. Phylogenies of developmentally important proteins do not support the hypothesis of two rounds of genome duplication early in vertebrate history. *J. Mol. Evol.* **48,** 565–576 (1999).
376. Lander, E. S. & Schork, N. J. Genetic dissection of complex traits. *Science* **265,** 2037–2048 (1994).
377. Horikawa, Y. *et al.* Genetic variability in the gene encoding calpain-10 is associated with type 2 diabetes mellitus. *Nature Genet.* **26,** 163–175 (2000).
378. Hastbacka, J. *et al.* The diastrophic dysplasia gene encodes a novel sulfate transporter: positional cloning by fine-structure linkage disequilibrium mapping. *Cell* **78,** 1073–1087 (1994).
379. Tischkoff, S. A. *et al.* Global patterns of linkage disequilibrium at the CD4 locus and modern human origins. *Science* **271,** 1380–1387 (1996).
380. Kidd, J. R. *et al.* Haplotypes and linkage disequilibrium at the phenylalanine hydroxylase locus PAH, in a global representation of populations. *Am. J. Hum. Genet.* **63,** 1882–1899 (2000).
381. Mateu, E. *et al.* Worldwide genetic analysis of the CFTR region. *Am. J. Hum. Genet.* **68,** 103–117 (2001).
382. Abecasis, G. R. *et al.* Extent and distribution of linkage disequilibrium in three genomic regions. *Am. J. Hum. Genet.* **68,** 191–197 (2001).
383. Taillon-Miller, P. *et al.* Juxtaposed regions of extensive and minimal linkage disequilibrium in Xq25 and Xq28. *Nature Genet.* **25,** 324–328 (2000).
384. Martin, E. R. *et al.* SNPing away at complex diseases: analysis of single-nucleotide polymorphisms around APOE in Alzheimer disease. *Am. J. Hum. Genet.* **67,** 383–394 (2000).
385. Collins, A., Lonjou, C. & Morton, N. E. Genetic epidemiology of single-nucleotide polymorphisms. *Proc. Natl Acad. Sci. USA* **96,** 15173–15177 (1999).
386. Dunning, A. M. *et al.* The extent of linkage disequilibrium in four populations with distinct demographic histories. *Am. J. Hum. Genet.* **67,** 1544–1554 (2000).
387. Rieder, M. J., Taylor, S. L., Clark, A. G. & Nickerson, D. A. Sequence variation in the human angiotensin converting enzyme. *Nature Genet.* **22,** 59–62 (1999).
388. Collins, F. S. Positional cloning moves from perditional to traditional. *Nature Genet.* **9,** 347–350 (1995).
389. Nagamine, K. *et al.* Positional cloning of the APECED gene. *Nature Genet.* **17,** 393–398 (1997).
390. Reuber, B. E. *et al.* Mutations in PEX1 are the most common cause of peroxisome biogenesis disorders. *Nature Genet.* **17,** 445–448 (1997).
391. Portsteffen, H. *et al.* Human PEX1 is mutated in complementation group 1 of the peroxisome biogenesis disorders. *Nature Genet.* **17,** 449–452 (1997).
392. Everett, L. A. *et al.* Pendred syndrome is caused by mutations in a putative sulphate transporter gene (PDS). *Nature Genet.* **17,** 411–422 (1997).
393. Coffey, A. J. *et al.* Host response to EBV infection in X-linked lymphoproliferative disease results from mutations in an SH2-domain encoding gene. *Nature Genet.* **20,** 129–135 (1998).
394. Van Laer, L. *et al.* Nonsyndromic hearing impairment is associated with a mutation in DFNA5. *Nature Genet.* **20,** 194–197 (1998).
395. Sakuntabhai, A. *et al.* Mutations in ATP2A2, encoding a Ca2+ pump, cause Darier disease. *Nature Genet.* **21,** 271–277 (1999).
396. Gedeon, A. K. *et al.* Identification of the gene (SEDL) causing X-linked spondyloepiphyseal dysplasia tarda. *Nature Genet.* **22,** 400–404 (1999).
397. Hurvitz, J. R. *et al.* Mutations in the CCN gene family member WISP3 cause progressive pseudorheumatoid dysplasia. *Nature Genet.* **23,** 94–98 (1999).
398. Laberge-le Couteulx, S. *et al.* Truncating mutations in CCM1, encoding KRIT1, cause hereditary cavernous angiomas. *Nature Genet.* **23,** 189–193 (1999).
399. Sahoo, T. *et al.* Mutations in the gene encoding KRIT1, a Krev-1/rap1a binding protein, cause cerebral cavernous malformations (CCM1). *Hum. Mol. Genet.* **8,** 2325–2333 (1999).
400. McGuirt, W. T. *et al.* Mutations in COL11A2 cause non-syndromic hearing loss (DFNA13). *Nature Genet.* **23,** 413–419 (1999).
401. Moreira, E. S. *et al.* Limb-girdle muscular dystrophy type 2G is caused by mutations in the gene encoding the sarcomeric protein telethonin. *Nature Genet.* **24,** 163–166 (2000).
402. Ruiz-Perez, V. L. *et al.* Mutations in a new gene in Ellis-van Creveld syndrome and Weyers acrodental dysostosis. *Nature Genet.* **24,** 283–286 (2000).
403. Kaplan, J. M. *et al.* Mutations in ACTN4, encoding alpha-actinin-4, cause familial focal segmental glomerulosclerosis. *Nature Genet.* **24,** 251–256 (2000).
404. Escayg, A. *et al.* Mutations of SCN1A, encoding a neuronal sodium channel, in two families with GEFS+2. *Nature Genet.* **24,** 343–345 (2000).
405. Sacksteder, K. A. *et al.* Identification of the alpha-aminoadipic semialdehyde synthase gene, which is defective in familial hyperlysinemia. *Am. J. Hum. Genet.* **66,** 1736–1743 (2000).
406. Kalaydjieva, L. *et al.* N-myc downstream-regulated gene 1 is mutated in hereditary motor and sensory neuropathy-Lom. *Am. J. Hum. Genet.* **67,** 47–58 (2000).
407. Sundin, O. H. *et al.* Genetic basis of total colourblindness among the Pingelapese islanders. *Nature Genet.* **25,** 289–293 (2000).
408. Kohl, S. *et al.* Mutations in the CNGB3 gene encoding the beta-subunit of the cone photoreceptor cGMP-gated channel are responsible for achromatopsia (ACHM3) linked to chromosome 8q21. *Hum. Mol. Genet.* **9,** 2107–2116 (2000).
409. Avela, K. *et al.* Gene encoding a new RING-B-box-coiled-coil protein is mutated in mulibrey nanism. *Nature Genet.* **25,** 298–301 (2000).
410. Verpy, E. *et al.* A defect in harmonin, a PDZ domain-containing protein expressed in the inner ear sensory hair cells, underlies usher syndrome type 1C. *Nature Genet.* **26,** 51–55 (2000).
411. Bitner-Glindzicz, M. *et al.* A recessive contiguous gene deletion causing infantile hyperinsulinism, enteropathy and deafness identifies the usher type 1C gene. *Nature Genet.* **26,** 56–60 (2000).
412. The May-Hegglin/Fetchner Syndrome Consortium. Mutations in MYH9 result in the May-Hegglin anomaly, and Fechtner and Sebastian syndromes. *Nature Genet.* **26,** 103–105 (2000).
413. Kelley, M. J., Jawien, W., Ortel, T. L. & Korczak, J. F. Mutation of MYH9, encoding non-muscle myosin heavy chain A, in May-Hegglin anomaly. *Nature Genet.* **26,** 106–108 (2000).
414. Kirschner, L. S. *et al.* Mutations of the gene encoding the protein kinase A type I-α regulatory subunit in patients with the Carney complex. *Nature Genet.* **26,** 89–92 (2000).
415. Lalwani, A. K. *et al.* Human nonsyndromic hereditary deafness DFNA17 is due to a mutation in non-muscle myosin MYH9. *Am. J. Hum. Genet.* **67,** 1121–1128 (2000).
416. Matsuura, T. *et al.* Large expansion of the ATTCT pentanucleotide repeat in spinocerebellar ataxia type 10. *Nature Genet.* **26,** 191–194 (2000).
417. Delettre, C. *et al.* Nuclear gene OPA1, encoding a mitochondrial dynamin-related protein, is mutated in dominant optic atrophy. *Nature Genet.* **26,** 207–210 (2000).
418. Pusch, C. M. *et al.* The complete form of X-linked congenital stationary night blindness is caused by mutations in a gene encoding a leucine-rich repeat protein. *Nature Genet.* **26,** 324–327 (2000).
419. The ADHR Consortium. Autosomal dominant hypophosphataemic rickets is associated with mutations in FGF23. *Nature Genet.* **26,** 345–348 (2000).
420. Bomont, P. *et al.* The gene encoding gigaxonin, a new member of the cytoskeletal BTB/kelch repeat family, is mutated in giant axonal neuropathy. *Nature Genet.* **26,** 370–374 (2000).
421. Tullio-Pelet, A. *et al.* Mutant WD-repeat protein in triple-A syndrome. *Nature Genet.* **26,** 332–335 (2000).
422. Nicole, S. *et al.* Perlecan, the major proteoglycan of basement membranes, is altered in patients with Schwartz-Jampel syndrome (chondrodystrophic myotonia). *Nature Genet.* **26,** 480–483 (2000).
423. Rogaev, E. I. *et al.* Familial Alzheimer's disease in kindreds with missense mutations in a gene on chromosome 1 related to the Alzheimer's disease type 3 gene. *Nature* **376,** 775–778 (1995).
424. Sherrington, R. *et al.* Cloning of a gene bearing missense mutations in early-onset familial Alzheimer's disease. *Nature* **375,** 754–760 (1995).
425. Olivieri, N. F. & Weatherall, D. J. The therapeutic reactivation of fetal haemoglobin. *Hum. Mol. Genet.* **7,** 1655–1658 (1998).
426. Drews, J. Research & development. Basic science and pharmaceutical innovation. *Nature Biotechnol.* **17,** 406 (1999).
427. Drews, J. Drug discovery: a historical perspective. *Science* **287,** 1960–1964 (2000).
428. Davies, P. A. *et al.* The 5-HT3B subunit is a major determinant of serotonin-receptor function. *Nature* **397,** 359–363 (1999).
429. Heise, C. E. *et al.* Characterization of the human cysteinyl leukotriene 2 receptor. *J. Biol. Chem.* **275,** 30531–30536 (2000).
430. Fan, W. *et al.* BACE maps to chromosome 11 and a BACE homolog, BACE2, reside in the obligate Down Syndrome region of chromosome 21. *Science* **286,** 1255a (1999).
431. Saunders, A. J., Kim, T. -W. & Tanzi, R. E. BACE maps to chromosome 11 and a BACE homolog, BACE2, reside in the obligate Down Syndrome region of chromosome 21. *Science* **286,** 1255a (1999).
432. Firestein, S. The good taste of genomics. *Nature* **404,** 552–553 (2000).
433. Matsunami, H., Montmayeur, J. P. & Buck, L. B. A family of candidate taste receptors in human and mouse. *Nature* **404,** 601–604 (2000).
434. Adler, E. *et al.* A novel family of mammalian taste receptors. *Cell* **100,** 693–702 (2000).
435. Chandrashekar, J. *et al.* T2Rs function as bitter taste receptors. *Cell* **100,** 703–711 (2000).
436. Hardison, R. C. Conserved non-coding sequences are reliable guides to regulatory elements. *Trends Genet.* **16,** 369–372 (2000).
437. Onyango, P. *et al.* Sequence and comparative analysis of the mouse 1-megabase region orthologous to the human 11p15 imprinted domain. *Genome Res.* **10,** 1697–1710 (2000).
438. Bouck, J. B., Metzker, M. L. & Gibbs, R. A. Shotgun sample sequence comparisons between mouse and human genomes. *Nature Genet.* **25,** 31–33 (2000).
439. Marshall, E. Public-private project to deliver mouse genome in 6 months. *Science* **290,** 242–243 (2000).
440. Wasserman, W. W., Palumbo, M., Thompson, W., Fickett, J. W. & Lawrence, C. E. Human-mouse genome comparisons to locate regulatory sites. *Nature Genet.* **26,** 225–228 (2000).
441. Tagle, D. A. *et al.* Embryonic epsilon and gamma globin genes of a prosimian primate (*Galago crassicaudatus*). Nucleotide and amino acid sequences, developmental regulation and phylogenetic footprints. *J. Mol. Biol.* **203,** 439–455 (1988).
442. McGuire, A. M., Hughes, J. D. & Church, G. M. Conservation of DNA regulatory motifs and discovery of new motifs in microbial genomes. *Genome Res.* **10,** 744–757 (2000).

443. Roth, F. P., Hughes, J. D., Estep, P. W. & Church, G. M. Finding DNA regulatory motifs within unaligned noncoding sequences clustered by whole-genome mRNA quantitation. *Nature Biotechnol.* **16**, 939–945 (1998).

444. Cheng, Y. & Church, G. M. Biclustering of expression data. *ISMB* **8**, 93–103 (2000).

445. Cohen, B. A., Mitra, R. D., Hughes, J. D. & Church, G. M. A computational analysis of whole-genome expression data reveals chromosomal domains of gene expression. *Nature Genet.* **26**, 183–186 (2000).

446. Feil, R. & Khosla, S. Genomic imprinting in mammals: an interplay between chromatin and DNA methylation? *Trends Genet.* **15**, 431–434 (1999).

447. Robertson, K. D. & Wolffe, A. P. DNA methylation in health and disease. *Nature Rev. Genet.* **1**, 11–19 (2000).

448. Beck, S., Olek, A. & Walter, J. From genomics to epigenomics: a loftier view of life. *Nature Biotechnol.* **17**, 1144–1144 (1999).

449. Hagmann, M. Mapping a subtext in our genetic book. *Science* **288**, 945–946 (2000).

450. Eliot, T. S. in *T. S. Eliot. Collected Poems 1909–1962* (Harcourt Brace, New York, 1963).

451. Soderland, C., Longden, I. & Mott, R. FPC: a system for building contigs from restriction fingerprinted clones. *Comput. Appl. Biosci.* **13**, 523–535 (1997).

452. Mott, R. & Tribe, R. Approximate statistics of gapped alignments. *J. Comp. Biol.* **6**, 91–112 (1999).

Supplementary Information is available on *Nature*'s World-Wide Web site (http://www.nature.com) or as paper copy from the London editorial office of *Nature*.

Acknowledgements

Beyond the authors, many people contributed to the success of this work. E. Jordan provided helpful advice throughout the sequencing effort. We thank D. Leja and J. Shehadeh for their expert assistance on the artwork in this paper, especially the foldout figure; K. Jegalian for editorial assistance; J. Schloss, E. Green and M. Seldin for comments on an earlier version of the manuscript; P. Green and F. Ouelette for critiques of the submitted version; C. Caulcott, A. Iglesias, S. Renfrey, B. Skene and J. Stewart of the Wellcome Trust, P. Whittington and T. Dougans of NHGRI and M. Meugnier of Genoscope for staff support for meetings of the international consortium; and the University of Pennsylvania for facilities for a meeting of the genome analysis group.

We thank Compaq Computer Corporations's High Performance Technical Computing Group for providing a Compaq Biocluster (a 27 node configuration of AlphaServer ES40s, containing 108 CPUs, serving as compute nodes and a file server with one terabyte of secondary storage) to assist in the annotation and analysis. Compaq provided the systems and implementation services to set up and manage the cluster for continuous use by members of the sequencing consortium. Platform Computing Ltd. provided its LSF scheduling and loadsharing software without license fee.

In addition to the data produced by the members of the International Human Genome Sequencing Consortium, the draft genome sequence includes published and unpublished human genomic sequence data from many other groups, all of whom gave permission to include their unpublished data. Four of the groups that contributed particularly significant amounts of data were: M. Adams *et al.* of the Institute for Genomic Research; E. Chen *et al.* of the Center for Genetic Medicine and Applied Biosystems; S.-F. Tsai of National Yang-Ming University, Institute of Genetics, Taipei, Taiwan, Republic of China; and Y. Nakamura, K. Koyama *et al.* of the Institute of Medical Science, University of Tokyo, Human Genome Center, Laboratory of Molecular Medicine, Minato-ku, Tokyo, Japan. Many other groups provided smaller numbers of database entries. We thank them all; a full list of the contributors of unpublished sequence is available as Supplementary Information.

This work was supported in part by the National Human Genome Research Institute of the US NIH; The Wellcome Trust; the US Department of Energy, Office of Biological and Environmental Research, Human Genome Program; the UK MRC; the Human Genome Sequencing Project from the Science and Technology Agency (STA) Japan; the Ministry of Education, Science, Sport and Culture, Japan; the French Ministry of Research; the Federal German Ministry of Education, Research and Technology (BMBF) through Projektträger DLR, in the framework of the German Human Genome Project; BEO, Projektträger Biologie, Energie, Umwelt des BMBF und BMWT; the Max-Planck-Society; DFG—Deutsche Forschungsgemeinschaft; TMWFK, Thüringer Ministerium für Wissenschaft, Forschung und Kunst; EC BIOMED2—European Commission, Directorate Science, Research and Development; Chinese Academy of Sciences (CAS), Ministry of Science and Technology (MOST), National Natural Science Foundation of China (NSFC); US National Science Foundation EPSCoR and The SNP Consortium Ltd. Additional support for members of the Genome Analysis group came, in part, from an ARCS Foundation Scholarship to T.S.F., a Burroughs Wellcome Foundation grant to C.B.B. and P.A.S., a DFG grant to P.B., DOE grants to D.H., E.E.E. and T.S.F., an EU grant to P.B., a Marie-Curie Fellowship to L.C., an NIH-NHGRI grant to S.R.E., an NIH grant to E.E.E., an NIH SBIR to D.K., an NSF grant to D.H., a Swiss National Science Foundation grant to L.C., the David and Lucille Packard Foundation, the Howard Hughes Medical Institute, the University of California at Santa Cruz and the W. M. Keck Foundation.

Correspondence and requests for materials should be addressed to E. S. Lander (e-mail: lander@genome.wi.mit.edu), R. H. Waterston (e-mail: bwaterst@watson.wustl.edu), J. Sulston (e-mail: jes@sanger.ac.uk) or F. S. Collins (e-mail: fc23a@nih.gov).

Affiliations for authors: 1, Whitehead Institute for Biomedical Research, Center for Genome Research, Nine Cambridge Center, Cambridge, Massachusetts 02142, USA; 2, The Sanger Centre, The Wellcome Trust Genome Campus, Hinxton, Cambridgeshire CB10 1RQ, United Kingdom; 3, Washington University Genome Sequencing Center, Box 8501, 4444 Forest Park Avenue, St. Louis, Missouri 63108, USA; 4, US DOE Joint Genome Institute, 2800 Mitchell Drive, Walnut Creek, California 94598, USA; 5, Baylor College of Medicine Human Genome Sequencing Center, Department of Molecular and Human Genetics, One Baylor Plaza, Houston, Texas 77030, USA; 6, Department of Cellular and Structural Biology, The University of Texas Health Science Center at San Antonio, 7703 Floyd Curl Drive, San Antonio, Texas 78229-3900, USA; 7, Department of Molecular Genetics, Albert Einstein College of Medicine, 1635 Poplar Street, Bronx, New York 10461, USA; 8, Baylor College of Medicine Human Genome Sequencing Center and the Department of Microbiology & Molecular Genetics, University of Texas Medical School, PO Box 20708, Houston, Texas 77225, USA; 9, RIKEN Genomic Sciences Center, 1-7-22 Suehiro-cho, Tsurumi-ku Yokohama-city, Kanagawa 230-0045, Japan; 10, Genoscope and CNRS UMR-8030, 2 Rue Gaston Cremieux, CP 5706, 91057 Evry Cedex, France; 11, GTC Sequencing Center, Genome Therapeutics Corporation, 100 Beaver Street, Waltham, Massachusetts 02453-8443, USA; 12, Department of Genome Analysis, Institute of Molecular Biotechnology, Beutenbergstrasse 11, D-07745 Jena, Germany; 13, Beijing Genomics Institute/Human Genome Center, Institute of Genetics, Chinese Academy of Sciences, Beijing 100101, China; 14, Southern China National Human Genome Research Center, Shanghai 201203, China; 15, Northern China National Human Genome Research Center, Beijing 100176, China; 16, Multimegabase Sequencing Center, The Institute for Systems Biology, 4225 Roosevelt Way, NE Suite 200, Seattle, Washington 98105, USA; 17, Stanford Genome Technology Center, 855 California Avenue, Palo Alto, California 94304, USA; 18, Stanford Human Genome Center and Department of Genetics, Stanford University School of Medicine, Stanford, California 94305-5120, USA; 19, University of Washington Genome Center, 225 Fluke Hall on Mason Road, Seattle, Washington 98195, USA; 20, Department of Molecular Biology, Keio University School of Medicine, 35 Shinanomachi, Shinjuku-ku, Tokyo 160-8582, Japan; 21, University of Texas Southwestern Medical Center at Dallas, 6000 Harry Hines Blvd., Dallas, Texas 75235-8591, USA; 22, University of Oklahoma's Advanced Center for Genome Technology, Dept. of Chemistry and Biochemistry, University of Oklahoma, 620 Parrington Oval, Rm 311, Norman, Oklahoma 73019, USA; 23, Max Planck Institute for Molecular Genetics, Ihnestrasse 73, 14195 Berlin, Germany; 24, Cold Spring Harbor Laboratory, Lita Annenberg Hazen Genome Center, 1 Bungtown Road, Cold Spring Harbor, New York 11724, USA; 25, GBF - German Research Centre for Biotechnology, Mascheroder Weg 1, D-38124 Braunschweig, Germany; 26, National Center for Biotechnology Information, National Library of Medicine, National Institutes of Health, Bldg. 38A, 8600 Rockville Pike, Bethesda, Maryland 20894, USA; 27, Department of Genetics, Case Western Reserve School of Medicine and University Hospitals of Cleveland, BRB 720, 10900 Euclid Ave., Cleveland, Ohio 44106, USA; 28, EMBL European Bioinformatics Institute, Wellcome Trust Genome Campus, Hinxton, Cambridge CB10 1SD, United Kingdom; 29, Max Delbrück Center for Molecular Medicine, Robert-Rossle-Strasse 10, 13125 Berlin-Buch, Germany; 30, EMBL, Meyerhofstrasse 1, 69012 Heidelberg, Germany; 31, Dept. of Biology, Massachusetts Institute of Technology, 77 Massachusetts Ave., Cambridge, Massachusetts 02139-4307, USA; 32, Howard Hughes Medical Institute, Dept. of Genetics, Washington University School of Medicine, Saint Louis, Missouri 63110, USA; 33, Dept. of Computer Science, University of California at Santa Cruz, Santa Cruz, California 95064, USA; 34, Affymetrix, Inc., 2612 8th St, Berkeley, California 94710, USA; 35, Genome Exploration Research Group, Genomic Sciences Center, RIKEN Yokohama Institute, 1-7-22 Suehiro-cho, Tsurumi-ku, Yokohama, Kanagawa 230-0045, Japan; 36, Howard Hughes Medical Institute, Department of Computer Science, University of California at Santa Cruz, California 95064, USA; 37, University of Dublin, Trinity College, Department of Genetics, Smurfit Institute, Dublin 2, Ireland; 38, Cambridge Research Laboratory, Compaq Computer Corporation and MIT Genome Center, 1 Cambridge Center, Cambridge, Massachusetts 02142, USA; 39, Dept. of Mathematics, University of California at Santa Cruz, Santa Cruz, California 95064, USA; 40, Dept. of Biology, University of California at Santa Cruz, Santa Cruz, California 95064, USA; 41, Crown Human Genetics Center and Department of Molecular Genetics, The Weizmann Institute of Science, Rehovot 71600, Israel; 42, Dept. of Genetics, Stanford University School of Medicine, Stanford, California 94305, USA; 43, The University of Michigan Medical School, Departments of Human Genetics and Internal Medicine, Ann Arbor, Michigan

48109, USA; 44, MRC Functional Genetics Unit, Department of Human Anatomy and Genetics, University of Oxford, South Parks Road, Oxford OX1 3QX, UK; 45, Institute for Systems Biology, 4225 Roosevelt Way NE, Seattle, WA 98105, USA; 46, National Human Genome Research Institute, US National Institutes of Health, 31 Center Drive, Bethesda, Maryland 20892, USA; 47, Office of Science, US Department of Energy, 19901 Germantown Road, Germantown, Maryland 20874, USA; 48, The Wellcome Trust, 183 Euston Road, London, NW1 2BE, UK.

† Present addresses: Genome Sequencing Project, Egea Biosciences, Inc., 4178 Sorrento Valley Blvd., Suite F, San Diego, CA 92121, USA (G.A.E.); INRA, Station d'Amélioration des Plantes, 63039 Clermont-Ferrand Cedex 2, France (L.C.).

DNA sequence databases

GenBank, National Center for Biotechnology Information, National Library of Medicine, National Institutes of Health, Bldg. 38A, 8600 Rockville Pike, Bethesda, Maryland 20894, USA

EMBL, European Bioinformatics Institute, Wellcome Trust Genome Campus, Hinxton, Cambridge CB10 1SD, UK

DNA Data Bank of Japan, Center for Information Biology, National Institute of Genetics, 1111 Yata, Mishima-shi, Shizuoka-ken 411-8540, Japan

Correction

International Human Genome Sequencing Consortium

Nature 409, 860–921 (2001)
Published 2 August 2001

Nature 409, 860–921 (2001).

We have identified several items requiring correction or clarification in our paper on the sequencing of the human genome.

• Six additional authors should have been included: Pieter de Jong, Joseph J. Catanese, and Kazutoyo Osoegawa (Department of Cancer Genetics, Roswell Park Cancer Institute, Buffalo, New York 14263, USA; present address: Children's Hospital Oakland Research Institute, 747 52nd Street Oakland, California 94609, USA) and Hiroaki Shizuya, Sangdun Choi and Yu-Juin Chen (Division of Biology, California Institute of Technology, Pasadena, California 91125, USA). These investigators and their laboratories constructed the high-quality BAC libraries that were crucial in sequencing the genome, as described in Table 1. These libraries were not previously published. We apologize to our colleagues for this omission.

• The Supplementary Information on *Nature*'s website has been revised. Changes to the original Supplementary Information are available in the Supplementary Information to this Correction. We have added 7 additional investigators to the full list of authors. We have also added 79 additional references, citing previously published sequences that were included in the draft genome sequence.

• Table 27 reported 18 instances of apparently novel paralogues of genes encoding drug targets. We have carefully reviewed these 18 cases and found that two are incorrect: a paralogue of an insulin-like growth factor-1 receptor gene and a paralogue of the calcitonin-related polypeptide alpha gene. In both cases, we had incorrectly recorded the chromosomal location sequence of the known gene, thereby erroneously giving rise to an apparent paralogue (the first instance was identified by J. Englebrecht and C. Kristensen (personal communication)). Of the 16 remaining apparent paralogues, two (calcium channel paralogue IGI_M1_ctg17137_10 and heparan *N*-deacetylase/*N*-sulphotransferase paralogue IGI_M1_ctg13263_18) have so far been confirmed as bona fide genes[1,2].

• Several correspondents have written to point out that a handful of clones listed as human sequence in the HTG division of GenBank (established to house 'unfinished' sequence data) are actually mouse sequence (about two dozen out of 30,000 clones). They asked whether these clones give rise to contamination in the human draft sequence. As noted in the paper, we used computer programs to identify and eliminate instances of such contamination (with mouse sequence, vector sequence, and so on) before assembling the draft genome sequence. In reviewing the work, we identified one mouse clone that slipped through the filter. This clone has been eliminated in subsequent assemblies (http://genome.cse.ucsc.edu/). Because the draft sequence remains an imperfect partial product, we welcome additional comments that could help in improving it.

• The discussion of possible horizontal gene transfer from bacterial genomes to vertebrate genomes has provoked considerable discussions[3–5]. We reported 113 instances of human genes that had reasonably close homologues in bacteria, but either had no homologue or only a weaker homologue in non-vertebrate eukaryotes for which extensive genomic sequence was available. We suggested two hypotheses to explain these data: horizontal gene transfer (HGT) from bacteria to human or gene loss in the other lineages. We had no data to distinguish between these hypotheses, although we suggested that the latter was a more "parsimonious" explanation as it involved fewer independent events. In the introduction we stated that this seemed "likely".

Several correspondents have undertaken more comprehensive analyses and have argued that a significant proportion of the cases can be explained by gene loss[3–5]. We agree. We believe that the two hypotheses cannot be distinguished on the basis of parsimony, because too little is known about the relative rates of HGT and gene loss in evolution. Instead, extensive sequence data from many additional organisms will be required to assess definitively the provenance of each gene.

We note that the process of HGT into the vertebrate genome from other organisms has clearly occurred on multiple occasions, as seen from the sudden arrival of many DNA transposons with strong similarities to other organisms. The most recent documented cases occurred subsequent to the eutherian radiation (see Fig. 19).

• A key reference concerning 3'-transduction by LINE elements was omitted on page 887. The sentence citing references 205 and 206 should also have cited Goodier *et al.*[6].

• In Fig. 33, the unit on the *y* axis should be bp, not kb. The legend should read "Sequence properties of segmental duplications. Distributions of length and per cent nucleotide identity are shown as a function of the number of aligned bp from the finished vs finished human genomic sequence dataset. Intrachromosomal (blue), interchromosomal (red)".

• In Fig. 41, the legend should begin: "For each of the 27 common domain families, the number of different Pfam domain types that co-occur with the family in each of the five eukaryotic proteomes. The 27 families were chosen to include the 10 most common domain families in each proteome. The data are ranked ..."

• In Table 22, the entry 81,126 should be 8,126.

• On page 898, line 31, the final phrase of the sentence ("... and the representativeness of currently 'known' human genes") should be deleted. The sentence should read: "Before discussing the gene predictions for the human genome, it is useful to consider background issues, including previous estimates of the number of human genes and lessons learned from worms and flies".

• On page 900, line 38, remove "(see above)".

• We failed to acknowledge the crucial role of sequence editing software, which has been widely used for inspection and subsequent finishing of the sequence assemblies. The two principal programs used were CONSED[7] and GAP4[8].

1. Burgess, D. L. *et al.* A cluster of three novel (Ca(2+)) channel gamma subunit genes on chromosome 19q13.4: Evolution and expression–profile of the gamma subunit gene family. *Genomics* **71**, 339–350 (2001).
2. Aikawa, J. *et al.* Multiple isozymes of heparan sulfate/heparin GlcNAc *N*-deacetylase/GlcN *N*-sulfotransferase. Structure and activity of the fourth member, NDST4. *J. Biol. Chem.* **276**, 5876–5882 (2001).
3. Salzberg, S. L *et al.* Microbial genes in the human genome: Lateral transfer or gene loss? *Science* **292**, 1903–1906 (2001).
4. Stanhope, M. J. *et al.* Phylogenetic analyses do not support horizontal gene transfers from bacteria to vertebrates. *Nature* **411**, 940–944 (2001).
5. Reelofs, J. & Van Haastert, P. J. M. Genes lost during evolution. *Nature* **411**, 1013–1014 (2001).
6. Goodier, J. L., Ostertag, E. M. & Kazazian, H. H. Transduction of 3'-flanking sequences is common in L1 retrotransposition. *Hum. Mol. Genet* **9**, 653–657 (2000).
7. Gordon, D., Abajian, C. & Green, P. Consend: a graphical tool for sequence finishing. *Genome Res.* **3**, 195–202 (1998).
8. Staden, R., Beal, K. F. & Bonfield, J. K. The Staden package. 1998. *Methods Mol. Biol.* **132**, 115–130 (2001).

Supplementary information (with changes to the original Supplementary Information) is available on *Nature*'s World-Wide Web site (http://www.nature.com).

Genome speak

Allele Alternative version of a particular gene. Humans carry two sets of most genes, one inherited from each parent, so a single allele for each locus is inherited separately from each parent.

Amino acid Any of a class of 20 molecules that are combined to form proteins in living things.

Annotation Identification of the locations and coding regions of genes in a genome and the prediction of functions for these regions.

Autosome A chromosome not involved in sex determination. The diploid human genome contains 22 pairs of autosomes, and 1 pair of sex chromosomes. Compare with **sex chromosome**.

BAC (bacterial artificial chromosome) A chromosome-like structure, constructed by genetic engineering, that is used as a vector to clone DNA fragments of genome (100 to 300 kb insert size) in cells of the bacterium *Escherichia coli.*

Base pair (bp) Two nitrogenous bases (adenine and thymine or guanine and cytosine) held together by weak bonds. See **DNA, nucleotide**.

Bioinformatics The study of genetic and other biological information using computer and statistical techniques. In genome projects, bioinformatics includes the development of methods to search databases, to analyse DNA sequence information, and to predict protein sequence and structure from DNA sequence data.

BLAST (Basic Local Alignment Search Tool) A computer-search program that searches for sequence similarity to identify homologous genes.

Centromere The compact region at the centre of a chromosome.

Chromosome A rod-shaped structure inside the nucleus of a cell which contains a densely packed continuous strand of DNA. Different organisms have different numbers of chromosomes. The diploid human genome consists of 23 pairs of chromosomes, 46 in all: 22 pairs

of autosomes and two sex chromosomes. See **autosome, sex chromosome**.

Clone An exact copy made of biological material, such as a DNA segment (a gene or other region).

Cloning The process of generating multiple, exact copies of a particular piece of DNA to allow it to be sequenced or studied in some other way.

cDNA (complementary DNA) A DNA sequence made from a messenger RNA molecule. cDNAs can be used experimentally to determine the sequence of messenger RNAs after their introns (non-protein-coding sections) have been spliced out.

Conserved sequence A sequence of DNA (or an amino-acid sequence in a protein) that has remained essentially unchanged throughout evolution, usually because of functional constraints.

Contig A contiguous sequence of DNA created by assembling overlapping sequenced fragments.

DNA (deoxyribonucleic acid) The molecule that encodes genetic information. The four nucleotides in DNA contain the bases: adenine (A), guanine (G), cytosine (C) and thymine (T). Two strands of DNA are held together in the shape of a double helix by bonds between base pairs of nucleotides, where A pairs with T and G with C. See **nucleotide, base pair**.

Diploid A full set of genetic material, consisting of paired chromosomes, one from each parental set. Most animal cells except the gametes have a diploid set of chromosomes. Compare with **haploid**.

Draft sequence DNA sequence in which the order of bases is sequenced at least four to five times (an accuracy of 99.9%), which enables the reassembling of DNA fragments in their original order. Some segments can be missing or in the wrong order or orientation. Compare with **finished sequence**.

Euchromatin The gene-rich regions of a

genome. Compare with **heterochromatin**.

Eukaryote An organism whose cells have a complex internal structure, including a nucleus. Animals, plants and fungi are all eukaryotes. Compare with **prokaryotes**.

Exon The protein-coding DNA sequence of a gene. Compare with **intron**.

EST (expressed sequence tag) A short sequence from a coding region of a gene that identifies the gene.

Finished sequence DNA sequence in which bases are identified to an accuracy of 99.99% and are placed in the right order and orientation along a chromosome with almost no gaps.

FISH (fluorescence in situ hybridization) A process that vividly paints chromosomes or portions of chromosomes with fluorescent molecules. This technique is useful for identifying chromosomal abnormalities and gene mapping.

Gamete Mature male or female reproductive cell (sperm or egg) with a haploid set of chromosomes.

Gene The fundamental physical and functional unit of heredity. A gene is an ordered sequence of nucleotides located in a particular position on a particular chromosome that encodes a specific functional product.

Gene mapping Determination of the relative positions of genes on a DNA molecule (chromosome or plasmid) and of the distance, in linkage units or physical units, between them.

Genetic code The sequence of nucleotides, coded in triplets (codons) along the messenger RNA, that determines the sequence of amino acids in protein synthesis. The DNA sequence of a gene can be used to predict the messenger RNA sequence, and the genetic code can in turn be used to predict the amino-acid sequence.

Genome The complete genetic material of an organism; the entire DNA sequence.

Genomic library A collection of clones made from a set of randomly generated overlapping DNA fragments representing the entire genome of an organism.

Genomics The study of genomes and their sets of genes.

Genotype The set of genes that an individual carries; usually refers to the particular pair of alleles (alternative forms of a gene) that a person has at a given region of the genome. Genotype refers to what is inherited (for example, an allele for brown eyes), whereas phenotype refers to what is expressed (brown eyes in this case).

Haploid A single set of chromosomes (half the full set of genetic material) present in the egg and sperm cells of animals and in the egg and pollen cells of plants. Compare with **diploid**.

Haplotype A particular combination of alleles (alternative forms of genes) or sequence variations that are closely linked – that is, are likely to be inherited together – on the same chromosome.

Heterochromatin Compact, gene-poor regions of a genome, which are enriched in simple sequence repeats. Compare with **euchromatin**.

Intron The DNA sequence interrupting the protein-coding sequence of a gene; this sequence is transcribed into RNA but is cut out before the RNA is transcribed. Compare with **exon**.

Karyotype A photomicrograph of an individual's chromosomes arranged in a standard format showing the number, size and shape of each chromosome type; used in low-resolution physical mapping to correlate gross chromosomal abnormalities with the characteristics of specific diseases.

Kilobase (kb) Unit of length of DNA fragments equal to 1000 nucleotides.

Library An unordered collection of clones whose relationship to each other can be established by physical mapping.

Linkage The proximity of two or more markers (for example, genes) on a chromosome.

Long and short arms The regions either side of the centromere, a compact part of a chromosome, are known as arms. As the centromere is not in the centre of the chromosome, one arm is longer than the other.

Marker An identifiable physical location or landmark on a chromosome (for example, a restriction enzyme cleavage site) whose inheritance through generations can be monitored.

Megabase (Mb) Unit of length for DNA frag-

ments equal to 1 million nucleotides.

Meiosis The process of two consecutive cell divisions in the diploid progenitors of sex cells, which results in four progeny cells, each with a haploid set of chromosomes.

mRNA (messenger RNA) RNA that serves as a template for protein synthesis.

Mitosis The process of nuclear division in cells that produces two daughter cells that are genetically identical to each other and to the parent cell.

Mutation An alteration in a genome compared to some reference state.

Nucleotide A subunit of DNA or RNA comprised of a nitrogenous base (adenine, guanine, thymine or cytosine in DNA; adenine, guanine, uracil or cytosine in RNA), a phosphate molecule and a sugar molecule (deoxyribose in DNA and ribose in RNA). Nucleotides are linked to form the strands of a DNA or RNA molecule. See **base pair**.

Phenotype The observable properties and physical characteristics of an organism.

Physical map The localization of identifiable landmarks on DNA.

Plasmid An autonomously replicating, extra-chromosomal circular DNA molecule, distinct from the normal bacterial genome and non-essential for cell survival under non-selective conditions. Artificially constructed plasmids are used as cloning vectors.

Polymorphism A difference in DNA sequence among individuals. To be called a polymorphism, a variant should be present in a significant number of people in the population.

Prokaryote A cell or organism lacking a membrane-bound, structurally discrete nucleus and other subcellular compartments. Bacteria are prokaryotes. Compare with **eukaryote**.

Protein A large molecule composed of one or more chains of amino acids in a specific order; the order is determined by the base sequence of nucleotides in the gene coding for the protein.

Proteome The complete set of proteins encoded by the genome.

Pseudogene A region of DNA that shows extensive similarity to a known gene, but does not function.

Recombinant DNA A combination of DNA molecules of different origin that are joined using recombinant DNA technologies.

Recombination The process by which DNA is exchanged between pairs of equivalent chromosomes during egg and sperm formation.

Recombination has the effect of making the chromosomes of the offspring distinct from those of the parents.

RNA (ribonucleic acid) A molecule with a similar structure to DNA that plays an important role in protein synthesis and other chemical activities of the cell. There are several types of RNA molecules, including messenger RNA, which acts as an intermediary molecule between DNA and protein. See **nucleotide, mRNA**.

STS (sequence tagged site) A short (200 to 500 base pairs) DNA sequence that has a single occurrence in the human genome and whose location and base sequence are known.

Sex chromosome The X or Y chromosome in humans. Determines the sex of an individual. Females have two X chromosomes in diploid cells; males have an X and a Y chromosome. The sex chromosomes comprise the 23rd chromosome pair of a human genome. Compare with **autosome**.

SNP (single-nucleotide polymorphism) A polymorphism caused by the change of a single nucleotide. Most genetic variation between individuals is due to SNPs.

Splicing The process that removes introns (non-protein-coding portions) from transcribed RNAs. Exons (protein-coding portions) can also be removed. Depending on which exons are removed, different proteins can be made from the same initial RNA or gene. Different proteins created in this way are 'splice variants' or 'alternatively spliced'.

Telomere The end of a chromosome. This specialized structure is involved in the replication and stability of linear DNA molecules.

Transcription The process of copying a gene into RNA. This is the first step in turning a gene into a protein, although not all transcripts generate proteins. Compare **translation**.

Transcriptome The complete set of RNAs transcribed from a genome in a particular tissue at a particular time.

Translation The process of using a messenger RNA sequence to build a protein. The messenger RNA serves as a template on which transfer RNA molecules, carrying amino acids, are lined up. The amino acids are then linked together to form a protein chain.

YAC (yeast artificial chromosome) A type of vector used to clone DNA fragments inside yeast cells. It is constructed from the telomeric, centromeric and replication origin sequences needed for replication in yeast cells.

Index